POLLUTION CONTROL IN THE ORGANIC CHEMICAL INDUSTRY

POLLUTION CONTROL IN THE ORGANIC CHEMICAL INDUSTRY

Marshall Sittig

NOYES DATA CORPORATION

Park Ridge, New Jersey London, England

1974

Library of Congress Catalog Card Number: 74-77220
ISBN: 0-8155-0536-1
Printed in the United States

Published in the United States of America by
Noyes Data Corporation
Noyes Building, Park Ridge, New Jersey 07656

FOREWORD

Waste products from organic chemical plants manufacturing identical compounds may be quite dissimilar because of the difference in processes and raw materials. Since the organic chemical industry is now largely petrochemical, preference has been given to waste treatment from such operations although basic coal tar processes have not been neglected. Detailed treatability evaluation of each waste stream has become a prerequisite for any profitable product.

Physical methods for waste and by-product removal include gravity separation, air flotation, filtration, evaporation and adsorption techniques. Chemical methods include precipitation, polyelectrolyte treatment, oxidation and other chemical conditioning such as neutralization. Biological treatments comprise activated sludge formation and its modifications, i.e., trickling filters, aerated lagoons and waste stabilization ponds.

This book intends to assist the chemical engineer in treating chemical waste products and effluents in conjunction with prudent raw material choices and methods, and thus bring about process economies. It presents condensed vital data from government and other sources of information that are scattered and difficult to pull together. Important pollution removal processes are interpreted and explained by examples from 72 U.S. patents. This condensed information will enable you to establish a sound background for action in combating pollution by the organic chemical industry.

Advanced composition and production methods developed by Noyes Data are employed to bring our durably bound books to you in a minimum of time. Special techniques are used to close the gap between "manuscript" and "completed book." Industrial technology is progressing so rapidly that time-honored, conventional typesetting, binding and shipping methods are no longer suitable. We have bypassed the delays in the conventional book publishing cycle and provide the user with an effective and convenient means of reviewing up-to-date information in depth.

The Table of Contents is organized in such a way as to serve as a subject index. Other indexes by company, inventor and patent number help in providing easy access to the information contained in this book.

CONTENTS AND SUBJECT INDEX

INTRODUCTION 1
Background and Economic Trends 1
Air Pollution 3
Water Pollution 4
Solid Wastes 8

WATER USAGE IN ORGANIC CHEMICALS MANUFACTURE 12

WASTEWATER SAMPLING AND SURVEYING 20
Sampling Station Selection 21
Case Studies of Wastewater Surveys 22

PROCESS CATEGORIES AND RAW WASTE LOADS 34

AQUEOUS EFFLUENT LIMITATIONS 39

SPECIFIC PRODUCTS - MANUFACTURE AND POLLUTION PROBLEMS 47
Acetaldehyde from Ethanol 47
Acetaldehyde from Ethylene 50
Acetic Acid from Acetaldehyde 52
Acetone from Isopropanol 55
Acetylene from Methane 57
Acrylates from Acrylic Acid 60
Acrylic Acid from Acetylene 62
Acrylonitrile from Propylene 64
Adipic Acid from Cyclohexane 67
Alachlor from Diethylaniline 71
Aldicarb from Isobutylene 72
Aldrin from Dicyclopentadiene 73
Allyl Chloride from Propylene 75
Aniline from Nitrobenzene 76
Atrazine from Cyanuric Chloride 80
BTX from Petroleum 82
Bisphenol-A from Phenol and Acetone 85
Butadiene from Butane 87
Caprolactam from Cyclohexane 93
Captan from Butadiene 96

Carbaryl from Naphthalene 98
Chlordane from Cyclopentadiene 100
Chlorobenzene from Benzene 103
Chloromethane from Various Raw Materials 104
  Chlorination of Methane 105
  Hydrochlorination of Methanol 105
  Thermal Chlorination of Propane 106
  Carbon Disulfide Chlorination 107
Coal Tar Distillation 108
Cresol from Toluene 113
Cyclohexane from Benzene 116
2,4-D from Dichlorophenol 116
DDT from Chlorobenzene 119
Dieldrin from Aldrin 121
Dimethyl Terephthalate from Terephthalic Acid 123
Dinitrotoluene from Nitrotoluene 126
Disulfoton from Ethanol 127
Dyes and Pigments from Aromatics 128
Epichlorohydrin from Allyl Chloride 131
Ethylbenzene from Benzene 132
Ethyl Chloride from Various Raw Materials 133
  Hydrochlorination of Ethylene 133
  Chlorination of Ethane 134
  Hydrochlorination of Ethanol 134
Ethylene and Propylene from Paraffins 135
Ethylene Dichloride from Ethylene 146
Ethylene Glycol from Ethylene Oxide 148
Ethylene Oxide from Ethylene 150
Formaldehyde from Methanol 152
Long Chain Alcohols from Ethylene Oligomers 154
Malathion from Methanol 155
Methanol from Natural Gas 158
Methylamines from Methanol and Ammonia 162
Methyl Bromide from Methanol 164
Methyl Methacrylate from Acetone Cyanohydrin 164
Nitrobenzene from Benzene 170
Nitrochlorobenzene from Chlorobenzene 172
Nitroparaffins from Paraffins 173
Oxo Products from Olefins 175
Parathion from p-Nitrophenol 177
Perchloroethylene from Propane 178
Phenol from Chlorobenzene 180
Phenol from Cumene 182
Phorate from Ethanol 185
Phosgene from CO and Chlorine 185
Phthalic Anhydride from Naphthalene or Xylene 186
Propylene Glycol from Propylene Oxide 188
Propylene Oxide from Propylene 189
Styrene from Ethylbenzene 191
Synthetic Natural Gas (SNG) from Crude Oil 195
2,4,5-T from Trichlorophenol 196
Terephthalic Acid from Xylene 198

Tetraethyl Lead from Ethyl Chloride 203
Trichloroethane from Vinyl Chloride 206
Trichloroethylene from Various Raw Materials 207
Chlorination, then Dehydrochlorination of Acetylene 208
Oxyhydrochlorination of Dichloroethane 208
Trifluralin from Perfluoromethyl Chlorobenzene 209
Vinyl Acetate from Ethylene and Acetic Acid 211
Vinyl Chloride from Acetylene 213
Vinyl Chloride from Ethylene Dichloride 214

AIR POLLUTION CONTROL PROCESSES 216
Halogen Acids 216
Halogenated Hydrocarbons 217
Halogens 218
Hydrocarbons 219
Particulates 220
Sulfur Compounds 221
Air Pollution Prevention in Coking of Coal 221

WATER POLLUTION CONTROL PROCESSES 226
Primary 226
Oil Separation 227
Equalization 228
Neutralization 228
Sedimentation 230
Flotation 231
Flocculation 232
Nutrient Addition 233
Secondary 234
Activated Sludge 238
Extended Aeration 240
Trickling Filters 241
Aerated Lagoons 242
Waste Stabilization Ponds 243
Chemical Oxidation 244
Nitrification-Denitrification 249
Tertiary 250
Chemical Precipitation 250
Gas Stripping 251
Microstraining 253
Carbon Adsorption 254
Electrodialysis 257
Ion Exchange 259
Evaporation 261
Reverse Osmosis 262
Chlorination 265
Rapid Sand Filtration 266
Sludge Handling 267
Aerobic Digestion 269
Anaerobic Digestion 270
Wet Oxidation 273
Thickening 274

Lagooning 274
Sand Drying Beds 275
Vacuum Filtration 276
Centrifugation 276
Filtration 277
Land Disposal 278
Incineration 279
Sea Disposal 280
Ultimate Disposal 280
Thermal Oxidation 281
Deep Well Disposal 281

OVERALL WATER POLLUTION CONTROL MODELS 283
Best Practicable Control Technology Currently Available (BPCTCA) 283
Best Available Technology Economically Available (BATEA) 291
Best Available Demonstrated Control Technology (BADCT) 296

POLLUTION CONTROL ECONOMICS 299

FUTURE TRENDS 300

COMPANY INDEX 302
INVENTOR INDEX 303
U.S. PATENT NUMBER INDEX 305

INTRODUCTION

The frontiers of the chemical industry which have produced most of the exciting new products of the past years in the forms of new drugs, plastics, fibers and pesticides are firmly based on industrial organic chemistry.

Organic chemicals were once almost wholly based on coal, and indeed much of the growth impetus for the early days of the industry was provided by the profits from coal tar dyes. Then as lower cost petroleum hydrocarbons became available in huge volumes, the term "petrochemical" became almost synonymous with "organic chemical." Now, with a new look at hydrocarbon sources spurred by energy shortages, a more balanced choice of coal and petroleum sources for organic chemicals is in evidence. In any event, this multibillion pound, multibillion dollar industry has its own pollution problems which must be addressed in consort with problems of raw material choices and process economics.

It is the purpose of this book to outline some of those pollution problems as well as some of the cures for those problems.

BACKGROUND AND ECONOMIC TRENDS

Synthetic organic chemicals are derivative products of naturally-occurring raw materials (petroleum, natural gas, and coal) which have undergone at least one chemical conversion. The organic chemicals industry was initially dependent upon coal as its sole source of raw materials. However, during the last two decades it has moved rapidly from coal to petroleum-based feedstocks. Although the cost of coal is less than half that of most liquid fuels, the handling and processing of liquids is much cheaper. In addition, the extraction of coal is much more labor-intensive than is the extraction of liquid fuels.

In recognition of the change in origin of raw materials, the term "petrochemical" has come into common usage. Although a precise definition of petrochemicals has yet to gain universal acceptance, it commonly refers to all organic chemical products derived from petroleum fractions and by-products or from natural gas constituents. From its modest beginnings in the 1920's with the manufacture of isopropanol from refinery off-gas propylene, petrochemistry has by now not only made possible the almost total elimination of coal and coal tar as sources of chemical raw materials, but has also gone a long way towards replacing such methods of obtaining organic chemicals as fermentation, extraction of compounds from materials occuring in nature, and chemical transformation of vegetable fats and oils.

Now in 1974, however, with the world energy crisis putting an entirely new perspective on hydrocarbon economics, coal chemicals may well be on the increase again. After having become accustomed to relatively cheap energy and plant feedstocks, chemical makers must now pay more for these materials as other demands crowd in on their traditional sources.

The alternate use for natural gas is as fuel. In the past, this alternate value as fuel set a base price of about 0.4¢/lb on chemical feedstocks such as ethane and propane. With chemical producers willing to pay 0.7¢/lb for these feedstocks, the natural gas industry found it advantageous to sell them for chemical usage.

However, recent drastic increases in demand for natural gas as a pollution-free fuel, coupled with a leveling off of gross gas production, have more than tripled the base fuel value for ethane and propane as chemical feedstocks. This has led most chemical producers to plan future production of chemicals such as ethylene on processes that use heavier feedstocks such as liquid crude oil distillates.

Light liquid distillates, however, have an alternate use and value as gasoline. A typical barrel of crude oil usually contains only about 20% light distillates in a boiling range suitable for use as gasoline. All of this must be processed at some expense, and in order to satisfy the automobile-oriented society in the United States, another 25% of the higher boiling crude oil distillates must be converted into the gasoline boiling range by cracking and other refinery processing.

With crude oil valued at 7.5¢/gal ($3.15/bbl), the final gasoline product, representing 45% of the barrel, must be valued at close to 12¢/gal. The light distillate fractions suitable either for manufacturing of chemicals or for processing into gasoline carry an intermediate value of 8.5¢/gal, or 1.4¢/lb.

This increasing cost for feedstock amounts to about 40% of the total 3¢/lb ethylene price prevailing for Gulf Coast markets during the past several years, as pointed out by the Environmental Protection Agency in *Report EPA 440/1-73/009* entitled, "Development Document for Proposed Effluent Limitations Guidelines and New Source Performance Standards for the Major Organic Products Segment of the Organic Chemicals Manufacturing Point Source Category," Washington, D.C., EPA (Dec. 1973).

Increasing gasoline demands and a limited supply of available crude oil will only cause the shortage of chemical feedstocks to become more severe. In economic terms, the expected increase in energy costs between now and 1980 will bring some dramatic changes in the cost of key petrochemicals. The brief tabulation on the following page illustrates the theoretical effect of an increase of 50¢/bbl in the cost of crude oil.

Cost Increase Associated with 50¢/Bbl

Petrochemical	Current Cost	Increase in Crude Oil
Ethylene	3.0¢/lb	+0.3¢/lb
Butadiene	5.0¢/lb	+0.8¢/lb
Benzene	23.0¢/lb	+1.7¢/lb
p-Xylene	5.5¢/lb	+0.7¢/lb

The above price differentials, predicated on a 50¢/bbl increase in crude prices now seem like a ludicrously conservative projection at a time when in late 1973 crude oil prices increased by magnitudes ten times greater—of the order of $5.00 per barrel.

Costs must be developed for the pollution control systems which may be utilized to comply with the effluent limitations. The pollution control costs may be put on the same cents-per-pound basis and added to the required realization to provide a meaningful assessment of the economic impact on specified products. Performing this calculation for several of the products within a subcategory or between subcategories will provide a basis for general conclusions relating to the industry as a whole.

AIR POLLUTION

While air pollution in the sense of emission of hydrocarbons, H_2S, HCl, HF, and Cl_2 constitute something of a problem for manufacturers of organic chemicals, they do not seem to constitute a problem of the magnitude of the industry's water pollution problem.

The organic chemical industry must, however, be concerned with two aspects of the air pollution problem: 1. The effect of discharges on the public ranging from really toxic fumes to soot emissions on washdays. 2. The effect of discharges on the utility of air as a chemical raw material. This is essentially the problem of industry fouling its own nest.

As pointed out by M. Sittig in *Combining Oxygen and Hydrocarbons for Profit,* Houston, Texas, Gulf Publishing Co. (1962), the requirements of many organic chemicals manufacturing processes for reaction air are considerable indeed. Some examples are as follows:

Process	Tons Air Required/Ton Product
Synthesis Gas	4.0
Hydrogen Cyanide	10.0
Phthalic Anhydride	5.0
Carbon Black	22.0

The presence of dust, coke breeze, nitrogen oxides, sulfur oxides or carbon oxides in such air intakes may cause a variety of process problems such as system plugging and catalyst poisoning. A recent discussion of atmospheric emissions from the petroleum industry, while not wholly applicable, gives some guidance as to emissions which may be encountered in petrochemical manufacture and their control. The source is "Atmospheric Emissions from the Petroleum Refining Industry," *Report PB 225,040,* Springfield, Va., National Technical Information Service (Aug. 1973).

Now that the coking of coal may be growing in importance as a source of organic chemicals, the particular air pollution problems presented by coke ovens come into focus as a problem related to organic chemicals manufacture. Some control techniques associated with coke oven operations will be detailed later in this volume.

WATER POLLUTION

The 1972 amendments to the Federal Water Pollution Control Act impose much more severe and specific strictures on the organic chemicals manufacturing industry than did the Clean Air Act of 1970. Thus, a substantial part of this volume will be devoted to aqueous wastes from this industry and possible techniques for their treatment.

The sources of wastes emanating from refinery and petrochemical operations can be divided into five general categories as noted by W.L. Jewell and D.L. Ford of Engineering-Science, Inc. in *Report 12020 EID 03/71,* Washington, D.C., EPA (Mar. 1971):

1. Wastes containing a principal raw material or product resulting from the stripping of the product from solution.
2. By-products produced during reactions.
3. Spills, slab washdowns, vessel cleanouts, sample point overflows, etc.
4. Cooling tower and boiler blowdown, steam condensate, water-treatment wastes, and general washing water.
5. Storm waters, the degree of contamination depending on the nature of the drainage area.

The principal contaminants in the wastewaters include organics from residual products and by-products, oils, suspended solids, acidity, heavy metals and other toxic materials, color, and taste and odor-producing compounds. The concentration of BOD_5 and COD, respectively, in untreated refinery effluents has averaged 108 and 204 lb/1,000 bbl of crude oil refined. The pH of refinery wastewaters is normally alkaline, but may vary considerably depending on disposal of spent acids, acid washes, etc.

The large variety of compounds produced within the petrochemical and refinery industries makes the task of treating wastewaters difficult and complex. Wastewaters from plants manufacturing similar or even the same compounds usually display dissimilar characteristics. This can be ascribed to the use of different manufacturing processes coupled with the fact that the by-product disposal pattern may occur in a number of different ways. Thus, a wastewater treatability study should be undertaken when the treatment of petrochemical and refining wastes is considered.

A detailed treatability evaluation of each waste stream is a prerequisite to determining the proper integration of unit processes which constitute an optimum waste treatment system. The waste treatment methods applicable to the refining and petrochemical industries can be categorized as follows: physical, chemical, biological, special in-plant methods, and ultimate disposal.

Physical methods include gravity separation, air flotation, filtration, centrifugation, vacuum filtration, evaporation, and carbon adsorption. Gravity separators and air flotation units, which are used extensively throughout the industry, are designed primarily for removal of free-floating oil and settleable solids. Filtration is used primarily as a pretreatment for deep well injection; and centrifugation and vacuum filtration are used for sludge dewatering. Evaporation ponds are often efficient, but are limited to areas where land is available and climatic conditions are favorable. Carbon adsorption is used to remove refractory organic substances.

Chemical treatment methods include coagulation-precipitation, chemical oxidation, ion exchange, and chemical pretreatment or sludge conditioning. These methods enhance oil and solids removal, particularly with respect to oil emulsions.

The biological treatment methods include activated sludge and its modifications, trickling filters, aerated lagoons, and waste stabilization ponds. Usually some form of wastewater pretreatment is required to remove oils, suspended solids, and toxic substances, and to provide neutralization, equalization, and surge or holding capacity.

The activated sludge process is generally considered the most effective biological process for removing organic materials with removal efficiencies in the range of 70 to 95% for BOD_5, 30 to 70% for COD, and 65 to 99% for phenols and cyanides. The conventional activated sludge process is the most widely applied. Contact stabilization is most applicable when a large fraction of the organic constituents is in suspended or colloidal form. Extended aeration is particularly adaptable to industrial applications as the longer detention periods allow the microorganisms more time to degrade the complex organics and, more importantly, sludge production is minimized.

Deep well injection of petrochemical and refinery wastewaters is used primarily as an ultimate disposal method for low-flow, highly concentrated wastes. The efficacy of disposing of these streams by deep well injection depends on the extent of pre-injection treatment requirements, the receiving formation, and the risks involved in contaminating overlying or underlying aquifers.

The more important in-plant treatment control methods include stripping and recovery operations, neutralization and oxidation of spent caustics, ballast water treatment, slop oil recovery, and temperature control. The practice of such in-plant treatment methods not only reduces the waste loadings to the treatment facility, but also enhances its performance. In some cases, in-plant control will show a cost credit in the form of product recovery.

An integral part of in-plant treatment procedures is adequate in-plant waste control practices. In-plant control techniques include salvage of unreacted chemicals, recovery of by-products, multiple reuse of water, good housekeeping techniques to reduce leaks and spills, and curbing and diking of drainage areas.

These controls can reduce both the volume and concentration of pollutants requiring treatment.

New production methods are being directed toward increases in product yield, often resulting in reduced amounts of by-products and unused raw materials in waste streams. A more general indication of pollution reduction by in-plant processing practices is the much lower pollutant loadings per unit of throughput for "newer" refineries as compared to "older" refineries.

To date, biological treatment methods have afforded the most economical secondary treatment processes for pollution abatement. In order to minimize biological treatment costs, several pretreatment processes have been adopted, depending on the characteristics of the waste streams being treated. Design of biological treatment units must consider the possibility of spills, storm runoff, ballast water handling, variation in flow and contaminants, and toxic or inhibitory substances.

Completely mixed activated sludge units, therefore, are applicable to dampen these fluctuations and inhibitory effects. Aerated lagoons are often used where the waste has a large volume but relatively low concentration of dissolved organics. Anaerobic lagoons, anaerobic contact units and high rate trickling filters are used individually or conjunctively when high strength wastes are involved.

Facilities for handling the treatment of the accumulated primary and biologically synthesized sludges often represent a major portion of the treatment facility costs. One inexpensive method of disposing of sludges practiced in many parts of the country is lagooning. However, dewatering the sludge to reduce the volume or direct disposal at sea or in remote areas may be more favorable alternatives where land is unavailable. Aerobic digestion is presently favored over anaerobic treatment in industrial waste applications and is generally followed by thickening, vacuum filtration, filter pressing, or centrifugation and ultimate disposal by incineration or trucking.

The problem areas encountered in the treatment of wastes discharged from the refining and petrochemical industries are generally those associated with pretreatment and biological treatment systems. Problems may be attributed to process changes or modifications, accidental spills resulting in the discharge of slugs of contaminants, or poor in-plant management.

Problems related to pretreatment processes may be caused by excess concentrations of free and emulsified oils, high or widely fluctuating temperatures, acidity or alkalinity, or other contaminants which adversely affect process operation. Waste oil usually occurs from leaks, spills, washing operations, and rainfall runoff from oily areas. When oily wastes and water commingle, there is always a good chance for emulsification, which sometimes is difficult and costly to break.

Wastes affecting pH in the refinery and petrochemical industries include strong acids and alkalies as well as dissolved solids which often buffer their effects. Spills, leaks, scrubbing operations, and point discharges of spent acid are the main sources of acid wastes while the main contributor of acidity is sulfuric acid. However, HCl, H_2S, HF, and CO_2 contribute to a lesser extent to the acidity of the wastewater. Alkalies from process neutralization, process operations, and kettle washes are the main sources of alkalinity. The oxygen demand of refinery and chemical wastewaters is probably the most important factor to be considered. The most concentrated oxygen-demanding wastes discharged from a refinery usually occur in the crude and cracking units. Although much of this oxygen demand in terms of BOD, TOC, COD, etc., is attributable to the oily fraction, a significant portion results from the presence of lighter hydrocarbon contaminants. Moreover, substances such as sulfides and various nitrogen compounds are responsible for an additional oxygen demand.

Toxicity to living organisms is a critical factor in evaluating the treatability of chemical and refinery wastes. Ammonium salts, sulfides and polysulfides resulting as wastes from the use of ammonia as an anticorrosive agent, as well as other boiler and cooling tower inhibitors are toxic. Heavy metals used as catalysts or preparants in process operations are other sources of pollutants which may be toxic to living organisms.

The taste and odor-producing wastes resulting from washing high sulfur crude oil, or cracked distillate from high sulfur stocks, may have highly noticeable and persistent odors. The odor is attributable to complex organic sulfur compounds and alkylated substances.

Acid sludges are the next most important group responsible for taste and odor problems. Sludges from the acid treatment of light distillates have a much more pronounced odor than those from the treatment of lubricating stocks. Much of the odor from light oil sludges is due to mercaptans, which are disagreeable and can be detected at concentrations of a fraction of a part per trillion.

Large concentrations of sulfides in biological treatment units impair the treatment efficiency of these units because of the oxygen required to satisfy the chemical oxygen demand or the possible inhibitory or toxicological effects on the system bacteria. Preaeration of sulfide wastes results in stripping the sulfide from solution when the wastewater is acidic, or oxidizing the sulfides to thiosulfate or sulfate at high pH values.

Because of the diversity of products, it is difficult to generalize regarding the reuse of petrochemical wastewaters. The American Petroleum Institute's *Manual on Disposal of Refinery Wastes* (1969) summarizes the treatment processes presently used to treat 80 different organic chemicals, 20 inorganic chemicals, and 11 overall plant wastes. All of these chemical waste streams present serious problems whenever the COD exceeds 1,000 mg/l, or when phenols or heavy metals exceed toxic levels. The quality of wastewater from the petrochemical manufacturing processes depends on the type of petroleum feedstock, products manufactured, in-plant control, maintenance, and types of processing equipment.

A major source of contaminated water in petrochemical industries is the cooling water and steam equipment. The total volume of this water will amount to 10 to 80% of the total wastewater and its degree of contamination depends on the following factors:

1. Process leaks
2. Water treatment additives
3. Input from air scrubbing
4. Blowdown and condensates

These waste streams have a high potential for reuse, but their degree of contamination must be considered.

There are now three treatment levels designed to ultimately achieve the U.S. national goal of zero discharge of aqueous pollutants. Listed below are these levels of treatment, the first two to be applied to existing plants and the third to new plants.

Best Practicable Control Technology Currently Available (BPCTCA) (by 7-1-77)
Best Available Technology Economically Achievable (BATEA) (by 7-1-83)
Best Available Demonstrated Control Technology (BADCT) (New Sources)

The EPA has formulated these as model systems which are considered capable of attaining the reduction factors associated with each technology. It should be noted and understood that these particular systems chosen for use in cost estimation and economic analysis are not the only systems which are capable of attaining the specified pollutant reductions. There exist many alternate systems which either taken singly or in combination are capable of attaining the recommended effluent limitations and standards. These alternate choices include:

1. Different types of end-of-pipe wastewater treatment.
2. Different in-process modifications and pollution control equipment.
3. Different integrated combinations of end-of-pipe and in-process technologies.

Some of the specifics of these various systems will be discussed later in this volume.

It is the intent of the EPA to allow the individual manufacturers within the organic chemicals industry to make the ultimate choice of what specific combination of pollution control measures is best suited to his situations in complying with the published limitations and standards presented.

SOLID WASTES

A historical survey of the trends in solids disposal at refineries and petrochemical plants indicated that this is a significant problem as reported by W. Jewell and D.L. Ford of Engineering-Science, Inc. in *Report 12020 EID 03/71,* Wash., D.C., EPA (March 1971). Landfill was found to be the most common method of disposal and was used primarily for separator and tank bottom sludges, sewer cleanings, water treating sludges, cooling tower bottoms, and biological treatment plant sludges. Incineration and open burning has been used for general refuse. Contract disposal services have been used for filter cakes, treating clays, and slop oil disposal.

Based on the results of this survey, it was decided that further information on the following areas would improve the disposal techniques available.

1. Determination of hydrocarbon content and calorific values of solid wastes.
2. Evaluation of incinerators to establish capabilities for handling different wastes.
3. Investigation of land spreading techniques to establish land and manpower requirements.
4. Identification of wastes that are amenable to de-oiling treatment, i.e., separation of solids and hydrocarbons.
5. Investigation of development of spent caustic disposal systems.

The topic of "Solid Wastes in the Petrochemical Industry" has also been addressed in a technical report to the U.S. Environmental Protection Agency by R.G. Makela and J.F. Malina, Jr. of the Center for Research in Water Resources, Austin, Univ. of Texas (Aug. 1972).

As discussed in that report, waste solids in the petrochemical industry occur in diverse forms, many of which would not be brought to mind by the term "solid wastes." These waste solids or solid wastes may occur as actual solids, such as scraps of wood, paper, or metal; as semi-solids, such as tars and resins; as suspended solids, such as monomers and polymers in a process waste stream; or as dissolved solids, such as inorganic or organic solids. These materials may be characterized by such terms as combustible or noncombustible, organic or inorganic, inert or biodegradable, dry or mixed with either aqueous or nonaqueous liquids, or as a combination of several of these terms.

Almost every plant in the petrochemical industry has some form of solid waste handling or disposal facility on the premises. Every plant with a liquid waste stream probably has a solids removal unit and some type of solids concentration unit. Most plants will have one or more disposal facilities or areas on the site. Many plants, however, find it more economical or expedient to contract for someone else to handle some or all of their solid wastes.

The solid wastes found in the petrochemical industry may be stored, handled, and disposed of by various combinations of many different methods. Materials may be stored in large open or closed bulk containers, in smaller bulk containers such as drums or barrels, or in open piles or pits. Plant personnel or private contractors collect and transport the wastes in flat bed trucks, stake-sided trucks, dump trucks, open or closed top vans, compacting trucks, tank trucks, or by pipeline to the disposal facility.

Unit processes for solids removal from liquid waste streams include sedimentation, flotation, coagulation, flocculation, precipitation, and filtration. Unit processes for concentration of solids include gravity and flotation thickening, centrifugation, vacuum and precoat filtration, filter press and heat treatment.

On-site disposal methods include incineration and land disposal. Incineration facilities include multiple hearth furnaces, rotary kilns, fluidized bed reactors, controlled air open-pit incinerators, and liquid burners. Land disposal methods include lagooning, spreading on land or "land-farming," and landfill.

Off-site disposal methods include land disposal by lagooning, land-farming, or landfill, incineration by various methods, and ocean disposal by pipeline to inshore waters, bulk dumping of barge transported wastes in offshore waters, or deep sea dumping of materials in sealed containers. The exact combination of methods chosen is dependent upon the particular situation. Factors of consideration include: (1) characteristics of the wastes, such as weight, volume, density, ease of handling, rate of production, combustibility, biodegradability, toxicity, etc.; (2) potential values of salvaged materials for recycle into the same process or into new or different processes at the plant or at other facilities; (3) adaptability of the disposal method to the waste in question; (4) the availability of land for land disposal, and plans for the land's future use.

Several types of solid wastes and disposal methods encountered in the industrial solid waste survey are presented in Table 1. Detailed discussion of this information is presented in the report cited above.

TABLE 1: PETROCHEMICAL SOLID WASTE SURVEY

Waste Type	Disposal Methods Used
Water treatment sludge	Land disposal (a) Lagoon (b) Spread on land (c) Sanitary landfill (d) Dump
Cafeteria and lunchroom	Land disposal (a) Sanitary landfill (b) Dump Incineration (a) Stationary hearth furnace (b) Multiple hearth furnace (c) Rotary kiln
Plant trash	Land disposal (a) Sanitary landfill (b) Dump Incineration (a) Stationary hearth (b) Multiple hearth (c) Rotary kiln (d) Open pit
Ashes, flyash, and incinerator residue	Land disposal (a) Sanitary landfill (b) Dump (c) Spread on land Salvage and recycle
Plastic	Salvage and recycle Land disposal (a) Sanitary landfill (b) Dump Incineration (a) Open pit (b) Rotary kiln (c) Multiple hearth (d) Stationary hearth
Ferrous and nonferrous metals	Salvage and recycle Land disposal (a) Dump

(continued)

TABLE 1: (continued)

Waste Type	Disposal Methods Used
Ferrous and nonferrous metals (cont.)	(b) Sanitary landfill
Catalysts	Salvage and recycle Land Disposal (a) Spread on land (b) Lagoon (c) Dump (d) Sanitary landfill
Organic chemicals	Chemical Treatment Biological Treatment Land Disposal (a) Lagoon (b) Spread on land Incineration (a) Liquid burner (b) Rotary kiln (c) Fluidized bed reactor Ocean disposal (a) Bulk dumping (b) Sealed container dumping
Inorganic chemicals	Chemical Treatment Land disposal (a) Lagoon (b) Spread on land (c) Dump Ocean disposal (a) Bulk dumping (b) Sealed container dumping Salvage and recycle Incineration (a) Liquid burner (b) Rotary kiln (c) Fluidized bed reactor
Wastewater treatment sludges, filter cakes, and viscous solids	Land disposal (a) Lagoon (b) Spread on land (c) Dump (d) Sanitary landfill Incineration (a) Multiple hearth (b) Rotary kiln (c) Fluidized bed

Source: R.C. Makela and J.F. Malina, Jr., University of Texas, Aug. 1972

WATER USAGE IN ORGANIC CHEMICALS MANUFACTURE

Process water is defined to be all water which comes in contact with chemicals within the process and includes the following.

- Water required or produced (in stoichiometric quantities) in the chemical reaction;
- Water used as a solvent or as an aqueous medium for the reactions;
- Water which enters the process with any of the reactants or which is used as a diluent (including steam);
- Water associated with the catalyst system, either during the reaction or during catalyst regeneration;
- Water used as an absorbent or as a scrubbing medium for separating certain chemicals from the reaction mixture;
- Water introduced as steam to strip certain chemicals from the reaction mixture;
- Water used to wash, remove, or separate chemicals from the reaction mixture;
- Water associated with mechanical devices such as steam-jet ejectors for drawing a vacuum on the process;
- Water used as a quench or direct contact coolant such as in a barometric condenser;
- Water used to clean or purge equipment used in batch type operations; and
- Runoff or washwater associated with battery limits process areas.

The type and quantity of process water usage are related to the specific unit operations and chemical conversions within a process. The term "unit operations" is defined to mean specific physical separations such as distillation, solvent extraction, crystallization, adsorption, etc. The term "chemical conversion" is defined to mean specific reactions such as oxidation, halogenation, neutralization, etc. Water leaves the process through another group of unit operations associated with the physical separation of water from hydrocarbons. Some of these are: liquid-liquid separation equipment, such as decant drums; vapor-liquid separation equipment, such as distillation columns or flash chambers; and solid-liquid separation equipment, such as distillation columns or flash chambers; and solid-liquid separation equipment, such as crystallizers and filters.

At first glance, an organic chemicals plant often appears to be a chaotic maze of equipment, piping, and buildings that is totally unlike any other facility, even those which manufacture the same product.

Nevertheless, there are certain basic components common to almost all chemical plants: a process area; storage and handling facilities for raw materials, intermediates, and finished products; electrical, steam, air, and water systems with associated sewers and effluent treatment facilities; and, in most cases, a laboratory, an office, control rooms, and service roads.

The process area is normally referred to as the "battery limit," while the remainder of the plant is called the "off-sites." The off-sites can be broken down into their components: the storage and handling facilities, the utilities, and the services. This four-area concept in plant layout is illustrated by the plot plan shown in Figure 1.

The storage facilities associated with any chemical plant obviously depend upon the physical state (i.e., solid, liquid, or gas) of the feedstocks and products. Storage equipment frequently utilized includes: cone roof tanks, with or without floating roofs, for storage of liquid hydrocarbons; cylindrical or spherical gas-holding tanks; and concrete pads or silos for storage of solids.

Wastewater emanating from this part of the plant normally results from storm runoff, tank washing, accidental spills, and aqueous bottoms periodically drawn from storage tanks. Although the generation rate is sporadic and the volume small, these wastewaters have in most cases contacted the chemicals which are present in this area.

For this reason, they are normally sent to a process sewer and given the same effluent treatment as contact process wastewaters. Utility functions such as the supply of steam and cooling water generally are set up to service several processes. Boiler feedwater is prepared and steam is generated in a single boiler house. Noncontact steam used for surface heating is circulated through a closed loop whereby varying quantities are made available for the specific requirements of the different processes.

The condensate is nearly always recycled to the boiler house, where a certain portion is discharged as blowdown. The three major uses of steam generated within a chemical plant are shown below.

(1) For noncontact process heating. In this application, the steam is normally generated at pressures of 125 to 650 psig.

(2) For power generation such as in steam driven turbines, compressors, and pumps associated with the process. In this application, the steam is normally generated at pressures

FIGURE 1: PLOT PLAN FOR CHEMICAL PLANT ILLUSTRATING FOUR-AREA LAYOUT

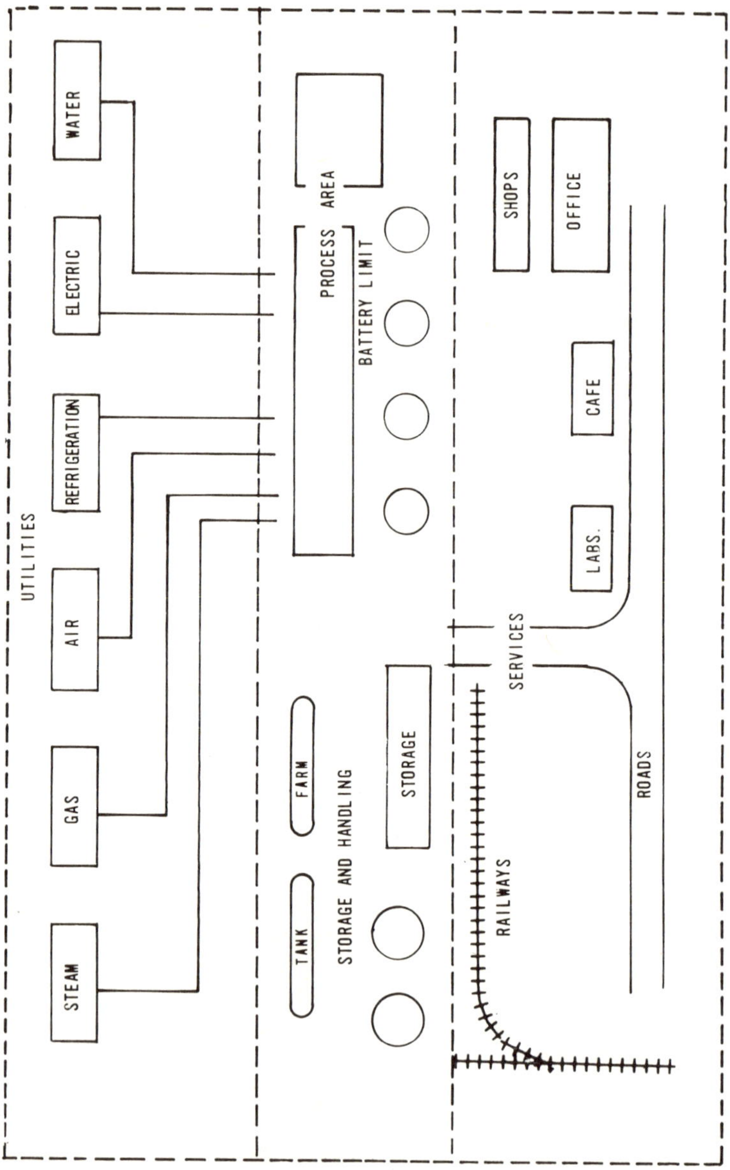

Source: Report EPA 440/1-73/009

of 650 to 1,500 psig and requires superheating.

(3) For use as a diluent, stripping medium, or source of vacuum through the use of steam jet ejectors. This steam actually contacts the hydrocarbons in the manufacturing processes and is a source of contact process wastewater when condensed. It is used at a substantially lower pressure than the foregoing and frequently is exhaust steam from one of the other uses.

Steam is supplied to the different users throughout the plant either by natural circulation, vapor phase systems, or by forced circulation liquid heat transfer systems. Both types of system discharge some condensate as blowdown and require the addition of boiler makeup water. The main areas of consideration in boiler operation are normally boiler efficiency, internal deposits, corrosion, and the required steam quality.

Boiler efficiency is dependent on many factors. One is the elimination of boiler tube deposition that impedes heat transfer. The main contributors to boiler deposits are calcium, magnesium, silicon, iron, copper, and aluminum.

Any of these can occur in natural waters, and some can result from condensate return line corrosion or even from makeup water pretreatment. Modern industrial boilers are designed with efficiencies on the order of 80%. A deposit of ⅛ inch in depth will cause a 2 to 3% drop in this efficiency, depending on the type of deposit.

Internal boiler water treatment methods have advanced to such a stage that corrosion in the steam generation equipment can be virtually eliminated. The control of caustic embrittlement in boiler tubes and drums is accomplished through the addition of sodium nitrate in the correct ratio to boiler water alkalinity.

Caustic corrosion in high heat transfer boilers can also be controlled by the addition of chelating agents. This type of solubilizing internal boiler water treatment has been shown to be more effective than previous precipitation treatment using phosphate.

Other factors influencing boiler efficiency include reduction of the amount of boiler blowdown by increased cycles of concentration of the boiler feedwater, efficiency of the blowdown heat recovery equipment, and the type of feed used.

Flash tanks are used in many plants to recover, as low pressure steam, as much as 50% of the heat lost from continuous boiler blowdown. The steam is then used for the boiler feedwater deaerator or other low pressure applications. Additional heat is recovered in some plants by installing heat exchangers following the blowdown flash tank. The blowdown is used to preheat the makeup boiler feedwater in these exchangers.

Steam purity is of prime importance if: the boilers are equipped with superheaters; the boilers supply power generation equipment; and the steam is used directly in a process where contamination could affect product quality or destroy some material, such as a catalyst, essential to the manufacture of the product. The minimum purity required for contact steam (or contact process water) varies from process to process. Limits for suspended solids, total solids, and alkalinity vary inversely with the steam pressure. The following tabulation summarizes boiler water concentration limits for a system providing a steam purity of 0.5 to 1.0 parts per million total solids, which is required for most noncontact steam used. It should be noted that the boiler operation must incorporate the use of anti-foam agents and steam separation equipment for the concentrations shown to be valid.

	Boiler Pressure (psig)			
Parameters	**0 to 300**	**301 to 450**	**451 to 600**	**601 to 750**
Total solids (mg/l)	6,000	5,000	4,000	2,500
Suspended solids (mg/l)	1,000	200	100	50
Total alkalinity (mg/l)	1,000	900	800	750

The concentrations of these contaminants found in actual boiler blowdowns were generally within the ranges shown above.

Water conditioning or pretreatment systems are normally part of the utilities section of most plants, and the required treatment may be quite extensive. Ion exchange demineralization systems are very widely employed, not only for conditioning water for high pressure boilers, but also for conditioning various process waters.

Clarification is also widely practiced and usually precedes the ion exchange operation. Noncontact cooling water also is normally supplied to several processes from the utilities area. The system is either a loop which utilizes one or more evaporative cooling towers, or a once through system with direct discharge. Cooling towers accomplish the cooling of water circulated over the tower by moving a predetermined flow of ambient air through the tower with large fans. The air-water contact causes a small amount of the water to be evaporated by the air. Thus, through latent heat transfer, the remainder of the circulated water is cooled.

Approximately 1,000 Btu are removed from the total water circulation by the evaporation of 1 lb of water. Therefore, if 100 lb of water are introduced at the tower inlet and 1 lb is evaporated to the moving air, the remaining 99 lb of water are reduced in total heat content by 1,000 Btu, or about 10 Btu/lb. The 99 lb of water leaving the tower have been cooled to 1°F/lb/Btu removed, and the exit temperature is reduced by about 10°F which leads to the common rule of thumb: 1% evaporation loss for each 10°F.

Since cooling is primarily by transfer of latent heat, cooling tower selection is based on the total heat content or enthalpy of the entering air. At any one enthalpy condition, the wet bulb temperature is constant. Therefore, cooling towers are selected and guaranteed to cool a specific volume of water from a hot water temperature to a cold water temperature while operating at a design wet bulb temperature. Design wet bulb temperatures vary from 60° to 85°F depending on the geographic area, and are usually equalled or exceeded only 2.5 to 5% of the total summer operating time. Hot water temperature minus cold water temperature is termed cooling range, and the difference between cold water and wet bulb temperature is called approach. A closed system is normally used when converting from once-through river cooling of plant processes. In the closed system, a cooling tower is used for cooling all of the hot water from the processes. Figure 2 illustrates this method. With the closed system, makeup water from the river is required to replace evaporation loss at the tower.

Two other water losses also occur. The first is drift, which is droplet carry-over in the air as contrasted to evaporative loss. The cooling tower industry has a standardized guarantee that drift loss will not exceed 0.2% of the water circulated. The second loss in the closed system is blowdown to sewer or river.

Although blowdown is usually taken off the hot water line, it may be removed from the cold water stream in order to comply with regulations that limit the temperature of water returned to the stream. Blowdown from a tower system will vary depending on the solids concentration in the makeup water, and on the occurrence of solids that may be harmful to equipment. Generally, blowdown will be about 0.3% per 10°F of cooling, in order to maintain a solids concentration in the recirculated water of three to four times that of the makeup water.

The quantity and quality of the blowdown from boilers and cooling towers depend on the design of the particular plant utility system. The heat content of these streams is purely a function of the heat recovery equipment associated with the utility system.

The amounts of waste brine and sludge produced by ion exchange and water treatment systems depend on both the plant water use function and the intake source. Usually none of these utility waste streams can be related directly to specific process units. Quantitative limitations on parameters such as dissolved solids, hardness, alkalinity, and temperature, therefore, cannot be allocated on a production basis.

The service area of the plant contains the buildings, shops, and laboratories in which most of the plant personnel work. The sanitary wastes from this area obviously depend on the number of persons employed.

It should be noted that most chemical plants run continuously and have three operating shifts per day. There are also wastes associated with the operation of the laboratory, machine shops, laundry, etc. The wastes from the service area normally are combined with the wastes from the process area prior to treatment.

As was mentioned previously, there are a large number of process combinations possible within the battery limits of the typical multiprocess plant. Choosing one of the many commercially viable processes for the manufacture of a specific chemical at a particular location or time is a decision based on a particular manufacturer's unique situation.

Each process is itself a series of unit operations which causes chemical and physical changes in the feedstock or products. In the commercial synthesis of a single product from a single feedstock, there generally are sections of the process associated with: the preparation of the feedstock, the chemical reaction, the

FIGURE 2: CLOSED SYSTEM

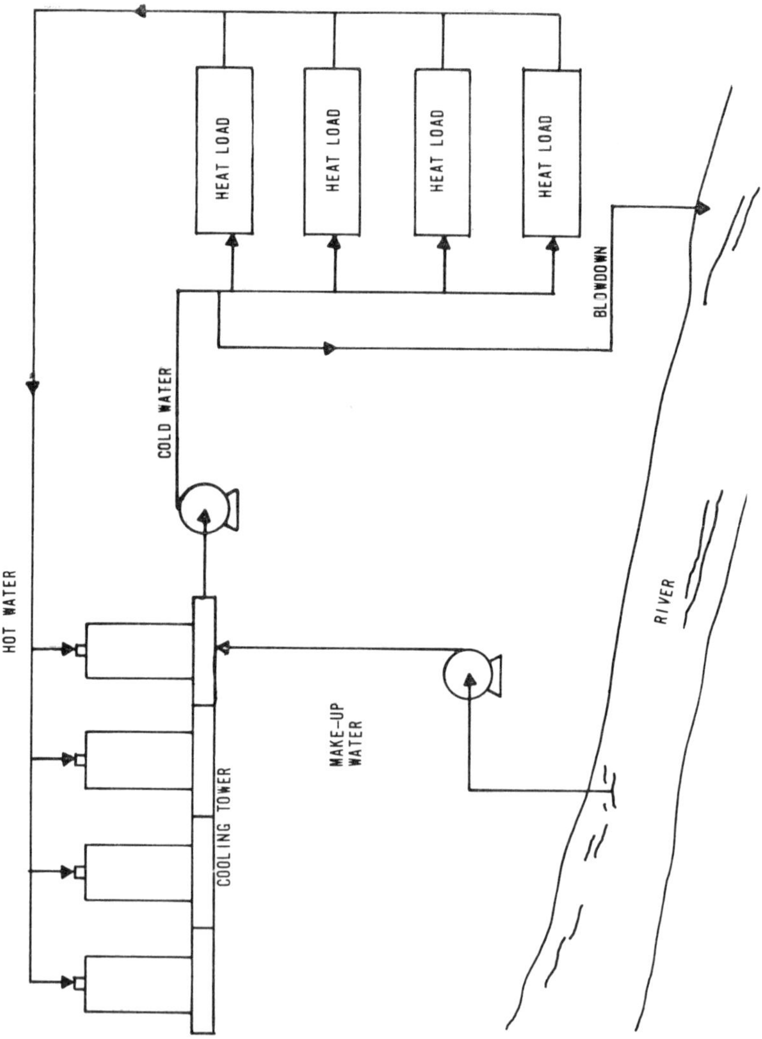

Source: Report EPA 440/1-73/009

separation of reaction products, and the final purification of the desired product. Each unit operation may have drastically different water usages associated with it. The type and quantity of contact wastewater are therefore directly related to the nature of the various processes. This in turn implies that the types and quantities of wastewater generated by each plant's total production mix are unique.

WASTEWATER SAMPLING AND SURVEYING

The general topic of pollutant sampling and analysis, is covered in a companion volume by the same author *Pollution Detection and Monitoring Handbook*, Park Ridge, N.J., Noyes Data Corp. (1974).

A wastewater survey must contain detailed information on waste flows, characteristics of the flows, and relevant environmental conditions in order that sound engineering decisions regarding the wastewater treatment can be made as noted by W.L. Jewell and D.L. Ford of Engineering-Science, Inc. in *Report 12020 EID* 03/71, Wash., D.C., E.P.A. (March 1971). Design data for wastewater treatment facilities should include but not be limited to the following:

(1) the source of all significant waste streams;
(2) samples at points where significant changes in waste characteristics or quantity of flow occur;
(3) samples which reflect the benefits from segregated sewer systems;
(4) consideration of water reuse and water saving process modifications;
(5) consideration of overloading of waste treatment systems resulting from unit shutdown, startups, turn-arounds, accidental dumping or spills, and other shock loadings;
(6) investigation of reducing waste streams by eliminating or decreasing flows by process modifications; and
(7) evaluation of wastewater sources related to variable occurrences such as storm flow drainage.

A complete audit of all water and contaminants entering and leaving a process or drainage area would yield sufficient information to satisfy the objectives of a wastewater survey. Practically, however, this mass balance is seldom possible to achieve due to unexpected and unknown variations in raw materials, chemical inputs, process operations, etc. Constraints placed on the industry such as required quality standards, limited financial resources, and pollution abatement policies often dictate the scope of a wastewater survey.

SAMPLING STATION SELECTION

Petrochemical and refinery liquid wastes may be classified as clean or highly contaminated process streams, clean and contaminated storm systems, and sanitary waste streams. The selection of wastewater sampling sites should be based on obtaining characterization data for each of the applicable categories, since segregation of wastewater conveyance systems gives more flexibility in solving pollution control problems.

"Clean" process streams normally are those discharged from cooling and boiler utility areas and from general washing operations. Under most operating conditions, 85 to 95% of the water used in petroleum refineries is for cooling purposes (API, 1968).

In "once-through systems" the major change in water quality will be only an increase in temperature. Recycled cooling water may have high salt concentrations due to evaporation loses as well as high concentrations of chromate or other toxic materials added for prevention of corrosion and/or growth of microorganisms. Other streams with relatively high concentrations of dissolved solids, i.e., boiler blowdown, neutralized demineralized resin rinses, etc., can be included. However, streams that may become contaminated with oil should be excluded. Thus, although organic contamination of these waters is usually not a problem, thermal considerations and the total dissolved solids concentration warrant careful attention.

Highly contaminated process streams include those streams containing concentrated organic materials, high concentrations of toxic substances, significant concentrations of oxygen-utilizing material (chemically and biologically oxidizable), extremely acidic and alkaline wastes, and oily streams containing high dissolved solids. These wastes usually originate from process operations, cleaning and washing, or accidental leaks or spills. Process discharges are caused by reaction products created by impurities in the feedstock, raw or partially reacted feed chemicals resulting from low process conversion efficiencies, and by-product formation in the chemical reactor.

Acidic wastes are among the most important wastes of petroleum refineries and originate principally in the treating plants, acid recovery areas, alkylation units, some water treating plants, and in some special product operations. They result from washing of intermediate petroleum products which have been acid treated, and from leakage and spills around these process areas and the leaching of acid wastes by rainwater.

Excessive alkalinity is normally not a problem although waste caustic or caustic sludges often merit special consideration. It varies in composition and appearance, depending on the crude being processed or the oils being washed. Waste causticity resulting from washing crudes and mixed paraffinic and naphthenic substances contain about 0.3% unreacted caustic soda, sodium sulfide, sulfate, sulfite, thiosulfate, phenol, mercaptans, and other organics.

Contaminated sewers usually contain "oily" wastewaters, although parallel sewers are provided occasionally to separately convey oily concentrated and nonoily concentrated waste streams. Oily sewers are routed to oil separation and recovery processes such as API gravity separators and air flotation units. This not only reduces the oil concentration of the unit process underflow flowing to subsequent treatment facilities, but also provides for oil recovery through skimming, emulsion breaking, storage, and reprocessing of recovered oil.

Sanitary waste streams emanating from administrative and laboratory areas are usually separated from industrial source discharges in order to prevent chlorinating the total combined flow. Consequently, these streams are usually separated from the industrial flow and treated separately.

CASE STUDIES OF WASTEWATER SURVEYS

The following wastewater surveys are presented as case histories to document typical wastewater surveys. Although the goals and objectives of the studies cited were different, they represent typical approaches utilized in conducting surveys to fulfill predetermined objectives.

Case History A: Petrochemical Industry — This survey was conducted in a large petrochemical complex having a total water use of 2.5 mgd. The primary products include butadiene, olefins, styrene, and nylon monomers.

At the time of the survey, the waste treatment system consisted of various in-plant recovery devices, oil separation facilities, neutralization systems, and deep injection wells with appropriate surface pretreatment units. The objectives of the survey were as follows:

(1) locate and classify major sources of pollution, organic and inorganic;
(2) categorize the major subcollection systems as to their amenability to biological treatment, and
(3) identify waste discharges with high toxicity potential on the proposed treatment system and receiving environment, and product recovery potential.

In order to fulfill these objectives, a comprehensive wastewater survey leading to the characterization and categorization of major waste streams was undertaken. Sampling points were selected in order to evaluate major sources of pollution. These sites are graphically illustrated in Figure 3 and are tabulated in Table 2.

After checking plant operational schedules and programming, it was determined that a 24 hour survey would indicate typical flows and characteristics, providing the sampling was sufficiently frequent and proper techniques of acquiring and compositing the sample were employed.

The flows of each of the major collection systems were estimated by measuring the depth and velocity of the partially filled gravity sewers, using the recorded slopes from construction drawings for verification. In some instances, the total flow from various locations within the complex was pumped and the flow charts were used as the basis of estimating flow. Grab samples were taken at each sampling station during the 24 hour period and composited according to flow at

FIGURE 3: EXAMPLE OF WASTEWATER SAMPLING STATION LOCATION IN A COMPLEX PETROCHEMICAL INDUSTRY

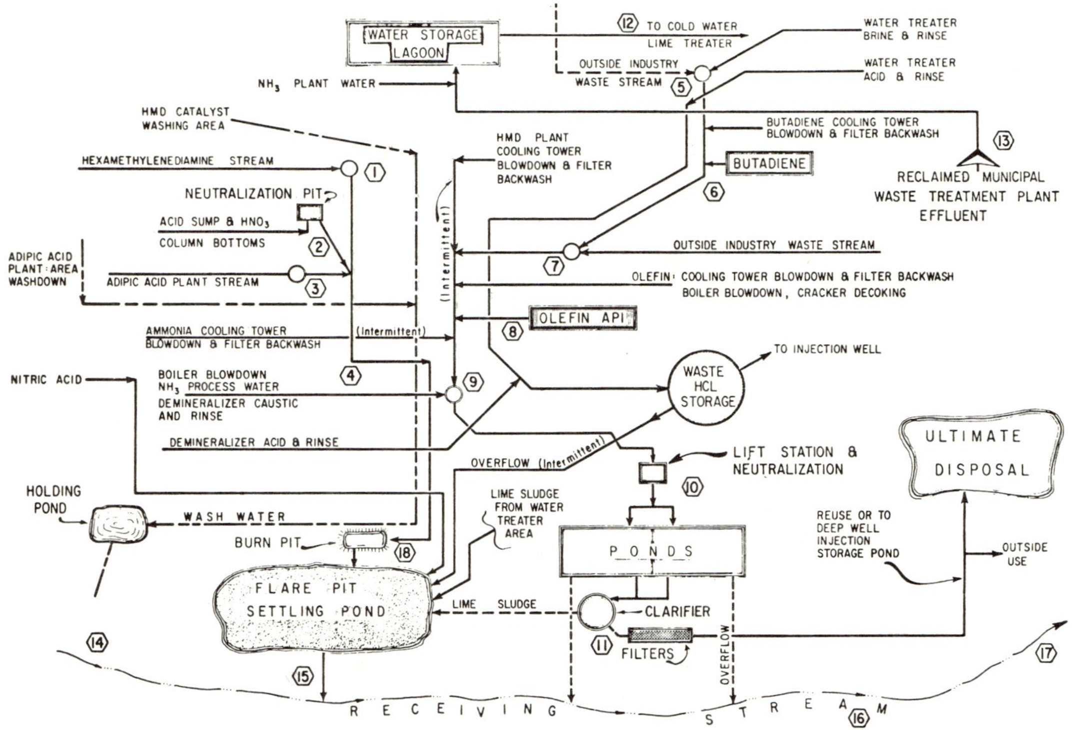

Source: EPA Report 12020 EID 03/71

TABLE 2: SAMPLING STATION DESCRIPTION

Sampling Point Designation	Description
NETWORK "A"	
1	Hexamethylene diamine area
2	Acid sump and column bottoms (neutralization)
3	Adipic acid production area
4	Total network "A" flow (API Separator effluent)
NETWORK "B" (Butadiene, Olefin Production Area, Outside Industry Stream, Blowdown Cooling Stream, Miscellaneous Streams)	
5	Outside industry waste stream
6	Butadiene production area (API Separator effluent)
7	Outside industry waste stream (manhole)
8	Olefin production area (API Separator effluent)
9	Four through Seven flow plus blowdown, ammonia process water
10	Total flow
11	Total flow
NETWORK "C" (Treated Sewage Effluent)	
12	Treated sewage effluent (sampling port)
13	Municipal treatment plant effluent
14	Catalyst washing area, surface runoff (open ditch)
15	Flare pit settling pond overflow
16	Upstream receiving water
17	Downstream receiving water
18	Nitric acid stream (sampled at Flare Pit Settling Pond)

Source: EPA Report 12020 EID 03/71

TABLE 3: ANALYSES COMPLETED IN A WASTEWATER CHARACTERIZATION STUDY OF A PETROCHEMICAL INDUSTRY

Sampling Point	Analyses Performed	Purpose
NETWORK "A", "B", and "C"	TOC COD	Degree of organic treatment required
	pH, alkalinity, acidity	Neutralization requirements
	Total Solids	
	Suspended	Pretreatment requirements
	Dissolved	Effect on treatment process; estimate of effluent quality with respect to that allowable
	Total Kjeldahl Nitrogen	Nutrients
	Total Phosphate	Nutrients
	Sulfates	Effect on proposed treatment process; effluent quality requirements
	Chlorides	Effect on proposed treatment processes; effluent quality requirements
	Oils	Efficiency of existing oil recovery systems; effect on proposed treatment processes; effluent quality requirements
	Temperature	Thermal pollution potential; basis for calculating temperature balances and effect on temperature dependent reactions

Source: EPA Report 12020 EID 03/71

two hour intervals. These composited samples were then cooled to 4°C and stored until the survey had been completed. All samples were then collected and transported to the laboratory for chemical analysis. The analyses selected for this particular survey are itemized in Table 3.

Once the analyses had been performed, the statistical distribution of the data was obtained and used in formulating design criteria. For example, the degree of variation was particularly significant when considering average values for design parameters. Additionally, the variation was descriptive of the nature of production process operations which is important when conceiving treatment system layout, in-plant modifications, or factors of safety to be incorporated into the sizing of the system.

The general conclusions resulting from this survey of the petrochemical plant as described herein can be summarized as follows.

(1) The wastewater from network "A" was found to have a low volume of flow, although high concentrations and variability of dissolved organic and inorganic constituents were observed.
(2) The network "A" stream analysis indicated the presence of some product spills and generally poor in-plant housekeeping. Concurrently, product recovery potential was indicated and steps were recommended to minimize this waste stream.
(3) Preliminary observations of the network "A" stream indicated difficulty in applying biological treatment methods for this source individually, without dilution water from other sources.
(4) The network "B" stream had a much higher volume of flow with much lower concentrations of organic and inorganic contaminants. The variation of these materials was also less. However, the suspended solids load from processes served in this network was higher.
(5) The network "B" stream appeared to be more amenable to conventional treatment applications than did the network "A" stream, although a preliminary step for removal of suspended solids would be required prior to subsequent secondary forms of treatment.

Case History B: Refinery-Petrochemical Complex — The objectives of this survey were different than the one previously cited in that the effects of the combined flow on an operating aerated lagoon and its performance characteristics were to be evaluated.

Generally, the wastewaters included those from refinery process operations, cooling and boiler units, and miscellaneous cleaning and washdown operations. Some forms of pretreatment were given to most streams within the complex.

For example, most oily streams such as those emanating from ballast dumps or in-plant separations were routed through a series of oil separation units prior to being pumped into the aerated lagoon.

It was determined that a 24 hour sampling and analysis period was sufficient to accurately characterize the performance of the lagoon. The approach in making this evaluation was to measure the level of organic and inorganic constituents for the influent and effluent, thus measuring organic removal efficiency as well as conversion levels of inorganic substances.

Concurrently, representative sampling profiles were established within the aerated lagoon basin in order to establish mixing and dispersion patterns necessary for the proper placement of mechanical surface aerators. The sampling and analysis schedules for the survey are itemized in Table 4 and 5.

TABLE 4: SAMPLING AND ANALYSIS SCHEDULE—INFLUENT AND EFFLUENT STATIONS

Analyses	Frequency	Purpose
BOD (five-day) (mg/l)	1/hr	organic load, plant effluent with respect to criteria
BOD (ultimate) (mg/l)	1/hr	organic load, plant effluent with respect to criteria
COD (mg/l)	1/hr	organic BOD comparison
Conductivity (mhos/cm)	1/hr	waste characterization
D. O. (mg/l)	continuous	process control
Flow (mgd)	continuous	organic and hydraulic load
Oils (mg/l)	1/hr	process control, toxicity, efficiency of existing oil recovery systems
ORP (mv)	1/hr	sulfide level; ratio of oxidants to reductants
pH	continuous	process control
NH_4 - N (mg/l)	1/hr	nutrients, waste characterization, potential toxicity
SS (mg/l)	1/hr	suspended solids loading
Sulfates (mg/l)	1/hr	sulfide oxidation, wastewater characterization; effect on subsequent waste stabilization ponds
TDS (mg/l)	2/hr	waste characterization
Temp (^oC)	1/hr	environmental, effect on temperature dependent reactions
VSS (mg/l)	1/hr	volatile solids loading

TABLE 5: SAMPLING AND ANALYSIS SCHEDULE—PROFILES WITHIN AERATED LAGOON BASIN

Analysis	Sampling Points: Horizontal	Sampling Points: Vertical	Purpose
BOD_5 (mg/l)	Every 20 feet	1/top 1/mid-depth 1/bottom	Mixing evaluation with respect to soluble organics
D. O. (mg/l)	"	"	O_2 profiles, anaerobic areas, zone of aerator influence evaluation
O_2 Uptake	One Per Profile	mid-depth	Evaluate O_2 demand of basin contents; aerator efficiency
ORP (mv)	"	"	Oxidant-reductant profile
TOC (mg/l)	"	"	Mixing evaluation with respect to soluble organics
TSS (mg/l)	"	"	Uniformity of TSS, effectiveness of aerators in maintaining solids in suspension
Velocity (fps)	"	"	Velocity profiles, mixing evaluation

Source: EPA Report 12020 EID 03/71

The general conclusions drawn from this aerated lagoon survey are listed as follows:

(1) The efficiency of the aerated lagoon in terms of biological removal of soluble organic materials was approximately equal to that predicted in preliminary bench scale treatability studies.
(2) There was an accumulation of heavy oily sludge in the bottom of the lagoon, indicative of the relatively high oily solids loading applied directly to the lagoon as well as the inability of the surface aerators to keep this sludge in suspension at the power level employed.
(3) The power level of 0.05 HP/1,000 gallons of basin volume was insufficient to keep all of the biological solids in suspension and pronounced gradients of DO, SS, and liquid velocities were noted in the lagoon. However, a small residual DO level was noted at most horizontal and vertical sampling stations within the aerated basin.
(4) The rapid increase of DO in the aerated lagoon when the wastewater inflow was temporarily diverted and the equally rapid decrease when the flow was returned to the lagoon indicated that the microbial floc was viable and active.

Case History C: Petrochemical Complex — A survey of this petrochemical facility producing polyester resins was conducted in order to determine the feasibility of treating the liquid wastes using biological methods. Although the plant produced a relatively low total waste flow in terms of volume, the wastewater organic concentration fluctuated greatly.

The sources of wastewater discharged from the facility were categorized as follows:

(1) Process wash and rinse – this wastewater results from periodic cleaning of reaction vessels;
(2) Base flow – a composite of wastewaters discharged from a multitude of sources exclusive of those derived from scrubber or process wash/rinse operations;
(3) Storm runoff – the portion of storm drainage which comes in contact with contaminated areas; and,
(4) Scrubber blowdown – from intermediate product washes.

The average COD and pH of these four waste streams are given in Table 6. The final recommendations of this survey included: segregation of the highly concentrated scrubber blowdown form the remaining waste streams using liquid incineration as the disposal process; and, altering the collection system for the remaining waste streams, including storm water, to be discharged to a single point and treating this flow biologically.

This concept was complicated to some extent because of the discontinuous pattern of flow. However, the impact of variable loading was sequestered by providing a completely mixed biological treatment basin sized for excess capacity. The results of continuous monitoring of all wastewater from this industry prior

FIGURE 4: FLOW VARIATIONS RESULTING FROM BATCH UNIT PROCESS OPERATION IN A PETROCHEMICAL INDUSTRY

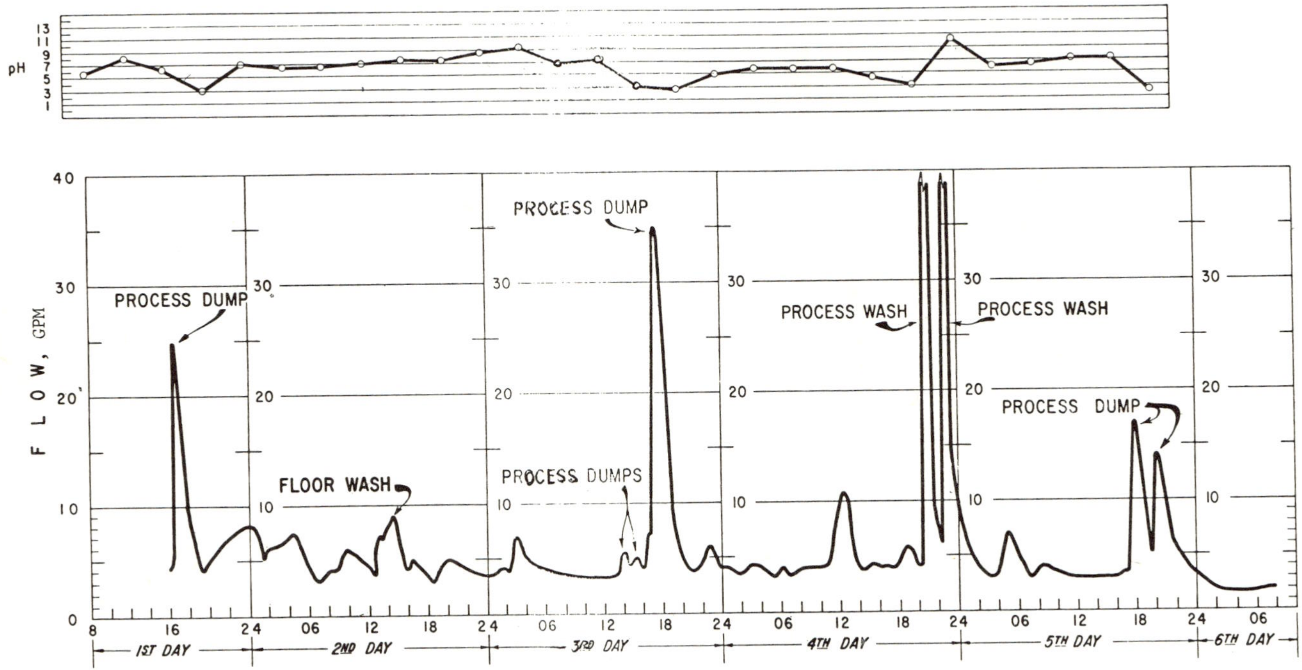

Source: EPA Report 12020 EID 03/71

FIGURE 5: OBSERVED FLOW VARIATIONS FOR PROCESS EFFLUENT SHOWN IN FIGURE 4 AFTER EQUALIZATION BASINS

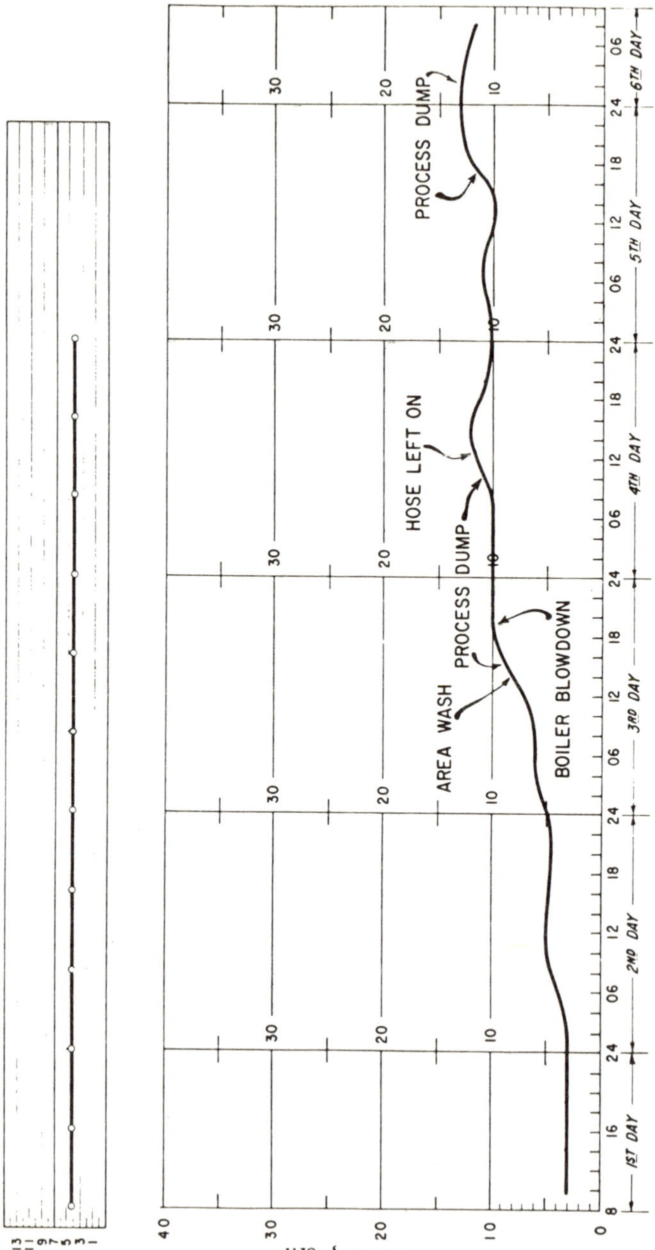

Source: EPA Report 12020 EID 03/71

to in-plant segregation and equalization is shown in Figure 3 while the effect of in-plant changes and equalization is shown in Figure 4.

TABLE 6: WASTEWATER CHARACTERISTICS FROM A SMALL PETRO-CHEMICAL INDUSTRY

Description of Waste Source	COD (mg/l)	pH
Intermittent Washing Operations	6,000	variable*
Base Flow From Process Operations	2,500	8.0
Storm Runoff (Design Basis = Two-Year Storm)	2,000	---
Off-gas Scrubber Blowdown	100,000	1.5

* Depends on Whether Process Wash or Process Rinse is being discharged.

Source: EPA Report 12020 EID 03/71

Case History D: Petrochemical Complex — This survey is cited to document the significance of storm runoff as a source of pollution. Rainfall information for this small petrochemical complex producing polyester resins was obtained from the Weather Bureau and the caluclated volume of runoff for several 24 hour design storms is given in Table 7. A runoff coefficient of 0.90 was used to estimate these volumes accumulating from the 13 acre contaminated area of the plant. Samples of runoff were taken during storm conditions and COD values up to 2,500 mg per liter were reported during the peak portion of the flow. Recommendations were made to provide retention facilities to store the first 15 minutes of runoff resulting from a five year storm, then discharging this contaminated volume of water to the biological facility over an extended period of time.

It is impractical in most cases to provide retention ponds when production and storage facilities encompass large areas. The calculated storage required to retain the contaminated fraction of a five year storm from one refinery complex occupying several thousand acres, for example, is 235,000,000 gallons.

Case History E: Chemical Complex — Storm runoff presented a different problem for a reprocessing chemical facility as shown in Figure 6. Samples were collected at a storm collection sump for each of the six contaminated areas. The varying magnitudes and patterns of runoff COD enabled the engineers to make decisions for collecting, storing and reprocessing the concentrated portion of the storm flow while using the less concentrated waters for boiler and cooling tower make-up. This approach not only provided credits in terms of recovering product and reducing fresh water make-up requirements, but also reduced the volume of contaminated storm runoff leaving the property of the chemical plant.

FIGURE 6: ORGANIC CONTAMINATION OF RUNOFF

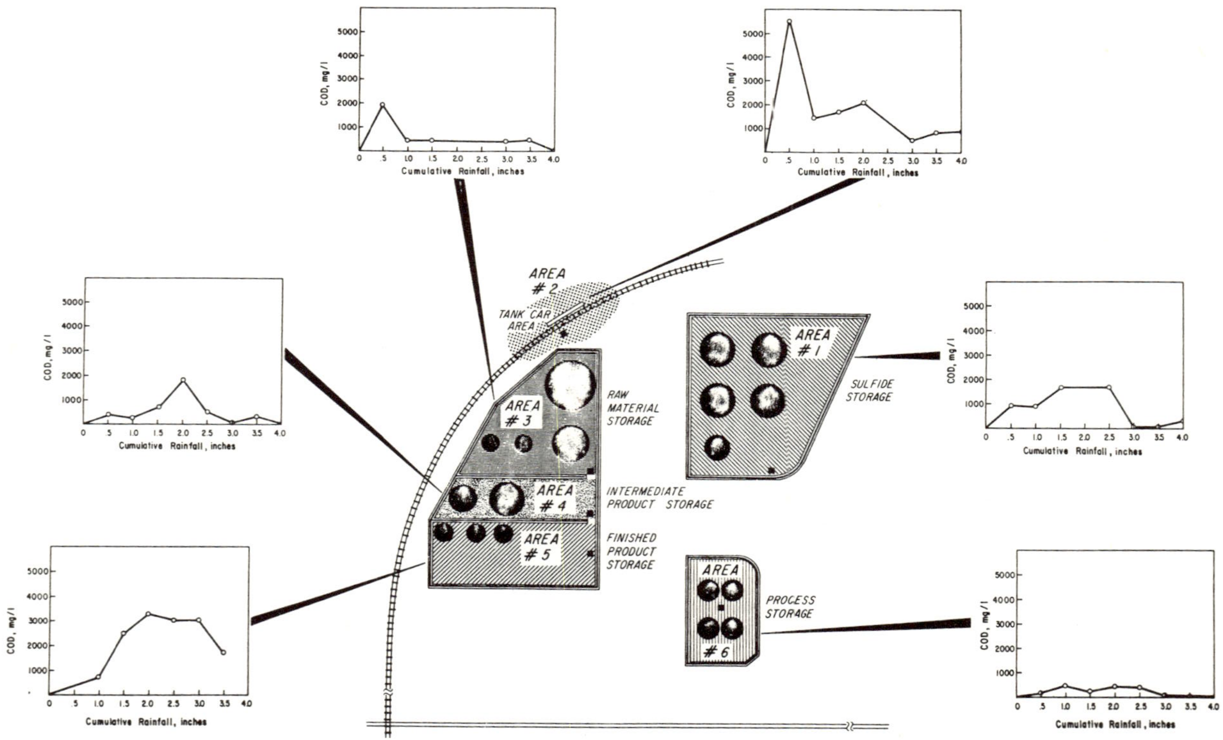

Source: EPA Report 12020 EID 03/71

TABLE 7: STORM OCCURRENCE AND QUANTITY OF RUNOFF FROM A PETROCHEMICAL INDUSTRIAL AREA*

Recurrence Interval (Years)	Total 24-Hour Rainfall (Inches)	Estimated Total Runoff (Gals.)
1	4.05	427,900
2	5.35	565,300
5	7.25	766,000
10	8.7	919,200
25	10.05	1,061,800

*Runoff Coefficient Taken To Be 0.95; Approximate Drainage Area is 13 Acres

Source: EPA Report 12020 EID 03/71

PROCESS CATEGORIES AND RAW WASTE LOADS

The organic chemicals manufacturing industry is a complex one in which interrelated chemicals compete for raw materials and markets via increasingly complex technologies. The water usage and subsequent waste water discharges are closely related to this mix of products and processes.

The diversity of products and manufacturing operations to be covered indicates the need for separate effluent limitations for different segments within the industry. To this end, process-oriented subcategories have been developed by the Environmental Protection Agency as noted in the *Development Document for Proposed Effluent Limitation Guidelines and New Source Performance Standards for the Major Organic Products Segment of the Organic Chemicals Manufacturing Point Source Category* issued as EPA 440/1-73/009 (December 1973). They are discussed in some detail in the pages which follow.

Subcategory A: Nonaqueous Processes – Contact between water and reactants or products is minimal. Water is not required as reactant or diluent, and is not formed as a reaction product. The only water usage stems from periodic washes or catalyst hydration.

To be considered within Subcategory A, the unit operations and chemical conversions within a process module must be essentially anhydrous. Contact water usage shall be only in the form of periodic washes or steaming used to treat nonaqueous catalysts or working solvents. The other sources of waste water are from external washing and maintenance operations within the process battery limits. External water sprays utilized to provide cooling on the outside of process pipes are contaminated through contact with chemicals present on the ground within the battery limits; consequently, they should be collected and discharged to a process sewer for subsequent treatment.

Subcategory B: Processes with Process Water Contact as Steam Diluent or Absorbent – Process water is in the form of dilution steam, direct product quench, or absorbent for effluent gases. Reactions are all vapor phase over solid catalysts. Most processes have an absorber coupled with steam stripping of chemicals for purification and recycle. Steam is also used for decoking of catalysts.

Subcategory B processes are characterized by unit operations and chemical conversions where the primary contact between water and chemicals is through vapor-liquid interfaces. Although final separation and discharge of water from the process may be as a liquid from a decant drum, contact within the process is normally: (1) through the mixing of steam with hydrocarbon vapors; (2) gaseous chemicals passing countercurrently through an aqueous absorption or quench medium; or (3) steam used to strip more volatile chemicals from liquid hydrocarbon mixtures. In all of these cases, the ultimate concentration of contaminants in the aqueous stream is governed by the specific vapor-liquid equilibria between the aqueous phases and the chemical phases. Hydrocarbon concentrations as total organic carbon (TOC) are generally less than 1 g/l or 1,000 mg/l in the aqueous streams associated with this type of processing.

Subcategory C: Aqueous Liguid-Phase Reaction Systems – Liquid-phase reactions where the catalyst is in an aqueous medium such as dissolved or emulsified mineral salt, or acid-caustic solution. Continuous regeneration of catalyst system requires extensive water usage. Substantial removal of spent inorganic salt by-products may also be required. Working aqueous catalyst solution is normally corrosive. Additional water may be required in final purification or neutralization of products.

The chemical conversions associated with Subcategory C processes are characterized by intimate contact between water and the reaction mixture or catalyst system. Water is used as a reaction medium in many of these systems because both the chemicals and the catalyst are infinitely soluble. The chemical conversions are generally multi-step reactions and generally more complicated than the vapor-phase reactions in Subcategory B. The Subcategory C reactions are also generally less selective in their yield to desired products and subsequently produce more by-products which must be removed from the system.

Typical unit operations involve liquid-solid interfaces where water is used to wash contaminants from solid chemical products. Because of the much larger quantities of chemicals and catalyst present in aqueous solution, most Subcategory C processes utilize many of the same unit operations as in Subcategory B, for the purpose of recovering these materials prior to discharging the water. There is also much more extensive internal recycling of aqueous process streams.

The hydrocarbon concentrations (as TOC) in the process waste waters which are ultimately discharged are in some cases 10-fold those for Subcategory B or approximately 10 g/l or 10,000 mg/l. The amount of contaminants, when expressed on a production basis, is also higher because of the required removal of by-products which are necessarily present in aqueous solutions.

Subcategory D: Batch and Semicontinuous Processes – Processes are carried out in reaction kettles equipped with agitators, scrapers, reflux condensers, etc. depending on the nature of the operation. Many reactions are liquid-phase with aqueous catalyst systems. Reactants and products are transferred from one piece of equipment to another by gravity flow, pumping, or pressurization with air or inert gas. Much of the material handling is manual with limited use of automatic process control. Filter presses and centrifuges are commonly used to separate solid products from liquid. Where drying is required, air or vacuum ovens are used. Cleaning of noncontinuous production equipment constitutes a major source of waste water. Waste loads from product separation and purification will be at least ten times those from continuous processes.

Subcategory D refers to batch processes. These operations are characterized by small production volumes and highly variable mixtures of products. A typical batch dye plant manufactures a wide variety of products at any specific point in time. This product mix itself may change completely on a schedule basis as short as one week. The segregation and characterization of process waste water associated with the production of any one specific dye is not possible, nor is it practical as a basis for establishing effluent limitations. Instead, the total waste water emanating from the batch plant is considered.

It is an economic necessity that equipment be transferred from the making of one chemical to that of another in multi-product batch plants. Although certain items may be used for only one type of chemical conversion, product purity requires that process lines and vessels be purged and cleaned between batches. Water is the most common cleaning solvent used in such applications, both because of the relatively low cost associated with its use and because other organic solvents cannot provide the required removal of contaminants. Waste water from cleaning is, therefore, a major contributor to the raw waste load RWL of Subcategory D processes.

Additional considerations include the fact that most of the chemical conversions are carried out in aqueous media and are generally much more complex than those done continuously. The reactions are generally less selective and produce greater quantities of waste by-products. They also frequently require rapid cooling which can be provided only through the direct additon of ice or refrigerated quench water.

The raw waste load (RWL) data obtained in field surveys of representative processes in the various categories surveyed by EPA in the study cited above are summarized in Table 8.

Subcategories B, C, and D were further subcategorized on the basis of raw waste loads. For Subcategories B and C, product processes were classified above and below the median raw waste load (BOD_5 and COD) for the subcategory. Those product processes with raw waste loads above the median subcategory raw waste load were designated B_2 and C_2, respectively, and those product processes with raw waste loads below the median value were designated B_1 and C_1, respectively, for each subcategory. Subcategory D includes one subcategory, azoic dyes and components.

The raw waste loads (RWL) associated with each of the continuous process Subcategories (A, B, and C) are based on contact process water only. Most continuous processes are able to achieve segregation and do not include noncontact cooling water or steam. Subcategory D includes all water usage associated

TABLE 8: MAJOR RWL'S OF POLLUTANTS BASED ON PROCESS WASTEWATER

Category	Flow RWL gals./1,000 lbs	BOD_5 RWL lbs/1,000 lbs	COD RWL lbs/1,000 lbs	TOC RWL lbs/1,000 lbs
A	0.25 - 2,000	0.1 - 0.13	0.3 - 3.7	0.034 - 0.9
Conc. Range (mg/l)		(400 - 1,000)	(200 - 10,000)	(50 - 2,000)
B	50 - 3,000	0.09 - 7.0	0.47 - 21.5	0.2 - 40
Conc. Range (mg/l)		(50 - 500)	(200 - 5,000)	(100 - 2,000)
C	30 - 3,000	1.3 - 125	1.9 - 385	1.5 - 150
Conc. Range (mg/l)		(3,000 - 10,000)	(10,000 - 50,000)	(3,000 - 15,000)
D	10,000 - 100,000	52 - 220	180 - 4,800	60 - 1,600
Conc. Range (mg/l)		(100 - 3,000)	(1,000 - 10,000)	(200 - 2,000)

Source: Report EPA 440/1-73/009.

with the process in that rapid cooling with direct contact is required in the manufacture of dyes.

The subcategorization assigns specific products to specific subcategories according to the manufacturing process by which they are produced. Where more than one process is commercially used to produce a specific chemical, it is possible that the chemical may be listed in more than one subcategory, because the unit operations and chemical conversions associated with different feedstocks may differ drastically in regard to process water usage and associated RWL. The most common chemical conversions used within the industry were examined and are themselves subcategorized in the tabulation below:

Subcategory A	Subcategory B	Subcategory C	Subcategory D
Acylation	Amination	Alcoholysis	Alkylation
Alkylation	Hydration	Ammonolysis	Amination
Aromatization	Dehydration	Dehydration	Condensation
Friedel-Crafts Reactions	Hydrogenation	Esterification	Nitration
Halogenation	Dehydrogenation	Hydroformulation	
	Oxidation	Hydration	
	Pyrolysis	Neutralization	
		Nitration	
		Oxidation	

AQUEOUS EFFLUENT LIMITATIONS

The effluent limitations proposed by EPA in Report EPA 440/1-73/009 cited earlier in this volume are based primarily on the dissolved organic pollutant contaminants in the process wastewaters associated with the processes listed in the various subcategories. No specific limitations are proposed for pollutants associated with noncontact wastewaters, such as boiler and cooling tower. These are primarily inorganic materials, and it is difficult to allocate such wastes among specific porcesses in many multiproduct plants. Since raw waste loads are based on process wastewaters, it is assumed that all noncontact waters be segregated from process wastewaters. Otherwise, combined wastewaters are subject to effluent limitations.

Separate limitations are presented for each of the four subcategories. The parameters involved are: biochemical oxygen demand (BOD_5), chemical oxygen demand (COD), total suspended solids (TSS) and phenols.

Other possible RWL parameters were considered during the study: total organic carbon, ammonia, cyanide, extractable oils and various metals, but were found to be in concentrations substantially lower than those which would require specialized end-of-pipe for the entire industry. Effluent limitations are not established for cyanide and cadmium pollutants although these have been designated as toxic substances. It is expected that the best practicable control technology currently available end-of-process treatment will substantially reduce these pollutants in the effluent stream. Effluent standards will be established for toxic pollutants wherever applicable.

Effluent limitations for the best practicable control technology currently available were based upon three significant pollutant parameters: BOD_5, total suspended solids (TSS) and phenols. The application of alternate oxygen demand parameters such as COD or TOC in lieu of the BOD_5 parameter may be possible, in situations where a direct correlation with BOD_5 has been satisfactorily established. For both New Source Performance Standards (BADCT) and Best Available Technology Economically Achievable, four significant pollutant parameters are specified: BOD_5, COD, TSS, and phenols.

End-of-process treatment for the 1977 standard is defined as biological treatment as typified by current exemplary processes: activated sludge, trickling filters, aerated lagoons, and anaerobic lagoons. These systems will be adequately equipped with pH control and equalization in order to control variable waste loads and clarification with the addition of chemicals to aid in removing suspended solids where this is required. Suspended solids are maintained at an average concentration of 30 mg/l.

These systems do not preclude the use of equivalent chemical-physical systems such as activated carbon in situations where necessary land area is not available. Additionally, suitable in-process controls are also applicable for the control of those pollutants which are biotoxic to the biological waste treatment system. Phenolic compounds are expected to be maintained at an average of 0.5 mg/l in the effluent by suitable combinations of in-process controls, (i.e., activated carbon, solvent extraction) and end-of-process treatment systems.

Best available technology economically achievable, BATEA, (1983 Standard) is based upon the addition of activated carbon to biological systems. This technology is based upon substantial reductions of dissolved organics compounds which are biorefractory as well as those which are biodegradable. Exemplary in-process systems are also applicable to this technology. End-of-process activated carbon treatment does not preclude the use of such treatment as an in-process technology. The following in-process controls are applicable to BATEA:

(1) the substitution of noncontact heat exchangers for direct contact water cooling;
(2) the use of nonaqueous quench media as a substitute for water where direct contact quench is required;
(3) the recycle of process water, such as between absorber and stripper;
(4) the reuse of process water (after treatment) as a make-up to evaporative cooling towers through which noncontact cooling water is circulated;
(5) the use of process water to produce low pressure steam by noncontact heat exchangers in reflux condensers of distillation columns;
(6) the recovery of spent acids or caustic solutions for reuse;
(7) the recovery and reuse of spent catalyst solutions; and
(8) the use of nonaqueous solvents for extraction of products.

End-of-process technology for new sources utilizing the best available demonstrated control technology (BADCT) is defined as biological treatment with suspended solids removal via clarification, sedimentation, sand, or dual-media filtration. In addition, exemplary in-process controls, as previously enumerated, are also assumed to be applicable, particularly where biotoxic pollutants must be controlled. This technology does not preclude the use of equivalent chemical-physical systems such as activated carbon as either an in-process or end-of-process treatment. This may be advantageous in areas where land availability is limited.

Substantial reductions of BOD_5 are expected from new sources as well as significant removals of dissolved organic compounds as measured by COD. Suspended solids and phenolic compounds are also maintained at average levels of 15 mg/l and 0.1 mg/l respectively.

Effluent limitations for BPCTCA, BATEA, and New Sources (BADCT) were developed on the basis of major subcategory median raw waste loads. Performance of exemplary treatment plants were considered in deriving the BPCTCA limitations for each category. Performance of BATEA and BADCT systems, together with in-process controls were considered in determining effluent limitations for each level of technology. It was determined that Subcategories B and C would be further subcategorized on the basis of raw waste loads. In these cases, median raw waste loads for subcategory groups were determined for B_1, B_2, C_1 and C_2 (as defined in the previous chapter.)

Finally, effluent limitations were derived on the basis of the maximum of any one day and the maximum average of daily values for any period of thirty consecutive days. The factors used in deriving these time based limitations were determined from long term performance (i.e., daily, weekly, monthly) from the best treatment systems evaluated. Time based limitations consider the normal variations of exemplary designed and operated waste treatment systems.

Effluent limitations commensurate with the best practicable control technology currently available are presented for each industrial subcategory of the organic chemicals manufacturing industry. Major product-process segments of the industry which are applicable to these limitations are listed in Table 9. Effluent limitations are presented in Table 10 for the 1977 standard (BPCTCA). It should be noted that process wastewaters subject to these limitations include all process water exclusive of auxiliary sources such as boiler and cooling water blowdown, water treatment back wash, laboratories and other similar sources.

End-of-process technology for BPCTCA involves the application of biological treatment systems as typified by activated sludge, trickling filters, aerated lagoons and anaerobic lagoons. Equalization with pH control and oil separation is provided in order to smooth out raw waste variations. Chemical flocculation aids, when necessary, are added to the clarification system to control suspended solids levels.

In-process controls as previously described are provided to remove those pollutants which interfere with biological waste treatment systems. Biological waste treatment does not preclude the use of equivalent chemical-physical systems. It may be advantageous to provide such systems within the process or at the end of process, especially where land availability is a limiting factor.

Effluent limitations to be attained by the application of the best available technology economically achievable are presented in Table 11 for the major product-process segments listed in Table 9. End-of-process treatment for BATEA include the addition of activated carbon systems to biological waste treatment processes. Exemplary in-process controls, as discussed in the previous section of this book, are also applicable to this technology. It is emphasized that the model treatment system does not preclude the use of activated carbon within the plant. Such systems are frequently employed for recovery of products, byproducts, and catalysts.

The best available demonstrated control technology for new sources includes the most exemplary process controls, as previously enumerated, with biological waste treatment and systems for removal of suspended solids. Effluent limitations for the major product-process segments are presented in Table 12. The most exemplary performance of the biological treatment system is assumed in terms of BOD_5 and COD reductions since in-process controls substantially reduce biotoxic pollutants.

TABLE 9: SUBCATEGORIES OF THE ORGANIC CHEMICALS MANUFACTURING INDUSTRY (PHASE I – MAJOR PRODUCT-PROCESSES)

Subcategory A	Process Descriptions
Cyclohexane	Hydrogenation of Benzene
Ethyl Benzene	Alkylation of Benzene with Ethylene
Vinyl Chloride	Addition of HCl to Acetylene
BTX Aromatics	Hydrotreating Pyrolysis Gasoline
BTX Aromatics	Solvent Extraction of Reformate
Subcategory B	
B1 Products	
Ethylene and Propylene	Pyrolysis of Naptha or liquid Petroleum Gas (LPG)
Butadiene	Co-product of Ethylene
Methanol	Steam Reforming of Natural Gas
Acetone	Dehydrogenation of Isopropanol
Acetaldehyde	Dehydrogenation of Ethanol
Vinyl Acetate	Synthesis of Ethylene and Acetic Acid
B2 Products	
Butadiene	Dehydrogenation of n-Butane
Butadiene	Oxidative dehydrogenation on n-Butane
Acetylene	Partial Oxidation of Methane
Ethylene Oxide	Oxidation of Ethylene
Formaldehyde	Oxidation of Methanol
Ethylene Dichloride	Direct Chlorination of Ethylene
Vinyl Chloride	Cracking of Ethylene Dichloride
Styrene	Dehydrogenation of Ethyl Benzene
Methyl Amines	Addition of Ammonia to Methanol
Subcategory C	
C1 Products	
Acetaldehyde	Oxidation of Ethylene
Acetic Acid	Oxidation of Acetaldehyde
Acrylic Acid	Synthesis with CO and Acetylene
Aniline	Nitration and Hydrogenation of Benzene
Bisphenol A	Condensation of Phenol and Acetone
Caprolactam	Oxidation of Cyclohexane
Coal Tar	Pitch forming and Distillation
Ethylene Glycol	Hydrogenation of Ethylene Oxide
Dimethyl Terephthalate	Esterification of Terephthalic acid (TPA)
Oxo Chemicals	Carbonylation and Condensation
Phenol	Oxidation of Cumene
Terephthalic Acid	Oxidation of p-Xylene
Terephthalic Acid (Polymer grade)	Purification of Crude Terephthalic Acid

(continued)

TABLE 9: (continued)

C2 Products	
Acrylates	Esterification of Acrylic Acid
P-Cresol	Sulfonation of Toluene
Methyl methacrylate	Acetone Cyanohydrin Process
Terephthalic Acid	Nitric Acid Process
Tetraethyl Lead	Addition of Ethyl Chloride to Lead Amalgam
Subcategory D	
Organic Dyes: Azoic Dyes and Components	Batch Processes

Source: Report EPA 440/1-73/009.

TABLE 10: EFFLUENT LIMITATIONS FOR BPCTCA ORGANIC CHEMICAL MANUFACTURING INDUSTRY

Effluent Characteristics	Effluent Limitations* kg/kkg Production	
	Maximum for Any One Day	Maximum Average of Daily Values for Any Period of Thirty Consecutive Days
Subcategory A		
BOD5	0.045	0.025
TSS	0.038	0.023
Phenols	0.00050	0.00025
Subcategory B		
B1 Product-Processes		
BOD5	0.10	0.06
TSS	0.10	0.06
Phenols	0.0013	0.00066
B2 Product-Processes		
BOD5	0.30	0.17
TSS	0.27	0.16
Phenol	0.0034	0.0017
Subcategory C		
C1 Product-Processes		
BOD5	0.30	0.17
TSS	0.27	0.16
Phenol	0.0034	0.0017
C2 Product-Processes		
BOD5	1.5	0.9
TSS	0.80	0.49
Phenols	0.011	0.0050

(continued)

TABLE 10: (continued)

Subcategory D		
BOD_5	15.0	9.0
TSS	13.0	7.88
Phenols	0.17	0.088

*kg/kkg production is equivalent to lb/1000 lb production.

Source: Report EPA 440/1-73/009.

TABLE 11: EFFLUENT LIMITATIONS FOR BATEA ORGANIC CHEMICALS MANUFACTURING INDUSTRY

Effluent Characteristics	Effluent Limitations* kg/kkg Production	
	Maximum for Any One Day	Maximum Average of Daily Values for Any Period of Thirty Consecutive Days
Subcategory A		
COD	0.04	0.02
BOD_5	0.004	0.002
TSS	0.006	0.004
Phenols	0.00005	0.000025
Subcategory B B1 Product-Processes		
COD	0.130	0.065
BOD_5	0.008	0.004
TSS	0.017	0.010
Phenols	0.00013	0.000065
B2 Product-Processes		
COD	0.74	0.37
BOD_5	0.02	0.01
TSS	0.0042	0.0025
Phenols	0.00034	0.00017
Subcategory C C1 Products-Processes		
COD	0.78	0.39
BOD_5	0.02	0.01
TSS	0.0083	0.005
Phenols	0.00068	0.00034
C2 Products-Processes		
COD	14.4	7.2
BOD_5	0.4	0.2
TSS	0.27	0.16
Phenols	0.0022	0.0011

(continued)

TABLE 11: (continued)

Subcategory D		
COD	130.0	65.0
BOD5	0.8	0.4
TSS	2.19	1.30
Phenols	0.017	0.0085

*kg/kkg production is equivalent to lb/1000 lb production.

Source: Report EPA 440/1-73/009.

TABLE 12: STANDARDS OF PERFORMANCE FOR NEW ORGANIC CHEMICALS MANUFACTURING SOURCES

Effluent Characteristics	Effluent Limitations* kg/kkg Production	
Subcategory A	Maximum for any One Day	Maximum Average of Daily Values for any Period of thirty Consecutive Days
BOD5	0.02	0.012
COD	0.15	0.10
TSS	0.012	0.0075
Phenol	0.0001	0.0005
Subcategory B B1 Product-Processes		
BOD5	0.06	0.035
COD	0.55	0.4
TSS	0.033	0.02
Phenol	0.00026	0.00013
B2 Product-Processes		
BOD5	0.15	0.085
COD	3.0	2.2
TSS	0.083	0.05
Phenol	0.00068	0.00034
Subcategory C C1 Product-Processes		
BOD5	0.15	0.085
COD	3.3	2.3
TSS	0.083	0.05
Phenol	0.00068	0.00034
C2 Product Processes		
BOD5	0.75	0.40
COD	60.0	40.0
TSS	0.27	0.16
Phenol	0.0022	0.0011

(continued)

TABLE 12: (continued)

Subcategory D		
BOD_5	1.5	0.85
COD	540.0	390.0
TSS	4.38	2.60
Phenol	0.034	0.017

*kg/kkg production is equivalent to lb/1000 lb production.

Source: Report EPA 440/1-73/009.

SPECIFIC PRODUCTS MANUFACTURE AND POLLUTION PROBLEMS

In the discussion of specific processes and associated pollution problems in the pages which follow, the major sources which have been quoted are as follows.

Report EPA 440/1-73/009 entitled *Development Document for Proposed Effluent Limitations Guidelines and New Source Performance Standards for the Major Organic Products Segment of the Organic Chemicals Manufacturing Point Source Category,* Washington, D.C., Environmental Protection Agency (December 1973)

Midwest Research Institute report entitled *The Pollution Potential in Pesticide Manufacturing,* Report PB 213,782, Springfield, Va., Nat. Tech. Information Service (June 1972).

Processes Research, Inc. report entitled *Air Pollution from Nitration Processes,* Report PB 217,106, Springfield, Virginia, Nat. Tech. Information Service (March 31, 1972).

Processes Research, Inc., report entitled *Air Pollution from Chlorination Processes,* Report PB 218,048, Springfield, Virginia, Nat. Tech. Information Services (March 31, 1972).

ACETALDEHYDE FROM ETHANOL

Acetaldehyde is produced in the United States by processes using ethylene, ethyl alcohol, or liquefied petroleum gas as feedstock. The breakdown of 1970 U.S. capacity for each route is shown below.

Feedstock	Process	Percent of 1970 U.S. Capacity
Ethylene	Oxidation	56
Ethyl alcohol	Oxidative dehydrogenation	36
LPG	Oxidation	8

The following discussion is of the ethyl alcohol route. The ethylene route will be discussed in the section which follows. In the oxidative-dehydrogenation process, ethanol and air enter an oxidation furnace. The primary reaction is given below:

$$C_2H_5OH + \frac{1}{2}O_2 \longrightarrow CH_3CHO + H_2O$$

The reaction is vapor phase and is carried out over a solid silver gauze catalyst at about 1000°F. The reactor effluent is condensed and is passed to a phase separator. The gaseous phase is absorbed in refrigerated water. Off-gases pass from the system, and the wash is combined with the liquid stream. The combined liquid stream is fractionated into product acetaldehyde, alcohol for recycle, and wastewater. Dehydrogenation of ethanol is based on the chemical reaction:

$$C_2H_5OH \longrightarrow CH_3CHO + H_2$$

This reaction is also vapor phase and is carried out over a solid copper catalyst promoted by cobalt or chromium on an asbestos support at 500°F. These processes yield from 85 to 95% of the stoichiometric amount of acetaldehyde. The only wastewater stream generated in these processes is either from the acetaldehyde flash column or the ethanol recovery still, and contains liquid byproducts such as acetic acid. Plant survey data are shown in the following tabulation.

	Plant 1	Plant 2
Flow	1,600 gal/1,000 lb	140 gal/1,000 lb
COD	186 mg/l	*
	2.48 lb/1,000 lb	*
BOD_5	84 mg/l	*
	1.12 lb/1,000 lb	
TOC	*	14,400 mg/l
		16.7 lb/1,000 lb

*Not available

Although direct comparison of COD, BOD_5 or TOC values between two plants was not possible, the magnitude of the above parameters and the general relations between COD and TOC, and between BOD_5 and TOC, show a significant difference in RWL for the two plants.

The difference is attributed mainly to differing efficiencies of acetaldehyde flash columns and ethanol recovery stills. By improving operating conditions of those stills, Plant 2 should be able to reduce its RWL to the representative RWL of Plant 1.

BADCT and BATEA control technology are defined by an in-process modification of the water-scrubbing system, entailing division of the scrubber into a fresh water-scrubbing portion and a bulk recycle water portion. Such a division can significantly reduce fresh water usage by permitting recycle of four-fifths of the existing waste stream. The amount of wastewater flow from the scrubber

FIGURE 7: ACETALDEHYDE BY OXIDATIVE DEHYDROGENATION OF ETHANOL

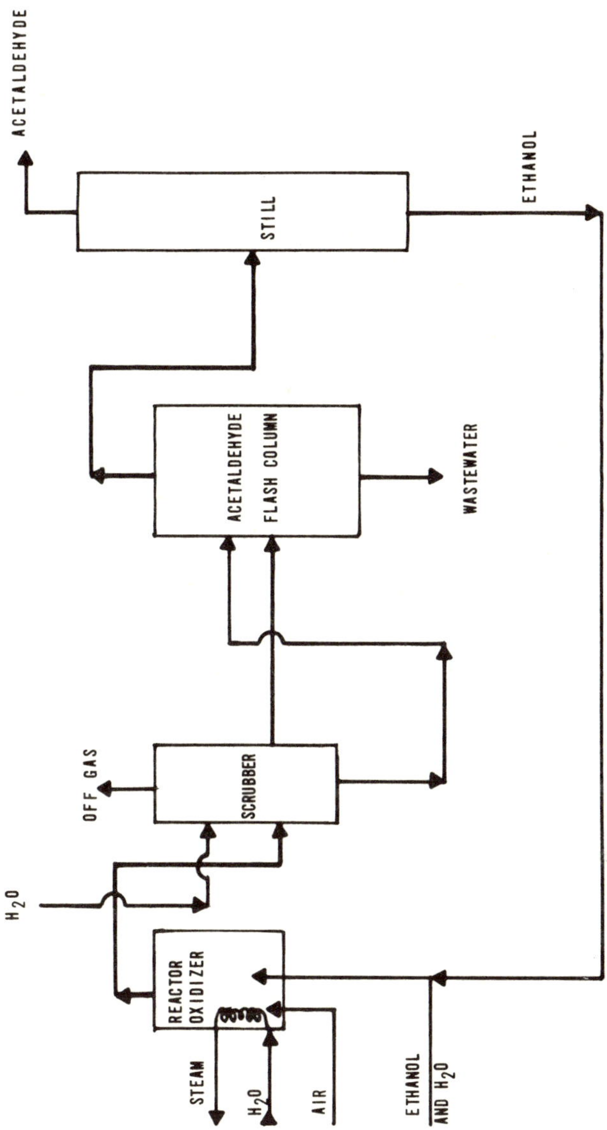

Source: Report EPA 440/1-73/009

is thus one-fifth of the original flow. However, because the concentrations of contaminants in this waste stream will increase proportionally, no reduction in RWL will occur. Total process water and cooling water usages of the two plants are summarized as follows.

Plant	Process Water (lb/lb product)	Cooling Water (lb/lb product)
1	13	104
2	1	100

Process water is applied mainly to the scrubber. The more water used, the larger the amount of waste flow.

ACETALDEHYDE FROM ETHYLENE

The Wacker process employs an aqueous catalyst solution of palladium chloride, promoted (for metal oxidation) by copper chloride. The chemistry involved in the process can be summarized by the reaction below.

$$C_2H_4 + \tfrac{1}{2}O_2 \longrightarrow CH_3CHO + \text{heat}$$

The catalyst acts as the oxygen carrier and causes selective conversion of ethylene to acetaldehyde. The reaction steps essentially are:

$$\text{Reaction: } C_2H_4 + 2CuCl_2 + H_2O \xrightarrow{PdCl_2} CH_3CHO + 2HCl + 2CuCl$$

$$\text{Regeneration: } 2CuCl + 2HCl + \tfrac{1}{2}O_2 \longrightarrow 2CuCl_2 + H_2O$$

There are two basic process variations, and choice depends upon such factors as oxygen cost, utilities prices, and available ethylene purity. In the single-stage process, pure oxygen is employed as the oxidant. The reactor effluent is condensed and water-scrubbed. Unreacted gas is recycled into the reactor. By-products and water are separated from the acetaldehyde product by distillation. Both the reaction and regeneration are effected at the same time.

In the two-stage process, the oxidant is air. The reaction is carried out with catalyst solution and ethylene in one reactor, and the regeneration is carried out with air in a separate reactor. Lower purity ethylene can be used with this version of the process. However, this process forms more by-products and requires high operating pressures.

The major wastewater sources in this process are the effluents from the scrubber that is required for removal of unreacted ethylene and uncondensed acetaldehyde vapor, and from the aqueous bottoms of the acetaldehyde still. The characteristics of the wastewater are shown in the tabulation on page 52.

FIGURE 8: ACETALDEHYDE BY SINGLE-STAGE WACKER PROCESS

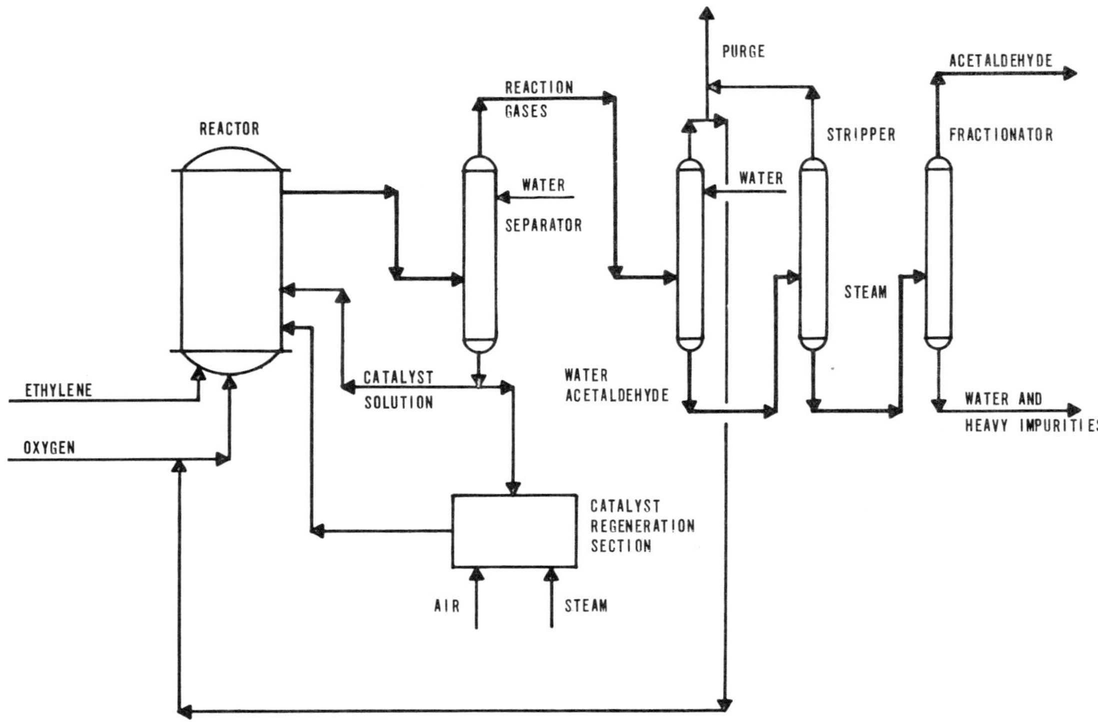

Source: Report EPA 440/1-73/009

	Plant 1	Plant 2	Plant 3
Flow	90 gal/1,000 lb	61 gal/1,000 lb	35 gal/1,000 lb
COD	58,718 mg/l	11,400 mg/l	20,240 mg/l
BOD_5	–	3,700 mg/l	11,500 mg/l
TOC	–	7,000 mg/l	12,500 mg/l

The foregoing data show the same order of magnitude of raw waste loads in Plants 2 and 3, and this level of RWL can be considered as standard for BPCTCA. The high RWL of Plant 1 is mainly due to sloppy operation of the acetaldehyde still. To define BADCT and BATEA control technology, it is required that a steam stripper be installed to recover and reuse the organic contaminants in the wastewater. A description of the steam stripper, as well as its estimated economics has been given in the section on aniline.

Because of the aqueous phase reaction, catalyst metals are present in the wastewater from the acetaldehyde still bottoms as a result of carry-over from the reactor. The aqueous catalyst solution is also quite acidic and corrosive. Survey data also shows that, in addition to metallic contaminants found in wastewater stream, sulfate and oil contaminants are found at concentrations in excess of general criteria for biological treatment processes. Pretreatment or dilution to reduce their concentrations is required.

Average process water usage for this process, including steam directly supplied to the process, is 0.92 lb/lb of acetaldehyde, while cooling water usage amounts to 330 lb/lb of product. Alternate routes for manufacturing of acetaldehyde, as well as U.S. production capacity, have been discussed under "Acetaldehyde from Ethanol" above.

ACETIC ACID FROM ACETALDEHYDE

Acetic acid is produced by the liquid phase oxidation of acetaldehyde, using either air or oxygen according to the reaction given below:

$$CH_3CHO + \tfrac{1}{2}O_2 \rightarrow CH_3COOH$$

The reaction is carried out in the liquid phase at 150°F and 60 psig, with manganese acetate dissolved in aqueous solvent as catalyst.

Acetaldehyde and solvent are fed to the oxidation reactors with a manganese acetate catalyst solution. The reactor effluent (containing unreacted oxygen, nitrogen, acetaldehyde, and solvent) is cooled, and the acetaldehyde and solvent are condensed and recycled back to the reactor. The noncondensibles are water-washed before being discharged into the atmosphere.

The degassed liquid stream as well as water from the scrubber are sent to a light-ends column, where the light ends are distilled overhead. The bottoms from these distillation columns are sent to a dehydration column in which water is removed overhead using benzene as the azeotropic agent. The aqueous phase

FIGURE 9: ACETIC ACID FROM OXIDATION OF ACETALDEHYDE

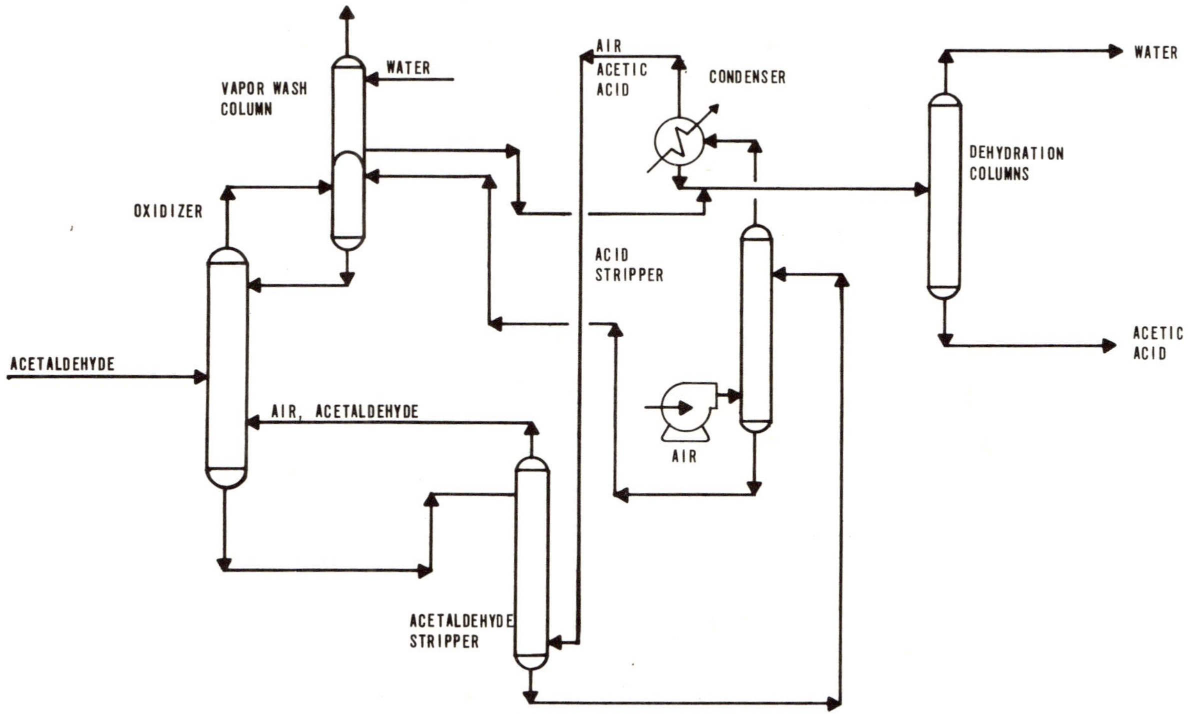

Source: Report EPA 440/1-73/009

in the distillate stream is sent to a solvent stripping column, where acetic acid is removed as distillate while the bottoms are sent to a waste disposal unit. The major wastewater source in this process is the water taken overhead from the dehydration column. The possible contaminants are unrecovered formic acid and acetic acid. The characteristics of the wastewater obtained from plant surveys is presented in the following tabulation.

	Plant 1	Plant 2
Flow	500 gal/1,000 lb	10.2 gal/1,000 lb
COD	186 mg/l	306,100 mg/l
	0.78 lb/1,000 lb	26.18 lb/1,000 lb
BOD_5	84 mg/l	64,000 mg/l
	0.35 lb/1,000 lb	5.44 lb/1,000 lb

The foregoing data show a significant variation in RWL between two plants. Examination of each process shows that the concentrated light ends and heavy ends from distillation columns are discharged into sewer lines by Plant 2 instead of being disposed of by incineration as commonly practiced. If these concentrated streams are excluded, the RWL of Plant 2 as shown below is comparable to RWL of Plant 1.

Flow	1.48 gal/1,000 lb
COD	7,500 mg/l
	0.925 lb/1,000 lb
BOD_5	26,700 mg/l
	0.33 lb/1,000 lb

There is a slight difference in manufacturing process between Plants 1 and 2. Plant 1 utilizes ethanol as part of its feedstock and generates at most 35 gal of reaction water per 1,000 lb of product, based on 100% ethanol feedstock. Also, instead of combining scrubber water with aqueous reactor effluent, Plant 1 sends scrubber water directly to an acetaldehyde recovery still and disposes of the bottom stream of the distillation column. This modification allows Plant 1 to use more scrubbing water in the scrubber and results in a high amount of wasteflow.

Based on the foregoing analysis, the RWL of Plant 1 can be considered the standard of BADCT and BATEA of this process. The standards of BADCT and BATEA should require recycling of scrubber water in Plant 1. This modification when implemented will reduce the flow of BPCTCA to one-tenth its current level, and the RWL by one-half.

Total process water usage of this process is directly proportional to the amounts of wastewater generated. The survey data show a variation from 4.2 lb of process water per pound of acetic acid at Plant 1 to 0.024 at Plant 2. The gross cooling water usages are 54 and 185 lb/lb of product for Plants 1 and 2, respectively. Several other process routes to acetic acid are also practiced commercially. The CO-methanol and petroleum gases (n-butane) processes are discussed briefly in the following paragraphs. Direct liquid phase oxidation of n-butane in petroleum gases is normally carried out at 300° to 350°F under a pressure of 700 to 800 psig, and the chemical reactions taking place are extremely involved. The reactor effluent is sent to a vapor-liquid separator, the gaseous products from this separator are scrubbed with a heavy hydrocarbon to recover unreacted n-butane, and the liquid product from the separator is split into an organic and aqueous phase. The organic phase is recycled, while the aqueous phase is fractionated to remove intermediate by-products.

The CO-methanol process is the most recent commercial route. Carbon monoxide and a liquid stream containing the catalyst system of cobalt iodide and cobalt carbonyl hydride are fed to a sparged reactor operating at 500°F and 10,000 psig. Product acetic acid is recovered by fractionation. The methanol feedstock is normally not introduced directly to the oxidizer, but rather is used to scrub the reactor off-gases, which contain catalyst in the form of methyl iodide vapor.

ACETONE FROM ISOPROPANOL

Acetone is produced by dehydrogenation of isopropanol. Fresh and recycle isopropanol are vaporized and fed to a tubular reactor at typical operating conditions of 5 psig and 450° to 550°C. A brass catalyst is commonly used. Conversion of isopropanol is about 90% per pass, and selectivities to acetone are above 95%.

The reactor effluent is passed into an absorption tower to clean up the hydrogen formed in the reaction. The water solution from the absorption tower is then purified by conventional fractionation techniques, and unconverted isopropanol is recycled back to the reactor. The chemical reaction is given below.

$$\begin{array}{c}CH_3CHCH_3\\ |\\ OH\end{array} \longrightarrow \begin{array}{c}CH_3CCH_3\\ \|\\ O\end{array} + H_2$$

In this process, water is used to absorb product acetone and unreacted isopropanol from the hydrogen produced. After fractionation, one or two wastewater streams are produced as bottoms from the isopropanol stripping still or as bottoms from the intermediate flash column. The wastewater contains acetone, isopropanol, and small quantities of heavier organic substances. RWL of this process from survey data in plant visits is summarized as follows.

	Plant 1	Plant 2
Flow	230 gal/1,000 lb	120 gal/1,000 lb
COD	246 mg/l	1,720 mg/l
	0.47 lb/1,000 lb	1.72 lb/1,000 lb
BOD_5	91 mg/l	338 mg/l
	0.18 lb/1,000 lb	6.34 lb/1,000 lb
TOC	132 mg/l	473 mg/l
	0.25 lb/1,000 lb	0.47 lb/1,000 lb

FIGURE 10: PRODUCTION OF ACETONE BY DEHYDROGENATION OF ISOPROPANOL

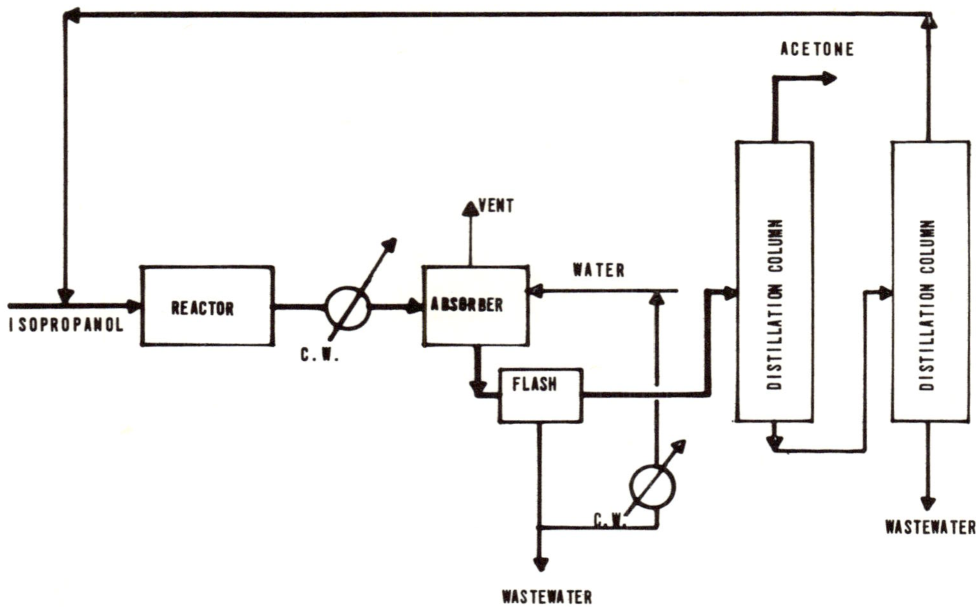

Source: Report EPA 440/1-73/009

Based on process information, it is concluded that the RWL of Plant 1 is representative of BPCTCA. Since the contaminants in the bottoms of the isopropanol stripping still have low concentrations of volatile organic compounds, this stream can be totally recycled to the absorber as absorbing water.

The wastewater stream from the intermediate flash column can also be recycled to reduce fresh water usage. Contaminant concentration in the intermediate flash column wastewater is a function of the column design. Recycle of the abovementioned streams will not change the characteristics of the existing waste stream. With this process modification, RWL of BADCT and BATEA can be expected to achieve the following values.

Flow	100 gal/1,000 lb
COD	103 mg/l
	0.086 lb/1,000 lb
BOD_5	53 mg/l
	0.044 lb/1,000 lb
TOC	52 mg/l
	0.043 lb/1,000 lb

The difference in RWL between the two plants is attributed to poor performance of the isopropanol stripping still of Plant 2. Average process water usage of this process is about 1.18 lb water per pound of acetone, while cooling water usage amounts to 119 lb water per pound of product.

Acetone can also be produced by several alternate routes. The most important recent development has been production of acetone as a coproduct in the cumene-to-phenol process. Another alternate process is the oxidation of isopropanol. The oxidation process is also vapor phase and is carried out with brass or copper catalysts. The economics of the acetone market are conditioned by the fact that the acetone produced as a coproduct in the cumene-to-phenol process can be credited at a low price.

ACETYLENE FROM METHANE

Acetylene is manufactured by burning preheated natural gas and pure oxygen in specially designed burners. The natural gas is partially oxidized with oxygen, and the evolved heat cracks the hydrocarbon to acetylene.

$$CH_4 + 2O_2 \longrightarrow CO_2 + 2H_2O$$

$$2CH_4 \longrightarrow HC{\equiv}CH + 3H_2$$

Cracking occurs at 1500°C with a residence time of 0.01 to 0.1 seconds. The resulting gases are rapidly quenched with water to prevent acetylene decomposition. A gas cooler and a series of distillation columns are then used to separate acetylene from by-products. Large quantities of carbon (coke) are produced by burning of the natural gas, and these fine particles are trapped in the quench stream. An air flotation unit or similar device must be provided to remove coke from the quench water before the water can be sent to a cooling tower and recycled. The solids removed can be dewatered and then disposed of by incineration.

The only wastewater stream from the process results from the cooling water system, which must be continually bled and replenished with fresh water to avoid buildup of dissolved substances. The results of survey data on cooling water draw-off stream are shown in the following tabulation.

Flow	561 gal/1,000 lb
COD	1,274 mg/l
	5.95 lb/1,000 lb
BOD_5	410 mg/l
	1.92 lb/1,000 lb
TOC	393 mg/l
	1.80 lb/1,000 lb

Comparing these data with those for the cooling water just prior to being discharged to the cooling tower inlet indicates that a significant amount of hydrocarbons are evaporated into atmosphere.

In order to further reduce the RWL and not sacrifice ambient air quality, a steam stripper can be installed to remove hydrocarbons from the waste stream before the wastewater is sent to the cooling tower. The collected hydrocarbons can then be disposed of by incineration. RWL of BADCT and BATEA will require this in-process modification to achieve lower waste loads.

Acetylene can also be produced by two other routes. The first is pyrolysis of a mixture of lime and coke at 2000°C in an electric furnace to form calcium carbide. The calcium carbide is ground under anhydrous conditions and then treated with a limited quantity of water to produce acetylene. Calcium hydroxide is a by-product.

The second, called the Wulff process, produces acetylene by pyrolysis of ethane, propane, naphtha, or similar material. An alternating cycle is used wherein hydrocarbons are heated by a hot tile checker work (1100°C) to produce acetylene.

Following this, there is a combustion step during which the bricks are heated in air to burn off tar deposits. The pyrolysis gases are contacted initially with dimethylformamide (DMF) to remove water, diacetylene, and other products. This is followed by absorption of the acetylene in DMF and final recovery of the acetylene by stripping.

Acetylene production had grown 10% annually from 1960 to 1965. This growth has been stimulated primarily by demand for vinyl chloride, vinyl acetate, and chloroprene. However, acetylene demand has exhibited a marked decline since then.

FIGURE 11: ACETYLENE MANUFACTURE

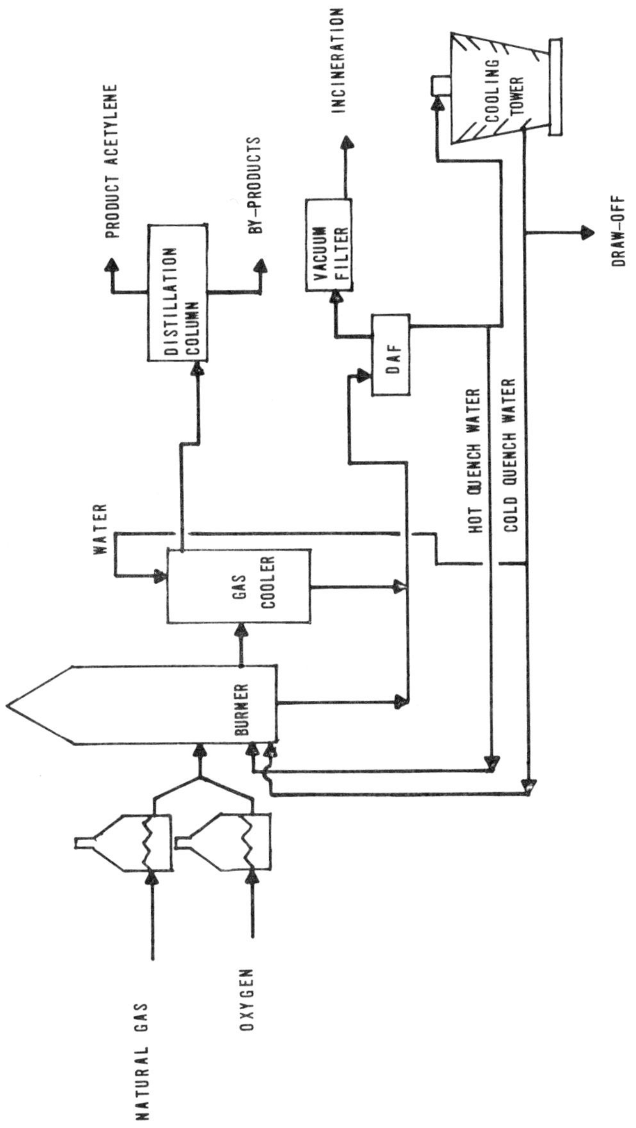

Source: Report EPA 440/1.73/009

ACRYLATES FROM ACRYLIC ACID

Acrylates are manufactured by esterification of acrylic acid. There are four main acrylates plus a large number of specialty, smaller volume derivatives. The main four are ethyl, 2-ethylhexyl, methyl and n-butyl in decreasing order of market share. The 2-ethylhexyl and butyl acrylates are produced in a separate facility from the methyl and ethyl esters due to their differences in volatility and solubility.

In the manufacture of methyl or ethyl acrylates, acrylic acid is reacted with an excess amount of methanol or ethanol in a concentrated sulfuric acid solution. The effluent from the reactor goes to an extraction column, where caustic removes the excess alcohol.

The effluent water stream is sent to a distillation column; alcohol is recovered overhead and recycles, while the acrylate stream is purified in two distillation columns by removal of light and heavy ends.

In the manufacture of butyl, 2-ethylhexyl, and higher acrylates, the esterification is conducted in the presence of cyclohexane, which is used to remove the water of reaction. The reactor effluent is first neutralized with caustic and then sent to a series of distillation columns. Acrylate is purified, while the excess alcohol is recovered and recycled. The chemical reaction can be expressed by the following equation.

$$C_2H_3COOH + R\text{–}OH \xrightarrow{H_2SO_4} C_2H_3COOR + H_2O$$

The two main sources of water pollution in acrylate manufacture are the bottoms of the alcohol recovery still and the effluent of the saponification kettle. The possible contaminants in the waste stream are acrylic acids, alcohols, and sodium salts of various acids. The results of the plant survey are presented in the following tabulation.

Flow	2,856 gal/1,000 lb
COD	4,870 mg/l
	117.5 lb/1,000 lb
BOD_5	1,942 mg/l
	47.1 lb/1,000 lb
TOC	3,290 mg/l
	79.5 lb/1,000 lb

Historical data over a period of two months show that total carbon in the waste stream ranges from 15.50 to 46.36 lb/1,000 lb of acrylate produced. Probability analysis of the data indicates that 50% occurrence is equivalent to 30.8 lb per 100 lb of product. From the data presented above, it is known that inefficient operation of distillation columns causes significant losses of organics such as alcohol, acrylic acid, and acrylates into the waste stream. Recovery of these organics can be achieved by modification of the distillation columns or by

FIGURE 12: MANUFACTURE OF ACRYLATES

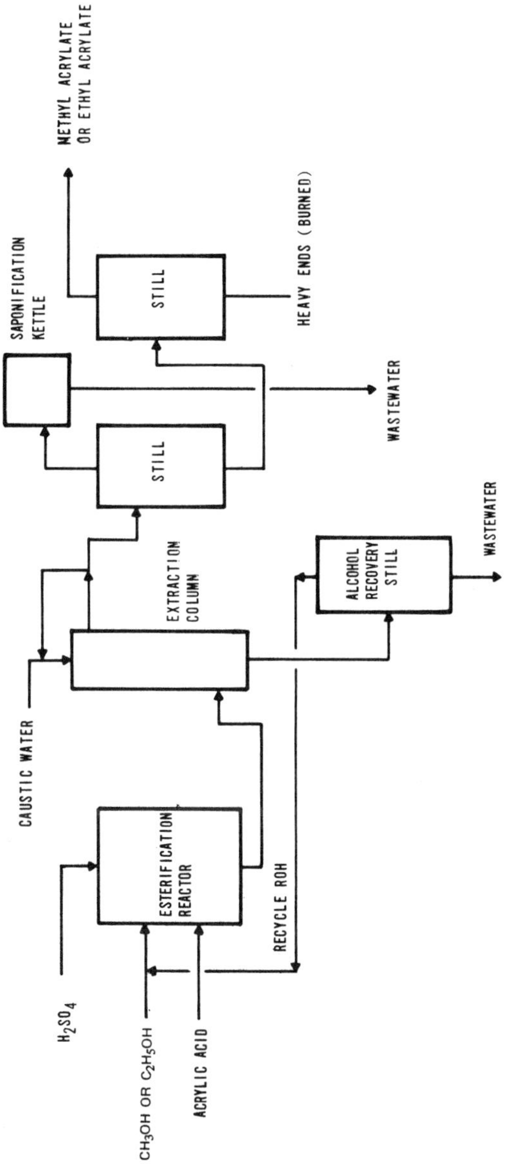

Source: Report EPA 440/1-73/009

installation of a steam stripper. The amount of waste flow can also be reduced by recycling the wastewater to an extraction column. BADCT and BATEA in-process controls should require a steam stripper to recover organic contaminants in the waste stream and thus achieve a low RWL.

ACRYLIC ACID FROM ACETYLENE

Acrylic acid is synthesized from acetylene and carbon monoxide in a catalytic solution. The chemistry can be represented by the following reaction.

$$C_2H_2 + H_2O + CO \longrightarrow C_2H_3COOH$$

The acetylene feedstock is first dissolved in THF (tetrahydrofuran) in an absorption tower. This solution and carbon monoxide are then mixed in a reactor, and the reaction is carried out at approximately 450°F and 1,500 psig in the presence of a nickel bromide and cupric bromide solution. The off-gas from the reactor is passed through a THF absorber to remove acrylic acid vapor and unreacted acetylene, and is then scrubbed by caustic water for further removal of THF and carbon monoxide from the gas stream.

The liquid reactor effluent, a mixture of acrylic acid, by-product acetaldehyde, and catalyst solution, is fed to a separation column. The overhead is extracted with water to recover THF and is distilled to yield purified acetaldehyde. The raffinate from the separation column is sent to a series of vacuum distillation and extraction columns. The THF and catalyst solution are recovered and recycled to the acid reactor. Technical grade glacial acrylic acid is produced in final distillation columns.

The major wastewater source is the caustic scrubber water. The contaminants are THF and Na_2CO_3. The characteristics of wastewater samples obtained during plant surveys are summarized in the following tabulation.

Flow	475 gal/1,000 lb
COD	414 mg/l
	1.64 lb/1,000 lb
BOD_5	186 mg/l
	0.737 lb/1,000 lb
TOC	387 mg/l
	1.53 lb/1,000 lb

Historical data over a period of two months show that TOC ranges from 1.73 to 6.92 lb/1,000 lb of acrylic acid and probability analysis indicates that 50% occurrence is equivalent to 3.08. The high wastewater flow rate is attributed to the utilization of steam jets used to produce a vacuum in the distillation columns.

Converting steam jets to vacuum pumps can certainly reduce the amount of wastewater generated, although the RWL in terms of COD, BOD_5, etc. will remain the same. Other than reducing wastewater flow rate, in-process modification is deemed unfeasible to further reduce RWL, and consequently, the data

FIGURE 13: ACRYLIC ACID FROM ACETYLENE

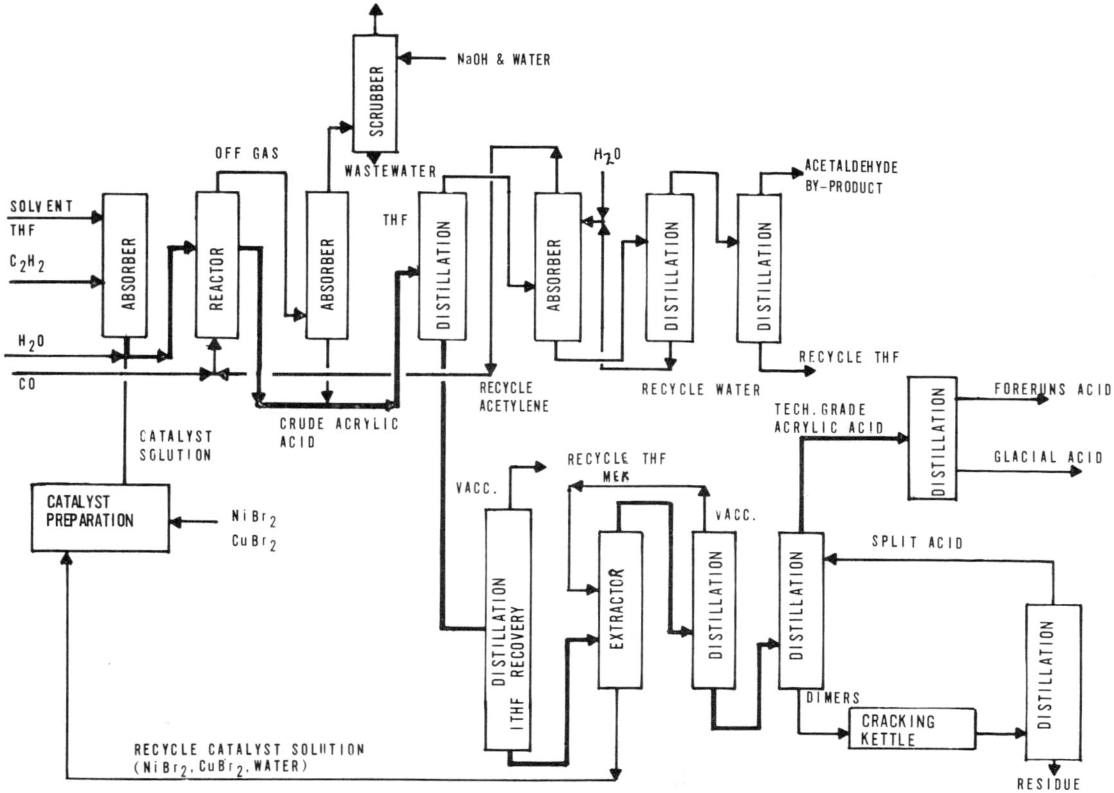

Source: EPA 440/1-73/009

presented can be considered as standards for BADCT and BATEA. A wide range of technology is used to produce acrylic acid. The other important route is based on propylene technology. A mixture of propylene, air, and steam is fed to two tubular catalytic reactors in series and cooled by circulation of molten salt.

Most of the acrylic acid is condensed and separated from the gaseous stream by quenching. The resulting aqueous solution is then subjected to an extraction with solvent, followed by distillation for purifying the product and recovering the solvent.

ACRYLONITRILE FROM PROPYLENE

About 16% of the acrylonitrile is produced by the following reaction, which will discussed in depth here.

$$4C_3H_6 + 6NO \longrightarrow 4C_3H_3N + N_2 + 6H_2O$$

About half of the acrylonitrile production is made by the Sohio process, which appears to be the process of the future.

$$C_3H_6 + NH_3 + \underset{\text{(air)}}{1\tfrac{1}{2}O_2} \longrightarrow C_3H_3N + 3H_2O$$

The remaining acrylonitrile is made from acetylene.

$$C_2H_2 + HCN \longrightarrow C_3H_3N$$

There is a fourth process, using ethylene oxide, which is obsolete. An annual growth rate of 13% is forecast. About 61% of the acrylonitrile is used for fibers. The next larger use, 17%, is for resins. The nitric oxide process is discussed below.

Nitric oxide is reacted with propylene in the vapor phase at 450° to 500°C and atmospheric pressure. There is a large excess of propylene in the feed. The reaction mixture is diluted with air or nitrogen. The catalyst is either lead titanate or silver/silica. A product yield of 80% is obtained at a conversion of 12%. The process produces the following by-products: acetone, 1,5-hexadiene, benzene, and hydrogen cyanide.

The reaction gas product stream passes through a water scrubber where all the reaction products except nitrogen and the dilution air are absorbed. The nitrogen and air are vented. The products and by-products are recovered and purified in a train of distillation columns. Unreacted propylene is recovered and recycled to the reaction step.

There is only one point where atmospheric emission occurs. This point is the vent from the scrubber. Normally, the emission will consist of nitrogen and air saturated with water vapor. The sole producer by the nitric acid oxidation process is DuPont with a plant at Beaumont, Texas. The manufacture of acrylonitrile (a subcategory C process) produces a harmful wastewater which is difficult to treat biologically. The harmful characteristics have been attributed to the presence of hydrogen cyanide in excessive quantities (500 to 1,800 mg/l). In addition, the wastewater is generally acidic (pH 4 to 6) and contains high concentrations of organic carbon. These wastewaters are generally segregated from other process wastes and disposed of by other means (e.g., incineration), and they are not generally discharged to municipal collection systems.

In a process described by *A.F. Miller and M.L. Salehar; U.S. Patent 3,468,624; September 23, 1969; assigned to The Standard Oil Company,* mineral acid ammonium salts are recovered from the wastewater streams produced in the wastewater distillation columns in the manufacture of acrylonitrile and methacrylonitrile, which streams also contain undesirable organic constituents.

The recovery process is carried out by adding to the waste stream a nitrogen-containing compound selected from the group consisting of ammonia, methylamine, ethylamine, n-propylamine, n-butylamine, n-amylamine, dimethylamine, di-n-propylamine, trimethylamine, ethylenediamine, trimethylenediamine, ethanolamine, diethanolamine and triethanolamine; maintaining the pH between 6.5 and 10; removing a portion of the water from the solution; cooling the solution and adding thereto additional amounts of the nitrogen-containing compound to effect precipitation therefrom of the ammonium salt.

A process has been developed by *A.T. Guttmann, M.G. Konicek and A.J. Grigsby; U.S. Patent 3,576,607; April 27, 1971; assigned to The Standard Oil Company* for the recovery of fertilizer grade ammonium salts in high yields from acrylonitrile wastewater process streams containing water-soluble organic tars by the removal of water and volatile organic matter from the wastewater process streams in the presence of a water-miscible solvent, recovering the solid ammonium salt from the residual ammonium salt-solvent-tar mixture and recovering the solvent for reuse.

A process has been developed by *J.-P. Zumbrunn; U.S. Patent 3,715,309; Feb. 6, 1973; assigned to L'Air Liquide, Societe Anonyme pour l'Etude de l'Exploitation des Procedes Georges Claude, France* for detoxicating wastewaters from acrylonitrile manufacture which contain organic cyanide impurities, particularly nitriles.

The detoxication process involves rapid degradation of the nitriles into amides and then into salts of organic acids by treatment with hydrogen peroxide in alkali medium, and the per compounds. This process, which has the advantage of saturating the effluents with oxygen, permits the total purification to a residual content lower than 0.5 ppm.

A process has been developed by *W.O. Fitzgibbons, E.M. Schwerko and A.H. Brainard; U.S. Patent 3,734,943; May 22, 1973; assigned to The Standard Oil Company* for the deep well disposal of a wastewater stream resulting from a catalytic ammoxidation process for the production of acrylonitrile comprising adding minor amounts of acrolein or a mixture of acrolein and ammonium sulfate to the wastewater stream prior to pumping the stream into the well.

Failure to treat the wet acetonitrile bottoms in this manner results in plugging of the disposal well when the material is pumped therein over a period of time. The mechanism whereby the addition of acrolein and ammonium sulfate to the

FIGURE 14: DEEP WELL DISPOSAL PROCESS FOR ACRYLONITRILE PROCESS WASTEWATER

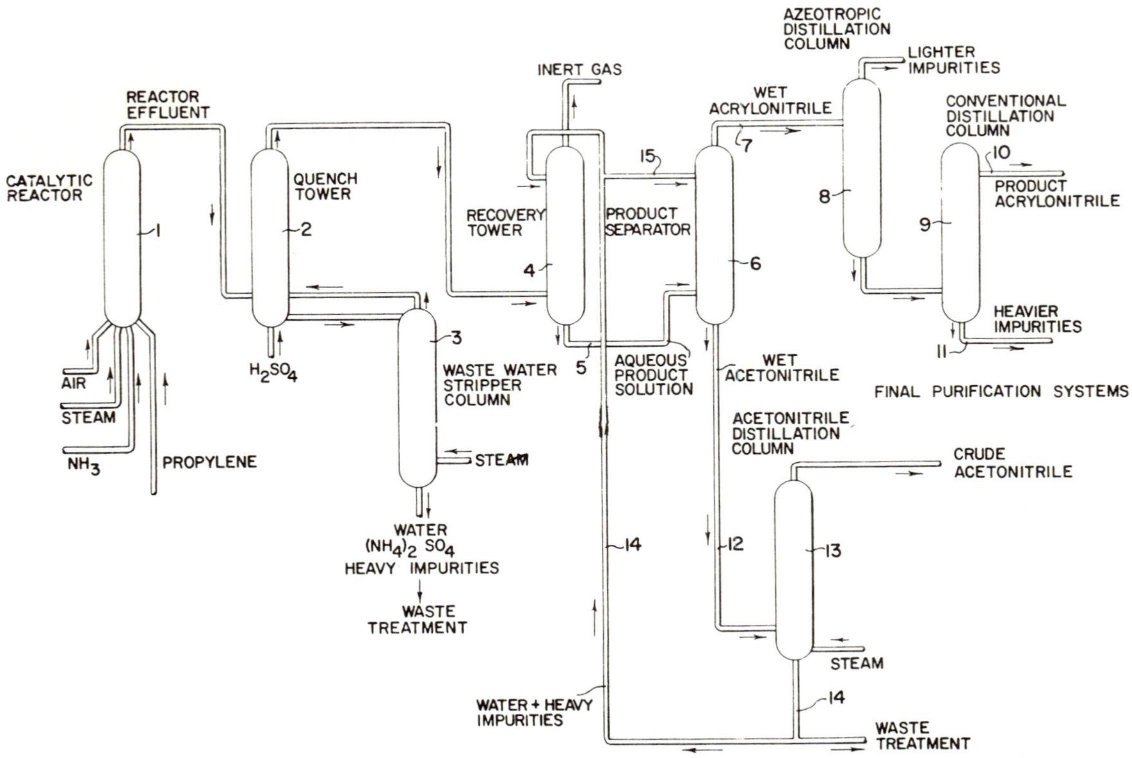

Source: W.O. Fitzgibbons, E.M. Schwerko and A.H. Brainard; U.S. Patent 3,734,943; May 22, 1973

wet acetonitrile bottoms stream eliminates plugging of the deep disposal well is not known. However, plugging is believed to be caused by suspended particles in the waste stream in the particle size range of about 0.45 to 5.0μ which cause a decrease in the permeability of the rock core and a corresponding increase in the injection pressure at the well.

It is preferred, and plugging of the well is virtually eliminated, when the wet acetonitrile bottoms stream is filtered through sand after it is treated with the acrolein or the acrolein-ammonium sulfate mixture.

A process has been developed by *Y. Fujii and T. Oshimi; U.S. Patent 3,756,947; September 4, 1973; assigned to Sumitomo Shipbuilding & Machinery Co., Ltd., Japan* in which a wastewater effluent containing nitriles and cyanides is treated by passing through an acclimated, activated sludge containing at least one of microorganisms capable of degrading nitriles and cyanides selected from the genera Alcaligenes and Achromobacter, for example, *Alcaligenes viscolactis* ATCC 21698 and *Achromobacter nitriloclastes* ATCC 21697 thereby to purify the wastewater effluent. The wastewater effluent containing 10 to 50 ppm of cyanide and 1,000 to 2,500 ppm of COD (Potassium Dichromate Method) can be purified with a high efficiency.

ADIPIC ACID FROM CYCLOHEXANE

Adipic acid is made by oxidizing crude cyclohexanol. The oxidizing agent can be nitric acid,

$$C_6H_{11}OH + 2HNO_3 \longrightarrow (COOH)_2(CH_2)_4 + N_2O + 2H_2O$$

by which over 90% of the adipic acid is made. Or it can be air,

$$C_6H_{11}OH + 2O_2 \longrightarrow (COOH)_2(CH_2)_4 + H_2O$$

In 1970, the production of adipic acid had exceeded 1.3 billion pounds. For the seventies, a 20% growth rate is expected. Adipic acid is a nylon intermediate.

The cyclohexanol feed is a crude mixture of cyclohexanol and cyclohexanone. This crude is usually obtained by the air oxidation of cyclohexane at 125° to 160°C and 50 to 250 psi in the presence of cobalt naphthenate catalyst. The reactor product is distilled to remove water and unreacted cyclohexane.

Another method for obtaining the cyclohexanol crude is from the hydrogenation of phenol at 300°F and 50 psig in the presence of a palladium catalyst. The cyclohexanol crude leaves the reactor as a vapor with unreacted hydrogen. The cyclohexanol crude is separated from the hydrogen to become the desired feed material for adipic acid production.

Cyclohexanol crude is oxidized to adipic acid using 40 to 50% nitric acid. The nitric acid fed is at least 1.5 times stoichiometric. The reaction is carried out at 50° to 150°C and 50 to 250 psig in the presence of a catalyst consisting of ammonium metavanadate and copper. Based on cyclohexane content of feed, a yield of 60 to 70% adipic acid is obtained.

FIGURE 15: NITRIC ACID OXIDATION OF CRUDE CYCLOHEXANOL

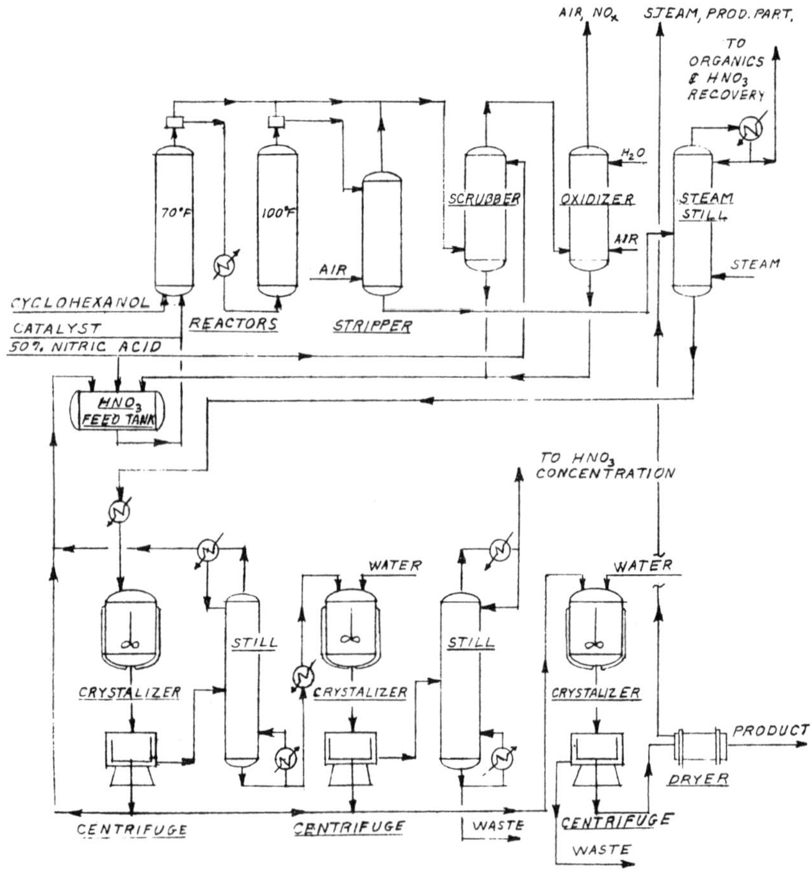

Source: Report PB 218,048; 3/31/72

The reactor liquid product is air stripped to remove nitrous oxide. It is then steam distilled to remove unreacted nitric acid and low boiling organics (as overheads). The liquid from the bottoms of the steam still are cooled to 40° to 50°C.

This crystallizes out the adipic acid. The crystals are separated from the mother liquor, redissolved in water, and recrystallized to remove impurities. The crystals are dried to produce the adipic acid.

The reaction produces nitrous oxide as a gas. This is oxidized with air, and scrubbed with water to produce nitric acid, which is recycled to the reactor. The overheads from the steam still contain nitric acid which is recovered and recycled to the reactor.

A portion of the mother liquor from the first crystallization step, which contains nitric acid and catalyst, is recycled to the reactor. Another portion is fed to a still. The overheads from the still, containing water and nitric acid, are recycled to the reactor. The bottoms are cooled to crystallize out adipic acid, which is combined with the crystals from the first crystallization step.

The mother liquor is distilled to recover any nitric acid present. This is concentrated and recovered. The bottoms is either a waste material or undergoes additional processing to recover the catalyst.

There are two sources of air pollutants. The first is the effluent from the oxidizer. This effluent will be air with any residual unoxidized and unabsorbed nitrous oxide or nitrogen dioxide. The other source is the vent from product drying which will contain product particulates as a contaminant.

Water pollution may be caused by both inorganic and organic materials from this process. The inorganic materials include metallic catalyst residues and the organic materials include by-product organic acids.

A process developed by *R. Johnson and C.R. Campbell; U.S. Patent 3,148,210; September 8, 1964; assigned to Monsanto Company* provides an improved method for recovering and reusing a copper-vanadium catalyst from waste liquor obtained in the manufacture of adipic acid by nitric acid oxidation of cyclohexanol and cyclohexanone mixtures.

This particular recovery technique is applied at the point where adipic acid is crystallized and separated from other organic dibasic acids such as glutaric and succinic acid by inducing the nitric acid in the liquors to evaporate to such an extent that the pH thereof is raised to within the range of 1.2 and 2.2 as measured after dilution of the evaporated residue with water.

Upon cooling the diluted residue, the vanadium component of the aforesaid mixed catalysts in the form of an organo-vanadium complex is selectively precipitated. In a suitable manner the vanadium precipitate is isolated from the mother liquor of the diluted residue for eventual reuse in the nitric acid oxidation of cyclohexanol and cyclohexanone. Thus, the vanadium precipitate can be dissolved in dilute nitric acid or the like and recycled to the feed material of the aforesaid nitric acid oxidation step. Preferably the mother liquor of the vanadium precipitation and separation is further diluted with water and cooled. This cooled, diluted mother liquor is brought into intimate contact with a polymerizate having cation exchange properties to effect substantially complete removal of the ionic copper in this mother liquor.

The polymerizate to which the copper ions are bound is separated from the liquid medium in a suitable manner. Finally, the copper ions are eluted from the ion exchange polymerizate, such as by contacting the polymerizate with dilute nitric acid or other suitable eluting agents. The eluted copper ions can be recycled to the feed material of the aforesaid nitric acid oxidation step.

A process developed by *C.R. Campbell, D.F. Danly and M.J. Mathews III; U.S. Patent 3,267,029; August 16, 1966; assigned to Monsanto Company* involves the removal of monobasic and dibasic acids, mineral acids, and other organic and inorganic material from aqueous mixtures prior to the discharge thereof to waste.

An example of a waste aqueous stream which is high in chemical oxygen demand and which cannot be discharged safely without treatment is the aqueous mixture which results from the overhead or make stream from the evaporation, either before or after crystallization of adipic acid, of the aqueous product of the nitric acid oxidation described above.

By this process, the aqueous mixture described above, or any similar aqueous mixture, is contacted either in single or multiple stages with a liquid ion-exchange material dissolved in a suitable solvent to remove the chemical oxygen demand causing components in the solvent phase, thereby leaving purified water suitable for safe, nonchemical polluting waste discharge to rivers or other natural waste areas.

The resulting solvent phase may be contacted with an inorganic base, anhydrous ammonia, an aqueous solution of an inorganic base or ammonia, or mixtures thereof to regenerate the solvent solution of ion-exchange material for recycle or other use, and the remaining concentrated aqueous solution of salts of the organic acids and other chemical oxygen demand causing components which remain in much reduced volume may be disposed of by incineration or other suitable nonpolluting means or recovered for profitable use.

A process developed by *W. Griehl and C. Suter-Homuth; U.S. Patent 3,523,064; August 4, 1970; assigned to Inventa AG für Forschung und Patentverwertung, Switzerland* involves the production of cellular protein from the waste liquors obtained in the cyclohexane oxidation.

Microorganisms of the family of Pseudomonadaceae convert these waste products to proteins in alkaline medium at ambient temperatures at 24° to 30°C and pH levels of substantially 6.7 to 9.2 within 50 to 90 hours. The process has the two-fold advantage of manufacturing a useful product from wastes and of preventing water pollution.

A process developed by *M. Seko, A. Yomiyama, T. Miyake and H. Iwashita; U.S. Patent 3,673,068; June 27, 1972; assigned to Asahi Kasei Kogyo KK, Japan* involves recovering a substantial portion of nitric acid and metal ions from an acidic waste liquid produced in the process for the production of adipic acid by oxidizing cyclohexanone or cyclohexanol in a liquid phase. More specifically, the recovery process comprises supplying the waste liquid to an electrodialysis apparatus including one or more electrodialyzers to recover the nitric acid and metal ions in a recovering liquid of a member selected from water and a diluted aqueous nitric acid.

ALACHLOR FROM DIETHYLANILINE

Alachlor is a selective herbicide based on acetanilide. It is a low-melting (40°C) solid material having a solubility in water of 140 ppm. It exhibits moderately low toxicity but is a nonpersistent skin irritant. Its formula and precise chemical name are shown below.

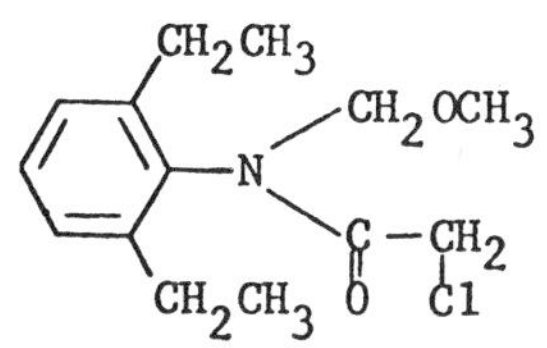

2-chloro-2',6'-diethyl-N-(methoxymethyl)acetanilide

It is produced by Monanto at Muscatini, Iowa, using the following production chemistry.

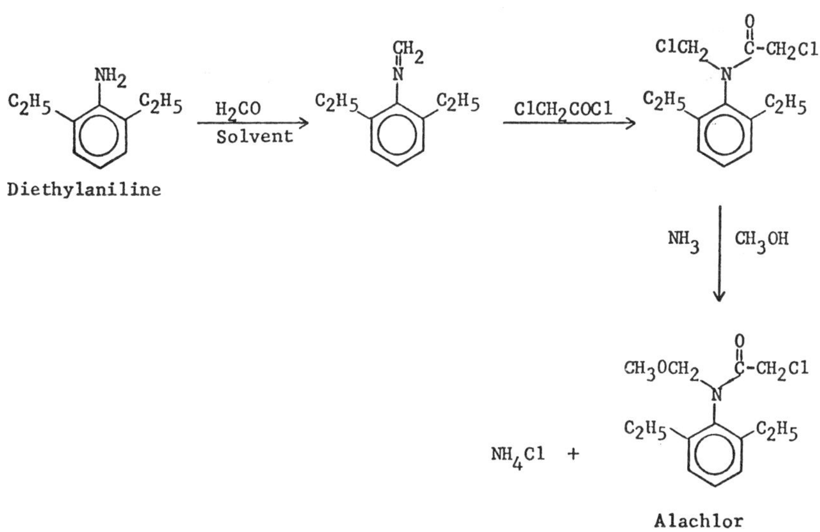

Process water is discharged to the Mississippi River after mechanical treatment only. Alachlor losses are about 300 lb/day. They are considering the recovery of by-product NH_4Cl. The process has no solid products and gives no off-gases. Production equipment requires cleanup only two or three times per year and presents no problem. The formulation equipment is used for some other products.

Packaging area has automatic bagging and filling equipment and is said to be very clean. The production and warehouse areas are not diked. Leaking containers have not been a serious problem: $\leqslant$1% on bags; $<$0.5% on cans. Used drums are a very small problem at Muscatine and would go to a landfill on site.

Wipe cloths, etc., are probably burned according to Midwest Research Institute in *Report PB 213,782,* Springfield, Virginia, Nat. Tech. Information Service (June 1972).

ALDICARB FROM ISOBUTYLENE

Aldicarb is an insecticide of the alkyl carbamate class. Its formula and chemical name are as follows.

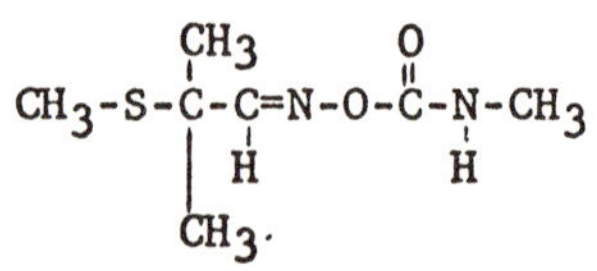

2-methyl-2-(methylthio)propionaldehyde-O-(methylcarbamoyl)oxime

It is a liquid of low volatility which is extremely toxic but nonpersistent. It is produced by Union Carbide at Institute, West Virginia, using the following production chemistry.

$$2(CH_3)_2C{=}CH_2 + 2NaNO_2 + 4HCl \longrightarrow 2NaCl + 2H_2O + [ClC(CH_3)_2CH_2NO]_2$$

2-chloro-2-methyl-1-nitrosopropane dimer

$$[ClC(CH_3)_2CH_2NO]_2 \xrightarrow{2CH_3SNa} 2NaCl + 2CH_3SC(CH_3)_2CH{=}NOH$$

2-methyl-2-(methylthio)-propionaldehyde oxime

$$CH_3SC(CH_3)_2CH{=}NOH \xrightarrow{CH_3NCO} CH_3SC(CH_3)_2CH{=}NO{-}C(O){-}NHCH_3$$

Aldicarb

All effluent streams from the unit are either incinerated or treated. No solid waste, as such, is generated from aldicarb production. NaCl goes to the liquid waste system. Organic losses are low, about 10 lb/100 lb aldicarb, exclusive of solvent loss. Overall, 1 to 2 lb of waste are generated per pound of product. It is anticipated that solvent losses will be significantly improved if a new plant is constructed.

Potentially toxic liquid streams are detoxified (with caustic) before they are sent to the plant waste treatment system. Other potentially toxic effluents are flared. Spills and leakers are handled per *Handling and Safety Guide for Temik 10G,* a publication of Union Carbide Corp.

One reactor in the process is vented to a NaOH scrubber; this waste then goes to the plant waste treatment system. For equipment cleanup steam washout is used in the aldicarb unit for decontamination.

The aldicarb area of the production facilities are diked so that all drainage goes to a caustic sump and then into the plant's secondary waste treatment plant. The solvent from the technical material is destroyed by flaring at the formulation site.

ALDRIN FROM DICYCLOPENTADIENE

Aldrin is a broad spectrum insecticide of the chlorinated hydrocarbon class. It is a solid which is insoluble in water. It is very toxic and is persistent in the form of dieldrin. Its chemical name and formula is shown below.

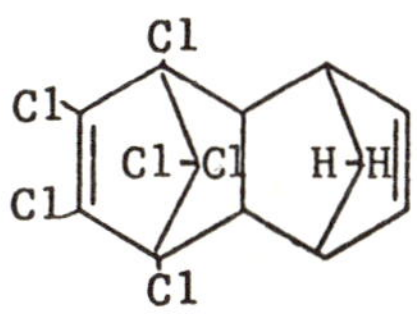

1,2,3,4,10,10-hexachloro-1,4,4a,5,8,8a-hexahydro-1,4-endo-exo-5,8-dimethanonaphthalene

It is produced by Shell at the Rocky Mountain Arsenal in Colorado using the following production chemistry.

FIGURE 16: PRODUCTION AND WASTE SCHEMATIC FOR ALDRIN

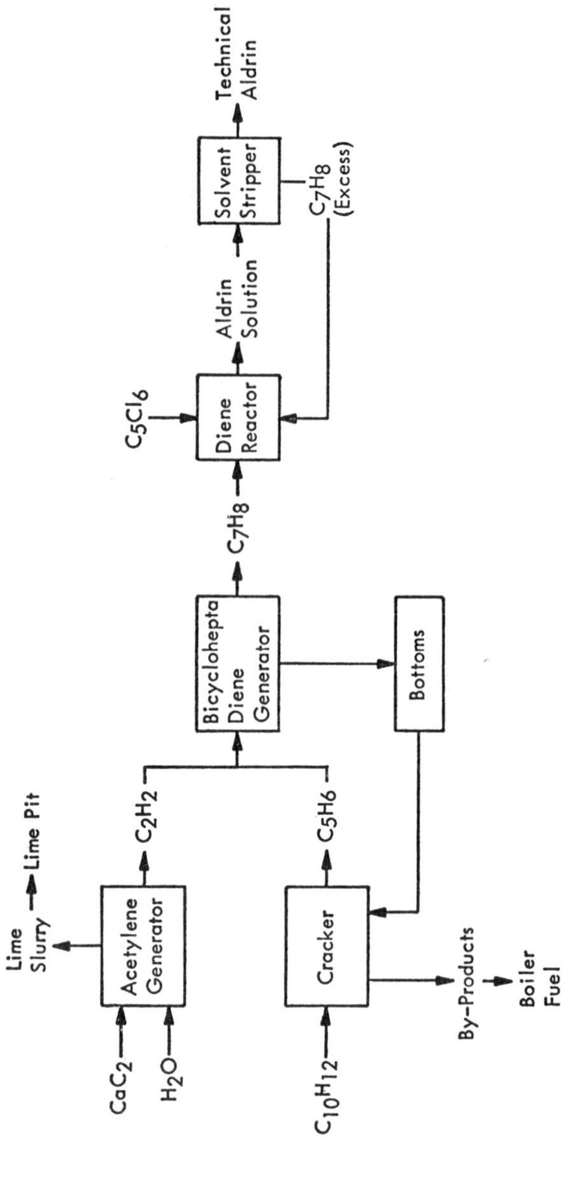

Source: Midwest Research Institute Report PB 213,782; 6/72

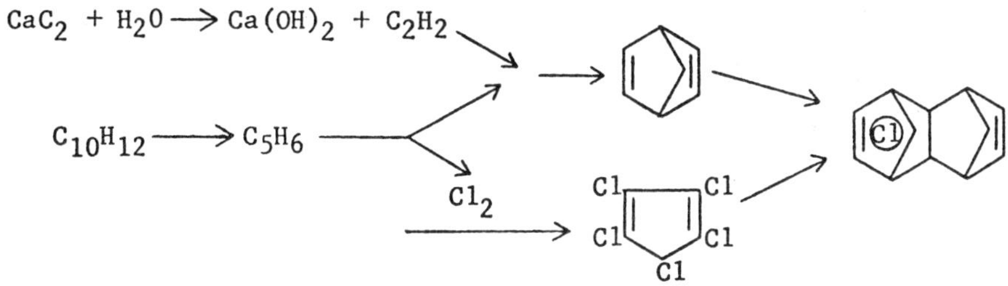

No liquid waste streams from the production plant contain aldrin. Small amounts of aldrin-containing liquid wastes from spill cleanup or floor washings would go to a 100-acre asphalt-lined evaporation basin on the arsenal. This basin is capable of evaporating 150 gal of water per minute year round and has shown no signs of leaks in its 16 years of use. The arsenal is also diked so that all surface waters go to unlined storage basins.

During shutdown the aldrin unit is washed with toluene and the washings go to dieldrin manufacture. Damaged drums containing aldrin are similarly cleaned, then incinerated, flattened and sold as scrap steel. Some wipe cloths are burned, some are cleaned. There have been no railroad accidents in shipping aldrin.

ALLYL CHLORIDE FROM PROPYLENE

Allyl chloride is made by chlorinating propylene at high temperatures (400° to 500°C). If lower temperatures (less than about 300°C) are used, the reaction becomes an addition reaction, yielding dichloropropane. For allyl chloride, the equation is shown below. The yields are about 85%.

$$Cl_2 + C_3H_6 \xrightarrow{500^\circ C} C_3H_5Cl + HCl$$

Allyl chloride is an important intermediate. Glycerol, allyl alcohol, and epichlorohydrin are made from allyl chloride. Industrial scale allyl chloride facilities perform three major processing operations: feed preparation, reaction, and product recovery. The propylene is purified by fractionation and drying. The chlorine is also dried.

The reaction is carried out in an adiabatic reactor designed to provide rapid and intimate mixing. Reaction temperature is controlled by balancing the mol ratio of the feed (usually 4 mols of propylene to one of chlorine) and propylene preheat temperature. The usual reaction temperature range is 500° to 510°C; pressures of 15 psig are used.

Chlorine utilization is in excess of 99%. Reactor products are then cooled to about 50°C. After initial removal of hydrogen chloride and propylene, the allyl chloride fraction is separated in a conventional two-step distillation. Yields of allyl chloride are about 85%. The raw materials are chlorine (dry gas) and propylene (dry gas). The by-products are coke, dichloropropane, hydrogen chloride, and chloropropenol (mostly 2-chloropropene).

There is one air pollution source from the allyl chloride process, the purge on the propylene recycle circuit to remove inerts, mainly impurities in the raw materials. The noncondensables from the stripping column are vented into the suction of the recycle compressor.

ANILINE FROM NITROBENZENE

Benzene is first converted to nitrobenzene in a mixture of nitric and sulfuric acids.

$$C_6H_6 + HNO_3 \xrightarrow{H_2SO_4} C_6H_5NO_2 + H_2O$$

The reactor effluent is decanted into a liquid/liquid separator, where crude nitrobenzene is separated from the acid solution. The acid solution is concentrated by steam stripping and recycled back to the reactor. Crude nitrobenzene is washed, vaporized, and fed to a fluidized bed reactor containing a copper-silica hydrogenation catalyst, where the following hydrogenation reaction occurs.

$$C_6H_5NO_2 + 3H_2 \longrightarrow C_6H_5NH_2 + 2H_2O$$

The unreacted hydrogen is recycled to the reactor. Reactor effluent goes to a separator, where two phases are formed. The organic phase contains water, and is fractionated in a two-tower system to remove heavy residue and water from the aniline product.

The aqueous layer, formed by the water of reaction, contains some aniline and is discharged into sewers. The major wastewater sources in this process are the crude nitrobenzene washwater and aniline water formed in the final separator. RWL survey data of this process are shown in the following tabulation

Flow	190 gal/1,000 lb
COD	13,400 mg/l
	21.2 lb/1,000 lb
BOD_5	15 mg/l
	0.02 lb/1,000 lb
TOC	12,150 mg/l
	19.2 lb/1,000 lb

Results of analyses indicate that, in addition to the parameters shown above, sulfate concentrations in wastewater streams are at levels inhibitory to biological treatment processes. The high RWL of this process is attributed to the high aniline concentration (3%) in aniline water from the final separator. It is a common practice to recover aniline by extraction either with incoming nitrobenzene or with benzene. However, such recovery was not practiced at the

FIGURE 17: ANILINE MANUFACTURE

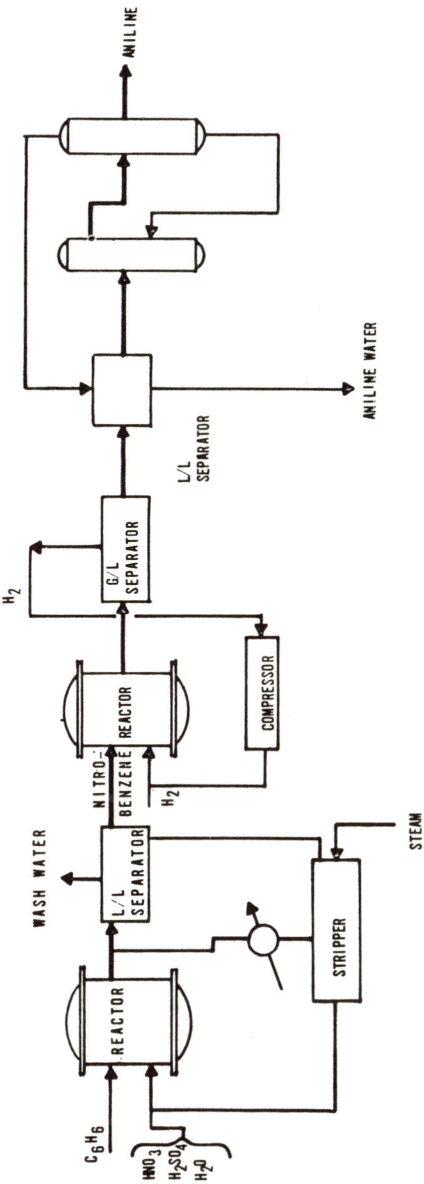

Source: Report EPA 440/1-73/009

plant visited during the survey. BADCT and BATEA in-process controls are defined by implementing an aniline recovery system to reduce process RWL. Instead of using a nitrobenzene extraction scheme, an effective steam-stripping system has been devised and the following is a description of the equipment and processing required.

Water from a 10^8 lb/yr aniline plant is steam stripped in a 2.5' x 40' tower. The feed to the stripper is 17 gpm containing 3.1% aniline by weight. The bottoms from the stripper will contain about 0.2% aniline. The overhead, essentially a 50/50 mixture of aniline and water is sent to incineration. Figure 18 is a process flowsheet of the proposed aniline stripper system.

FIGURE 18: ANILINE STRIPPER

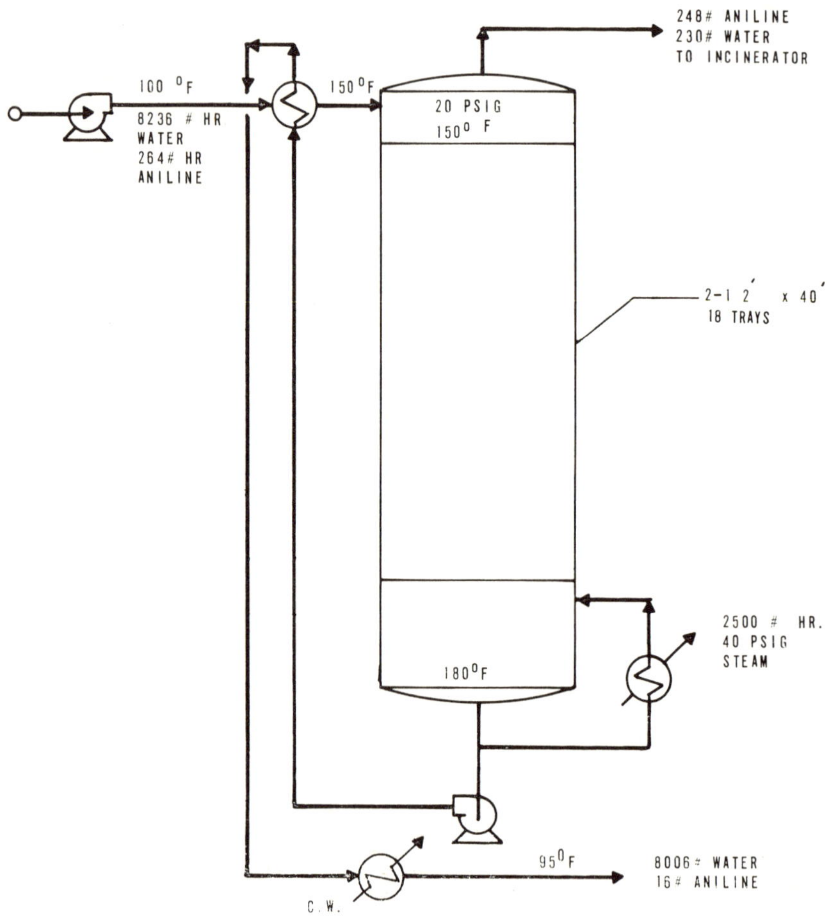

Source: Report EPA 440/1-73/009

With this modification, RWL can be expected to achieve the following values.

Flow	184 gal/1,000 lb
COD	1,390 mg/l
	2.13 lb/1,000 lb
TOC	1,490 mg/l
	2.29 lb/1,000 lb

The totally installed cost for the stripper, including heat exchange, pumps, instrumentation, piping, foundations, electrical wiring, structures, etc. is $115,000. The total annual operating cost, including depreciation, is about $45,000. For the 10^8 lb/yr aniline plant, this adds about $0.05 per pound to the cost of the aniline. Table 13 presents the economics of the proposed aniline stripper.

The alternate routes in manufacturing aniline are the traditional technique of nitrobenzene liquid phase reduction with iron filings and the liquid phase nitrobenzene hydrogenation technique.

TABLE 13: ANILINE STRIPPER ECONOMICS

Investment

Tower Cost, including trays, pumps, exchanges, instruments, piping, foundations, etc.	=	$115,000 Totally Inst.

Operating Costs

Utilities

		$/Yr
Steam: 2500 #/hr. x $.55/M# x 8000	=	$11,000
Power: 800,000 kwh x $.01/kwh	=	8,000
Cooling Water: 20 X 10^6 Gals. x $.25/M Gals	=	5,000
		$24,000

Investment Related

Maintenance Material & Labor 4%	=	$ 4,600
Plant Overhead 65% of Maintenance	=	3,000
Insurance, Taxes 1.5%	=	1,700
Depreciation 10% on BLCC	=	11,500
Total Expenses		$44,800

¢/Gal Handled 0.55 ¢/Gal.

¢/lb Aniline removed 2.36¢/lb

In 100 MM #/yr Facility 0.045¢/lb Product Aniline

Source: Report EPA 440/1-73/009

ATRAZINE FROM CYANURIC CHLORIDE

Atrazine is a selective herbicide of the triazine class. Its chemical name and formula are as follows.

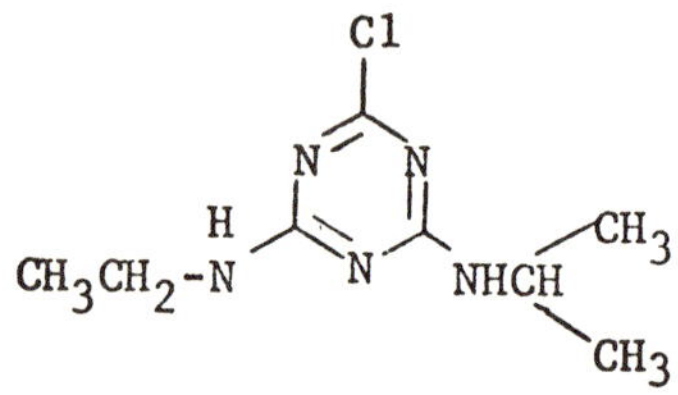

2-chloro-4-ethylamino-6-isopropylamino-s-triazine

It is a solid, nearly insoluble in water. It has low toxicity and is nonpersistent. It is produced by Geigy at St. Gabriel, Louisiana using the following production chemistry.

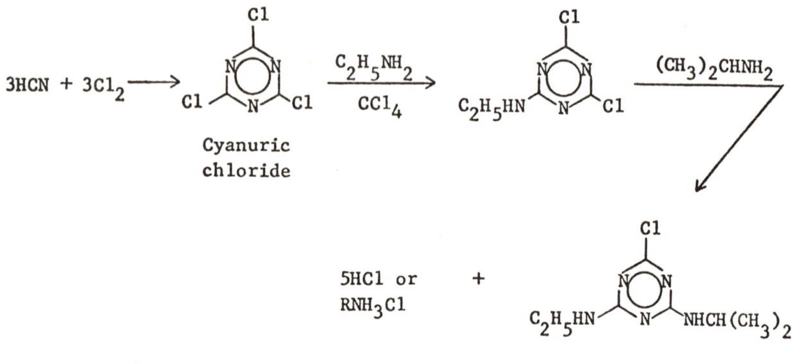

About 1 lb of effluent is generated per pound of atrazine produced, mostly NaCl. Liquid wastes from the cyanuric chloride production unit ordinarily go to a 6,000 ft deep disposal well, after receiving a preliminary polishing (pH adjustment and filtration).

The larger amount of liquid wastes from the remainder of the plant are discharged to the river. Sanitary wastes from the plant are chlorinated before they are discharged. The BOD of the waste going to the river is 500 lb/day at a 100 million pound per year production rate. Solid waste is primarily bag wrappers, car lining material, etc. This waste is disposed of by a commercial operator by landfill not located on the plant site. The formulation and packaging areas

FIGURE 19: PRODUCTION AND WASTE SCHEMATIC FOR ATRAZINE

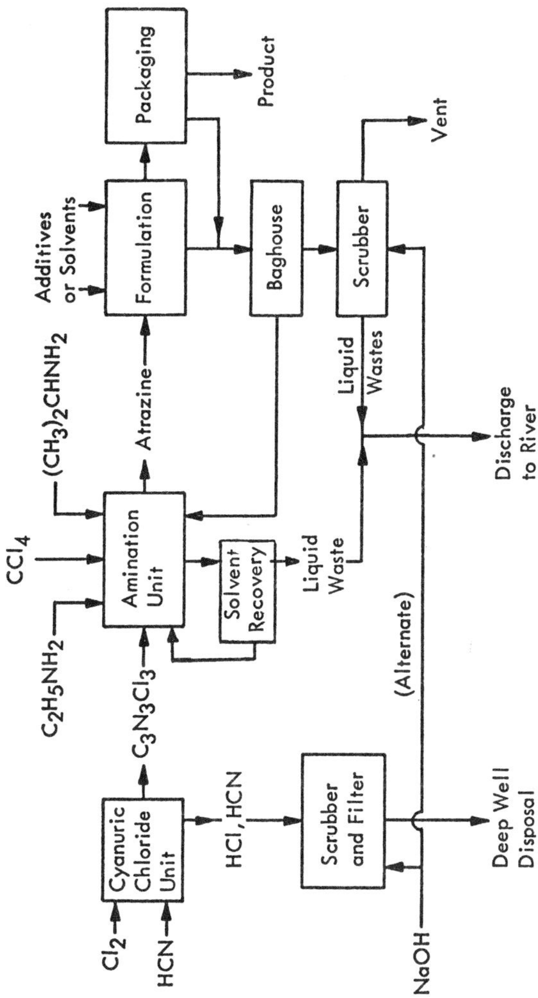

Source: Midwest Research Institute Report PB 213,782; 6/72

are controlled by baghouses and wet scrubbers and atmospheric monitors are used. Losses are said to be substantially $<1\%$. Breakage and leakers have not been a major problem. Returns have been $<1\%$. Overall repackaging is $<2\%$. Package disposal is a problem. They can be burned, but what happens to the atrazine is not known.

Within the next two years, an additional investment equal to 10% of the new plant's cost will be made for spill protection and the treatment of organic waste. Fifty percent of the next five years capital budget will be for environmental control. Chlorine recovery may have to be considered instead of deep well in the future, although Louisiana encourages the use of deep wells for waste disposal, and 20 wells within four miles of the atrazine plant have operated for over 10 years without major difficulty.

BTX FROM PETROLEUM

A mixture of BTX aromatics and saturates may be obtained as a by-product of ethylene manufacturing (by pyrolysis of naphtha feedstocks). This mixture must first be stabilized by hydrotreating prior to the recovery of BTX aromatics by solvent extraction.

The first stage hydrotreating of pyrolysis gasoline differs from conventional hydrotreating of virgin stocks in that the feedstock is difficult to handle and cannot be heated to the 500° to 700°F temperatures needed for conventional hydrotreating. Water injection is not required and the process itself should be nonpolluting.

In the low temperature processes, diolefins and other reactive compounds are hydrogenated to yield a product which can be stored or handled in conventional refinery and petrochemical processing.

The second stage hydrotreater is similar to the conventional refinery hydrotreater with a cobalt-molybdenum catalyst. Under certain conditions it may be necessary to prefractionate the feed to these processes as heavy polymers can rapidly deactivate catalyst. Frequently steam stripping, or steam injection with the fractionator feed is used in these operations. The condensate must then be disposed of.

The catalysts used for the first stage hydrotreating operation contain either nickel or a noble metal. They require more frequent regeneration than most refinery catalysts. Once in about four months may be a reasonable number. The steam-air decoking may result in an air pollution problem.

Some plants contain a provision to inject water to wash out ammonium bisulfite salts which might be formed in the reaction. In this case, a coalescer, or water separator, would be used to separate this water. This operation would only be performed intermittently, and would not be a major source of wastewater.

The feedstock and recycled hydrogen gas are preheated and passed through a series of hydrotreating reactors containing platinum catalyst. The reactor effluent is cooled and then discharged into a separator, where the gas stream taken

FIGURE 20: BTX PRODUCED BY HYDROTREATING PYROLYSIS GASOLINE WITH SUBSEQUENT SOLVENT EXTRACTION

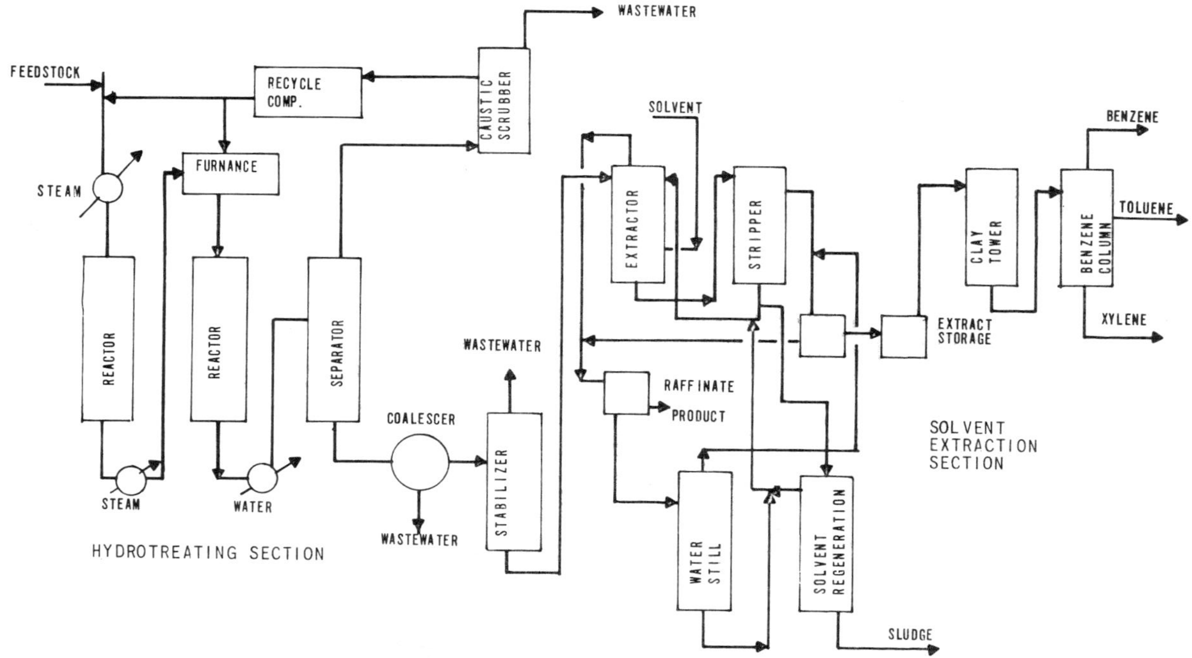

Source: Report EPA 440/1-73/009

overhead is recycled back to the reactor after being scrubbed with caustic solution. The liquid phase from the reactor is passed through a coalescer (where water is used to trap coke particles formed in the pyrolysis reactor) and a stabilizer (where light hydrocarbons are removed).

The survey data obtained from a plant with pyrolysis gasoline as feedstock are shown in the following tabulation. The data presented can be considered as the standard for all levels of control technology of this process. The only sources of wastewater are in the hydrotreating section of the process.

Flow	13.6 gal/1,000 lb
COD	2,755 mg/l
	0.31 lb/1,000 lb
BOD_5	914 mg/l
	0.104 lb/1,000 lb
TOC	306 mg/l
	0.034 lb/1,000 lb

The stabilized liquid is then extracted with a solvent (di- and triethylene glycol) to recover the aromatics, and the raffinate (containing paraffins) is sent to a cracking furnace to produce olefins. The dissolved aromatics (BTX) are separated from the solvent by distillation, and the solvent-free aromatics are waterwashed and then separated into the individual components: benzene, toluene, and xylene.

The separated solvent is regenerated and recycled to the process, while the sludge produced is disposed by landfill. There are many solvents that will extract aromatics from naphthas at high recoveries and purities, but for many reasons only a few are used commercially.

In recent years, several refiners have switched from di- and triethylene glycol to tetraethylene glycol. A change to tetraethylene glycol can usually be achieved with minor equipment modifications and no change in royalty status. The most recent development shows that sulfolane has established itself in the U.S. as the preferred solvent for BTX extraction.

Although the results and economics of using sulfolane have not been published, it is known that drastic changes in process conditions and a relatively high solvent cost cause a large capital investment for this solvent change. The sulfolane system differs from other solvent systems in that the solvent regeneration is under vacuum.

If steam ejectors with barometric condenser are used to produce the vacuum, the resultant oily water will be a significant water pollution source. Therefore, BPCTCA and BATEA levels of control technology for the sulfolane system will require vacuum pumps with surface condensers to produce vacuum for process needs, and the resulting oily stream should be disposed of by incineration.

Alternatively, solvent extraction may be employed on C_6 to C_8 reformate cuts to extract aromatics from low octane paraffins. The raffinate could be fed to synthetic natural gas (SNG) generation or a petrochemical facility or, alternatively, recycled to the catalytic reformer. The extraction unit here might be of different design than a unit for chemical production in that high purities and recoveries are not required. As the refinery picture is complex, similarly it is difficult to predict the growth of extraction processes. Sulfolane, or other solvent, becomes a loss in three ways which affect waste disposal: solvent degradation, units have provision for solvent regeneration but periodically heavy materials must be purged; losses to products due to incomplete waterwash; and leaks.

An estimate in a sulfolane unit was that 30 lb of sulfolane makeup were required for 1,000,000 lb of feed. Published information on Udex operation indicated losses about three times as great. (For Udex, 0.031 lb/barrel of feed, *Oil and Gas Journal*, May 1962.) These then represent the material, either as solvent, or degradation products, which can go to waste streams.

The sulfolane system differs from other solvents in that the solvent regeneration is under vacuum. In a 1,000 BPSD feed plant, about 5,000 lb/hr of 100 lb steam might be required to maintain the necessary vacuum for one design, and this quantity of oily condensate must be disposed of. If surface condensing is not used, but barometric condensers used, the quantity of oily water to be disposed of increases by a factor of 10 to 50 (water required to condense the steam).

It is difficult to project the future requirements for new solvent extraction units. With high severity reforming required for low-lead or no-lead gasoline, there should be ample benzene in reformate so that, there should be no need to extract toluene for conversion to benzene. Furthermore, for other chemical purposes toluene requirements are small and consequently toluene will be left in the gasoline pool.

In high severity reforming the C_8 aromatic fraction contains very little paraffins and para-xylene can be produced by crystallization without solvent extraction. With this argument, solvent extraction will only be used on reformate for benzene where the solvent/feed requirements are minimal. Consequently, existing facilities may be adequate and no new units may be required for refineries. RWL data for the Udex solvent extraction process are summarized below.

Flow RWL	60.4 (gal/1,000 BTX extract)
TOC RWL	0.144 (lb/1,000 lb BTX extract)

BISPHENOL-A FROM PHENOL AND ACETONE

Di(p-hydroxyphenyl)propane, also known as bisphenol-A, is produced by reacting phenol with acetone in the presence of acid catalyst, and the chemical reaction is given below.

$$2C_6H_5OH + CH_3COCH_3 \longrightarrow CH_3C(C_6H_4OH)_2CH_3 + H_2O$$

A number of by-products are formed in conjunction with the main reaction. The earlier processes eliminated these impurities by batchwise crystallization, while the new process, the Hooker process, employs a continuous distillation and extractive crystallization under pressure to purify the product.

FIGURE 21: BISPHENOL-A FROM PHENOL AND ACETONE

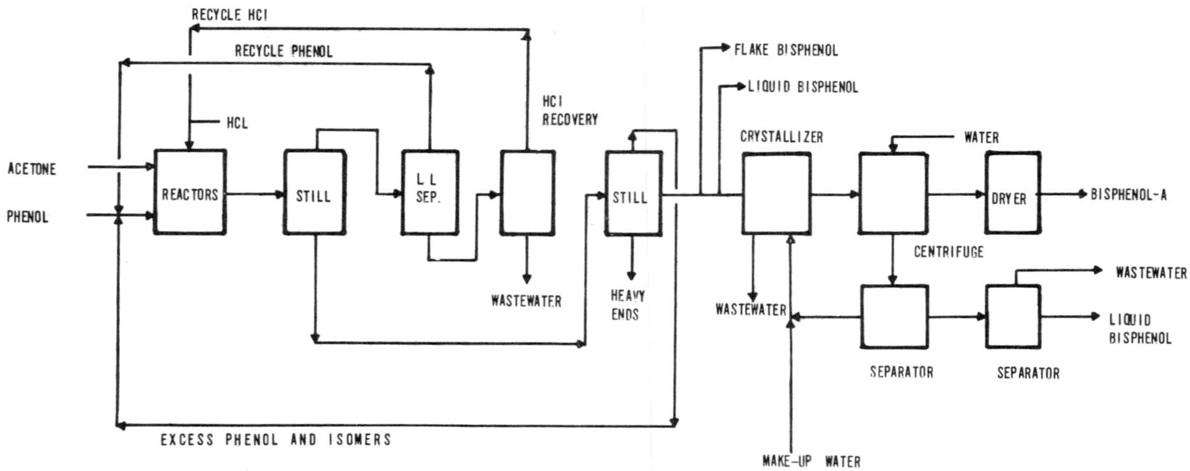

Source: Report EPA 440/1-73/009

Phenol and acetone at a molar ratio of approximately 3 to 1 are mixed, saturated with hydrogen chloride gas, and sent to the reaction vessel. Reaction conditions are about 40°C, close to atmospheric pressure, with a mercaptan used as a catalyst. The crude product is stripped of HCl and water of reaction. The overhead is decanted into an organic phase (consisting mainly of phenol which is recycled) and an aqueous phase. The latter goes on to an HCl recovery unit, and water is sent to disposal.

Bottoms from the stripper are sent to a series of purification distillation chambers, where excess phenol, isomers, and heavy ends are removed from the system for either recycle or disposal. Distillate from the last chamber is sent to the extraction operation, which produces a slurry of pure crystals. The filtrate from the centrifuge is partially recycled to the crystallizer, and the remainder is concentrated in an evaporator to produce liquid bisphenol-A.

The water separated from the HCl recovery unit, the extracted aqueous phase from the crystallizer, and the condensate from the final evaporator are the major wastewater sources. The characteristics of the wastewater obtained from survey data are presented in the following tabulation.

Flow	66.8 gal/1,000 lb
COD	30,699 mg/l
	17.11 lb/1,000 lb
TOC	9,216 mg/l
	5.13 lb/1,000 lb
Phenol	12,713 mg/l
	7.1 lb/1,000 lb

The high concentration of phenol produces an inhibitory effect and interferes with the BOD_5 measurement. The organic contaminants in the wastewater are mainly phenol, bisphenol, and organic solvent. Incomplete separation of the aqueous and organic phases in the decanter causes the high loss of organics into the wastewater. Organic vapor escaping from the final evaporator also contributes a significant amount of contaminants.

To define BADCT and BATEA, a steam stripper should be required to recover and recycle these organic contaminants in the two major waste streams. The specification and the estimated economics of a steam stripper have been presented in the discussion of Aniline. The total process water usage of this process is approximately 0.25 lb/lb of bisphenol-A, while the gross cooling water usage is about 197 lb/lb of product.

BUTADIENE FROM BUTANE

Butadiene is produced as the by-product of the cracking of hydrocarbons or by dehydrogenating of C_4 hydrocarbons, such as n-butane or butylenes, or as a coproduct of ethylene manufacture.

Dehydrogenation of n-Butane: The one-step catalytic dehydrogenation of n-butane is carried out in the vapor phase with solid chromium-on-alumina catalyst.

FIGURE 22: BUTADIENE FROM DEHYDROGENATION OF n-BUTANE

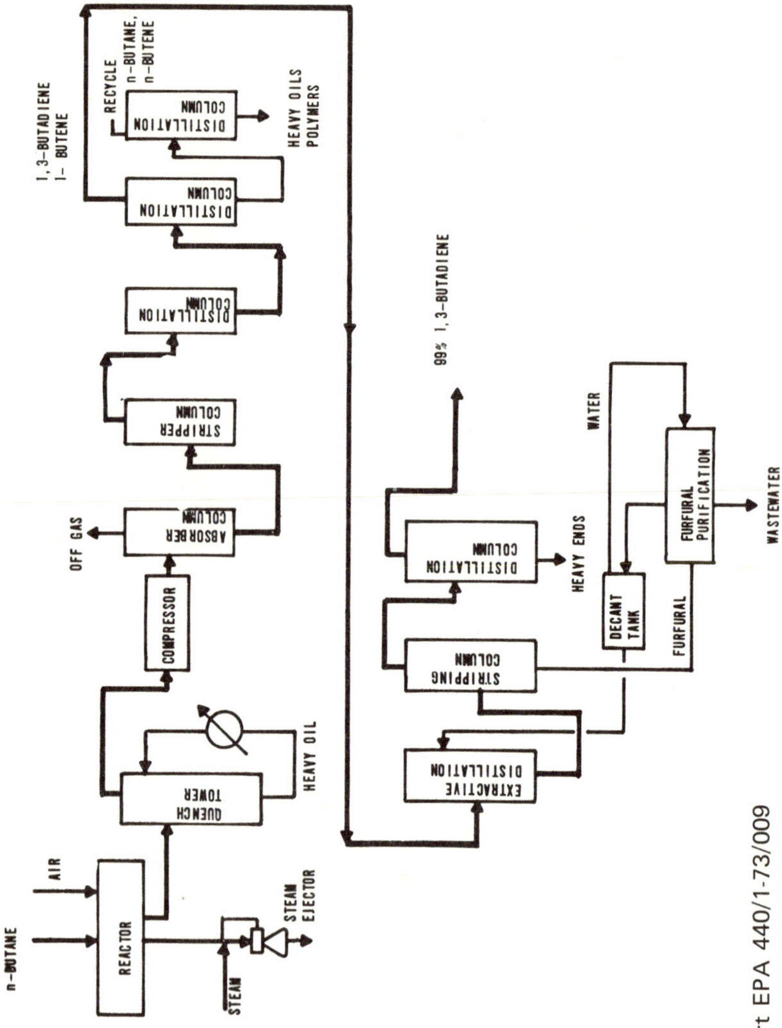

Source: Report EPA 440/1-73/009

The reactors operate under vacuum at approximately 3 lb per square inch absolute (psia) to obtain low hydrocarbon partial pressures. This prevents excessive coking on the catalyst. The chemical reaction is given below.

$$CH_3CH_2CH_2CH_3 \longrightarrow CH_2{=}CHCH{=}CH_2 + 2H_2$$

The feedstock (n-butane) and recycled butane and butenes are fed into a battery (n-butane) and recycled butane and butenes are fed into a battery of fixed bed reactors. The reactor effluent is oil-quenched, compressed, and sent to an absorption column, where hydrocarbon vapor is absorbed with light oil. The effluent from the absorber is then passed through a series of distillations where unreacted butane and butene are separated for recycle to the dehydrogenation reactors.

Butadiene is separated from the butene splitter overhead by extractive distillation with furfural or cuprous ammonium acetate (CAA) extraction.

A different process for producing butadiene from C_4 hydrocarbons was employed by one plant visited. The process dehydrogenates butane to butylenes using superheated steam as a diluent. After separation from light and heavy by-products, the butylenes are converted to butadiene by oxidative dehydrogenation. This reaction is illustrated below.

$$C_4H_8 + \tfrac{1}{2}O_2 \longrightarrow C_4H_6 + H_2O$$

Coproduct of Ethylene Manufacture: Butadiene is also produced by extraction from the C_4 and heavier residue produced in ethylene manufacture. The C_4 residue goes to an extractive distillation with furfural or cuprous ammonium acetate (CAA) extraction. The effluent is then sent to a steam stripper and fractionator, where heavy ends are removed. The stream taken overhead from the fractionator is further washed with water to remove solvents.

The dehydrogenation of n-butane produces wastewaters from scrubbing the gases used to periodically burn coke from the catalyst surface, and from the steam ejector-barometric condenser systems used to obtain vacuum in the reactors. The process also produces wastewaters from the final recovery of butadiene product.

Wastewaters generated by the process wherein butadiene is formed as a coproduct of ethylene manufacture are essentially the same as those from the final recovery unit in the dehydrogenation process. Survey data obtained from plant visits are summarized in the following tabulation.

Plant No.	Process	Flow gal/1,000 lb (mg/l)	COD lb/1,000 lb (mg/l)	BOD_5 lb/1,000 lb (mg/l)	TOC lb/1,000 lb (mg/l)
1	Dehydrogenation, extractive distillation	1,160	3.23 (334)	2.96 (306)	– –
2	Dehydrogenation, extrative distillation	1,451	245 (20,200)	72 (5,960)	– –

(continued)

Plant No.	Process	Flow gal/1,000 lb (mg/l)	COD lb/1,000 lb (mg/l)	BOD_5 lb/1,000 lb (mg/l)	TOC lb/1,000 lb (mg/l)
3	Coproduct of ethylene, extractive distillation	88	1.120 (1,525)	0.547 (745)	0.554 (755)
4	Coproduct of ethylene, extractive distillation	339	3.899 (1,378)	1.183 (418)	1.545 (546)
5	Coproduct of ethylene, extractive distillation	183	1.042 (683)	0.165 (102)	0.313 (205)

Since furfural has a relatively high boiling point, it can be easily separated from product butadiene by distillation. Furfural loss in the waterwashing step is minimal. The process (Plant 4) utilizing extraction with cuprous ammonium acetate must be equipped with a water scrubber to remove CAA from the final product and therefore produces considerably more wastewater as well as a higher RWL.

The data presented by Plants 3, 4 and 5 confirm the above argument and can be considered, respectively, as representatives of BPCTCA for the ethylene co-product process with extractive distillation and extraction. However, to define BPCTCA and BATEA control technology, a steam stripper should be used to recover solvents, furfural, or CAA from the scrubber water, which would then be recycled. With this in-process can be reduced to the following levels.

Flow	65 gal/1,000 lb
COD	0.43 lb/1,000 lb
BOD_5	0.18 lb/1,000 lb
TOC	0.21 lb/1,000 lb

As indicated in the data tabulation, RWL's presented by the two dehydrogenation plants show a significant variation. Although these two plants represent two different dehydrogenation routes, as described in the preceding discussion the variation is due mainly to less effective operation of wash columns and strippers rather than to differences in the processes. Consequently, the data presented by Plant 1 should be considered as representative of BPCTCA.

Since the dehydrogenation reactor is operated at approximately 3 psia, steam ejectors with barometric condensers are used to produce the vacuum, and these generate an excessive amount of wastewater. A vacuum pump system, such as that described in the Styrene section can substantially reduce the amount of wastewater and also eliminate organic losses in the exhaust stream.

To define BADCT and BATEA control technology for the dehydrogenation process, steam ejectors should be replaced with vacuum pumps and a steam stripper should be installed to recover organic solvent, as described in the ethylene co-product processes. An alternate route for butadiene manufacturing is the dehydrogenation of n-butene. It is a vapor phase reaction with a catalyst of iron oxide promoted by chromium oxide, magnesium-iron oxide, or calcium nickel phosphate. Butadiene is produced through several dehydrogenation reactors (in parallel) containing fixed bed catalysts.

The reaction gases are quenched and cooled in a series of quench towers. The condensate containing C_4 hydrocarbons is charged to a fractionating column where it is stabilized. The stabilized condensate is then treated for the removal of polymerized materials. Finally, the crude butadiene is purified by absorption or extractive distillation.

A process developed by *R.C. Woerner; U.S. Patent 3,336,414; August 15, 1967; assigned to Petro-Tex Chemical Corporation* involves the simultaneous purification of butadiene from undesirable carbonyl compounds and the disposal of boiler blowdown water from an adjacent boiler plant.

The oxygenated compounds are a serious contaminant in the unsaturated hydrocarbon product and must be essentially completely removed in order to have a product of suitable purity, e.g., a product having on the order of a few parts per million carbonyl compounds. The essentially complete removal of the oxygenated compounds is quite difficult for several reasons.

In the first place, the oxygenated compounds constitute only a very minor percentage of the gaseous stream to be purified. Normally the carbonyl compounds will only constitute less than 5 mol percent of the gaseous stream to be purified and more usually may constitute such as less than or up to 2.5 mol percent of the gaseous stream. The oxygenated compounds are therefore quite difficult to remove because of their low concentrations in the gaseous stream.

In addition, the oxygenated compounds may be difficult to separate from the hydrocarbons regardless of their relative concentration. Azeotropes may form between the oxygenated compounds and various hydrocarbons. For instance, an azeotrope is formed between acetaldehyde and butadiene-1,3.

Boiler blowdown waters are formed as high pH waste products from the operation of water boilers. In most states there are restrictions preventing discharge of compositions of this nature into the stream and water supply. For example, the high pH of the water must be neutralized. Consequently, it would be advantageous to have a method for the treatment of these waste products at low cost.

Boiler blowdown waters also have a high heat content. If these compositions are simply treated as such the heat content may be lost. Moreover, it may be necessary to utilize cooling water to bring these streams to a suitable temperature for treatment.

Ordinarily wastewater having a high heat content may be used as a heating medium such as in heat exchangers. However, if the wastewater has a high content of dissolved solids it may be unsatisfactory for this use because it will plug the tubes of the heat exchangers. Therefore, another object is to provide a method for utilizing the heat content of aqueous compositions containing a high level of dissolved solids.

In the development of the process, it was found that boiler blowdown water can be effectively utilized as a treating composition for the removal of oxygenated compounds from organic compounds such as butadiene contaminated with oxygenated compounds. This discovery was surprising in view of the high concentration of dissolved solids in the boiler blowdown.

It might have been thought that the use of boiler blowdown as a contacting medium in apparatus such as bubble cap columns would plug the contacting apparatus due to the high concentration of solids. It has been found, however, that this is not the case and that there are even some improved results in the operation of the apparatus when boiler blowdown is used as the contacting medium.

A process developed by *S.A. Cunningham; U.S. Patent 3,557,238; January 19, 1971; assigned to Phillips Petroleum Company* provides an alternative technique for handling oxygenated by-products from butadiene production by oxydehydrogenation. The process involves the following steps:

(1) condensing the oxygenated hydrocarbons along with steam in the reactor effluent;

(2) separating the thus condensed steam and oxygenated hydrocarbons from the hydrocarbon portion of the effluent;

(3) recycling the oxygenated hydrocarbons and condensed steam after vaporization to the reaction zone during dehydrogenation whereby there is an accumulation of oxygenated hydrocarbons in the condensed steam; and

(4) recycling the condensed steam containing accumulated oxygenated hydrocarbons after vaporization together with air to the reaction zone during regeneration of the dehydrogenation catalyst whereby, in the absence of fresh hydrocarbon feed, the oxygenated hydrocarbons are converted to carbon oxides and water thereby removing the necessity of purging the accumulated oxygenated hydrocarbons from the system as environmental pollutants.

Under normal plant operating conditions, these oxygenated by-products might be ultimately vented to the atmosphere and/or discharged with wastewater from the process depending upon the separation and recovery processes employed and their operating conditions.

However, it has been found that these by-products are toxic and result in damage to property, particularly crops and foliage and are probable contributors to photochemical smog and haze, especially when vented as aerosols. It is not only desirable to eliminate or at least reduce this source of air and water pollution, but such control is essential in many locations.

A process developed by *T. Hutson, Jr. and R.E. Riter; U.S. Patent 3,646,239; February 29, 1972; assigned to Phillips Petroleum Company* is one in which excess purge water containing oxygenated hydrocarbons resulting from hydrocarbon dehydrogenation processes is rendered nontoxic by the conversion of the oxygenated hydrocarbons to water and carbon oxides in the presence of microorganisms, e.g., saprophytic bacteria, protozoa, yeast and fungi, under aerobic conditions. In a preferred mode of operation the purge water prior to treatment with microorganisms is reboiled or stripped with a hot gas to reduce the concentration of oxygenated hydrocarbons.

A process developed by *R.A. Hinton and J.E. Cottle; U.S. Patent 3,679,764; July 25, 1972; assigned to Phillips Petroleum Company* is one in which water containing oxygenated hydrocarbons resulting from hydrocarbon oxidative dehydrogenation processes is rendered substantially nontoxic by stripping with steam to remove the oxygenated hydrocarbons including carbonyls and returning to the oxidative dehydrogenation process to suppress the formation of additional oxygenated hydrocarbons and the conversion of a portion of the steam stripped water to steam for use as stripping medium. The steam stripped water phase can be neutralized with a base prior to reboiling and converted to steam for use as the stripping medium.

CAPROLACTAM FROM CYCLOHEXANE

Caprolactam is produced in the Beckman process by the addition of hydroxylamine sulfate to cyclohexanone, which is derived from cyclohexane. The basic chemical equations are given below.

$$C_6H_{12} + O_2 \xrightarrow{H_3BO_3} C_6H_{10}O \xrightarrow{H_2NOH \cdot HSO_4} C_6H_{10}NOH$$

$$C_6H_{10}NOH \xrightarrow{H_2SO_4} \overline{CH_2(CH_2)_4NHCO} + (NH_4)_2SO_4$$

Feed and recycled cyclohexane are mixed with air in an oxidation reactor in the presence of boric acid, which minimizes adipic acid production. The oxidation is carried out at approximately 150 psig and 160°C. The gaseous effluent is scrubbed to separate unreacted cyclohexane from what is essentially nitrogen. The liquid effluent is flashed with water and separated into an organic phase and an aqueous catalyst phase, which is then sent to a catalyst recovery unit.

The organic phase is essentially a mixture of unreacted cyclohexane, cyclohexanone, and cyclohexanol. This mixture is first distilled to recover unreacted cyclohexane and followed by saponification and fractionation to separate cyclohexanone from cyclohexanol, which is then converted to cyclohexanone by dehydrogenation.

The hydroxylamine sulfate is obtained from ammonium nitrates and sulfur dioxide. Ammonia gas and air are fed to a converter where ammonia is burned at about 700°C in the presence of a catalyst and converted to disulfonate by contacting with ammonium carbonate and sulfur dioxide in series. The disulfonate is then hydrolyzed to hydroxylamine.

By addition of cyclohexanone to hydroxylamine sulfate, cyclohexanone oxime is first produced and rearranged in nearly quantitative yield to caprolactam in the presence of concentrated sulfuric acid. The product is neutralized, and the ammonium sulfate solution is extracted with benzene to recover the lactam product and discharged to a concentration and recovery step. The major water pollution sources in this process are the draw-offs from catalyst recovery unit, saponification and wash tower, and the final product purification step. The contaminants in the waste stream are small amounts of diacids formed during the oxidation step, sodium salts, and unrecovered intermediate products. The characteristics of the wastewater obtained from the plant survey are summarized in the following tabulation.

	Plant 1	Plant 2
Flow	1,334 gal/1,000 lb	2,500 gal/1,000 lb
COD	358 mg/l	*
	4.0 lb/1,000 lb	*
BOD_5	147 mg/l	
	1.64 lb/1,000 lb	11.2 lb/1,000 lb
TOC	109 mg/l	
	1.22 lb/1,000 lb	*

*Not available

Since it is deemed unfeasible to reduce RWL of this process by any in-process modification, the RWL presented in the preceding tabulation can be considered as standard for BADCT and BATEA. Several other commercial routes to caprolactam are available, and process highlights of each route are summarized in the following paragraphs.

In the Toyo Rayon process, nitrosyl chloride is first manufactured by reacting ammonia gas with air at 700°C and atmospheric pressure using platinum-rhodium gauze as a catalyst, then with concentrated sulfuric acid, and finally with hydrogen chloride. The nitrosyl chloride gas mixture is then reacted with cyclohexane to give the cyclohexane oxime hydrochloride.

The reaction is carried out in the liquid phase, using the visible light emitted by mercury lamps to induce the photonitrosation. Subsequently, cyclohexanone oxime hydrochloride is treated with oleum to produce a sulfuric acid solution of caprolactam, which is then purified by a series of purification steps.

The Snia Viscosa process is based on the nitrosation of hexahydrobenzoic acid with sulfuric acid in oleum. The feed toluene is oxidized with air and then hydrogenated over a palladium catalyst to form hexahydrobenzoic acid. Caprolactam is then formed by reacting hexahydrobenzoic acid with nitrosylsulfuric acid, which is prepared by bubbling N_2O_3 into the cyclohexanecarboxylic acid dissolved in oleum.

The other route (referred to as the Caprolactone process) produces caprolactam without any ammonium sulfate by-product. Caprolactone is first produced by oxidation of cyclohexanone with peracetic acid, which is produced by acetaldehyde oxidation.

The resulting caprolactone is distilled under vacuum and reacted with ammonia at high pressure to form caprolactàm, which is purified using conventional distillation techniques.

FIGURE 23: CAPROLACTAM FROM CYCLOHEXANE

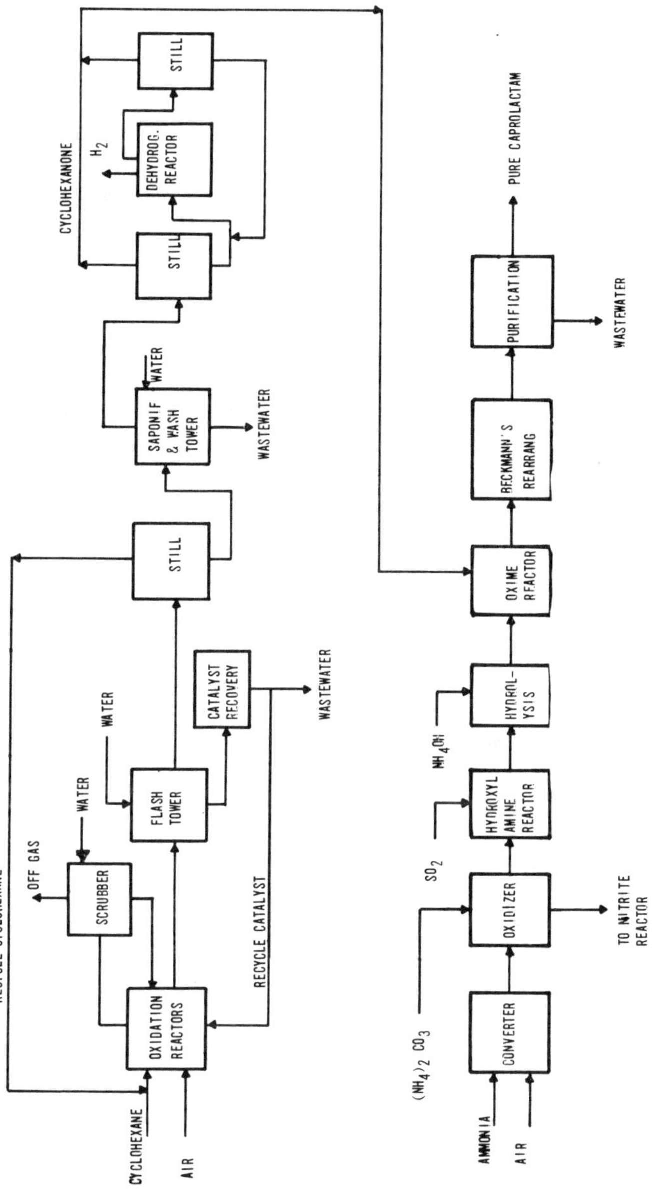

Source: Report EPA 440/1-73/009

CAPTAN FROM BUTADIENE

Captan is a fungicide of the chlorinated organosulfur compound type. Its chemical name and formula are as follows.

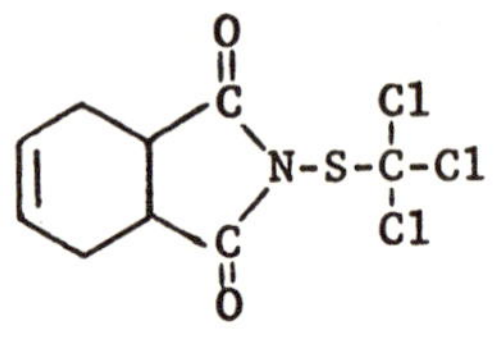

N-trichloromethylthio-4-cyclohexene-1,2-dicarboximide

It is a solid material, very nearly insoluble in water. It has very low toxicity and is nonpersistent. It is produced by Calhio Chemical using the following production chemistry.

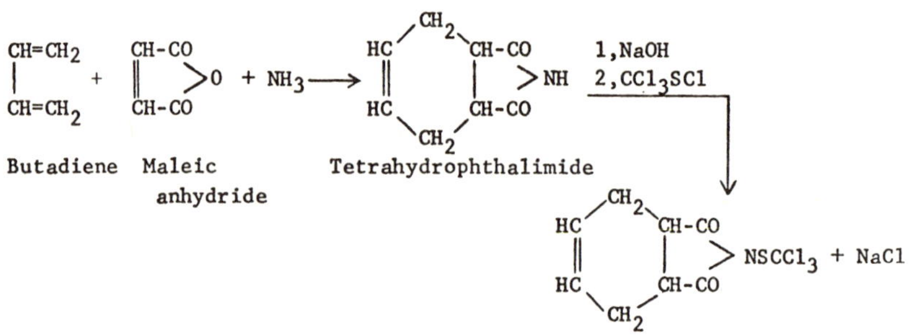

$$CS_2 + 3Cl_2 \xrightarrow{I_2} CCl_3SCl + SCl_2$$

Perchloromethyl mercaptan

All liquid effluents from the production facility (including 70 to 80 gpm dilute saline from the captan unit and that from the scrubbers) go to an asphalt-lined settling pond; retention time is about 6 or 7 days. The effluent from the holding pond (averaging 75 to 80 gpm) goes to a river that feeds into Lake Erie.

This effluent is monitored weekly. Captan levels in the discharge are said to be nil. This system has been in operation since the 1950s and no sludge has been removed to date. Depth ranges from 3½ to 4½ ft. Sanitary waste goes to a septic tank. Calhio is testing a deep well (6,000 ft deep) to be used for liquid wastes. Over 1 million gallons of clean water are being pumped down the well before waste injection can be initiated. This will be the only well in this part of the state; five or six wells are located in the western part of Ohio.

FIGURE 24: PRODUCTION AND WASTE SCHEMATIC FOR CAPTAN

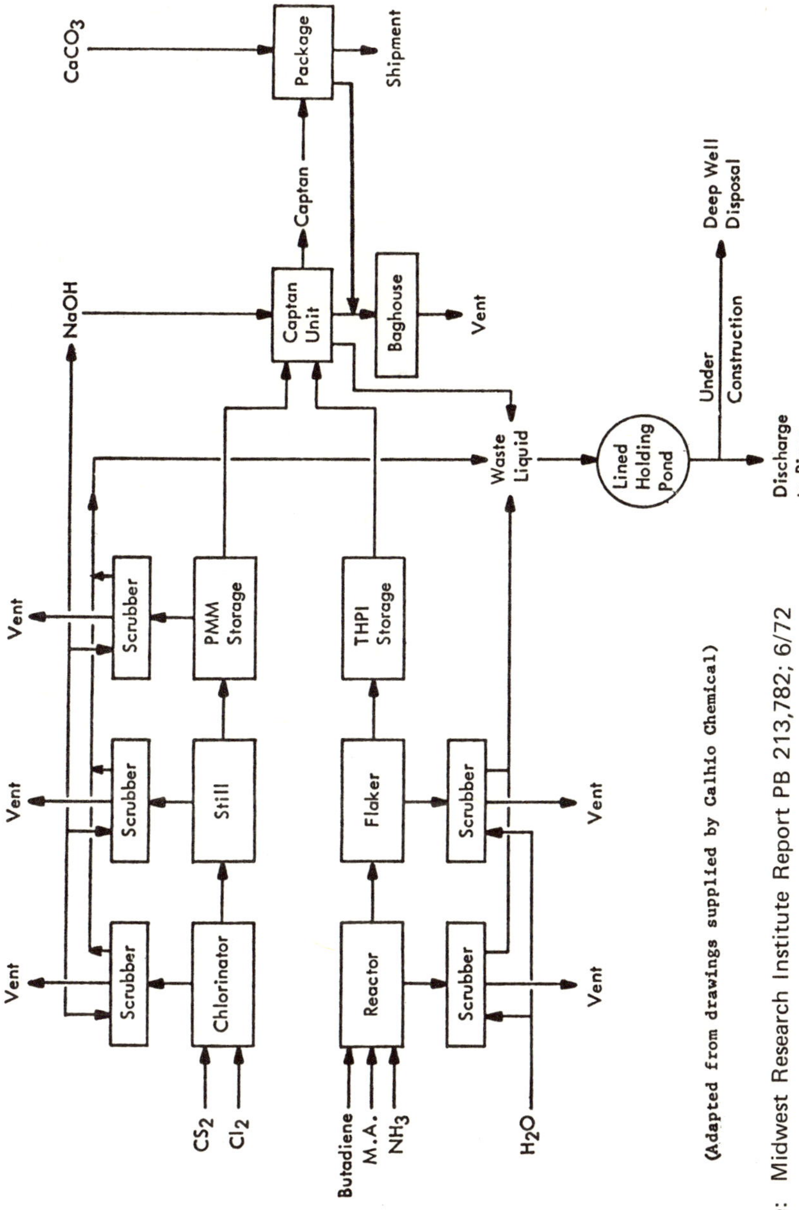

(Adapted from drawings supplied by Calhio Chemical)

Source: Midwest Research Institute Report PB 213,782; 6/72

Surface facilities for the well have not been designed yet, pending results from the tests. Capital investment at this plant for environmental protection, including the deep well, is between $1 and $2 million.

Solid Waste Disposal – Solid chemical waste (1,200 lb/yr) from spills, contaminated waste from floors, etc. are buried on plant property. Trash paper and cardboard (10 tons/yr) is moved out by a local collector. Metal scrap (25 tons per year) is sold to local scrap dealers.

Air Pollution Controls –

Baghouse stacks – 8,000 cfm
Particulates – captan, 3.43 lb/day (0.048 grains per cubic foot)

Filter hood exhaust – 6,500 cfm
Steam and particulates – captan and water, quantities unknown, less than above

Packer baghouse stack – 3,500 cfm
Particulates – quantities unknown, less than above

Product changeover is about once per year. Cleanup is required of the imide and captan processes only. All the material from the cleanup of these units is salvaged. Dust is returned to the broken bag handling system; lumps are slurried and refiltered; and water from the cleanup goes into the waste system.

CARBARYL FROM NAPHTHALENE

Carbaryl is an insecticide of the alkyl carbamate variety. Its chemical name and formula are as follows.

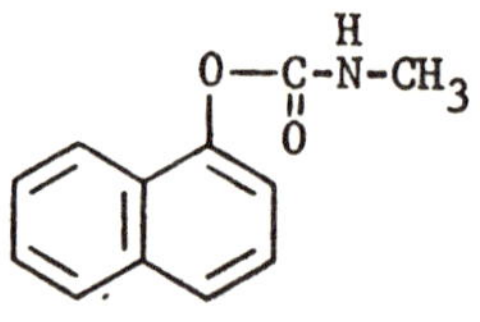

1-naphthyl-N-methylcarbamate

It is a solid with low (99 parts per million) solubility in water. It is moderately toxic but nonpersistent.

It is produced by Union Carbide at Institute, West Virginia according to the production chemistry shown on the following page.

FIGURE 25: PRODUCTION AND WASTE SCHEMATIC FOR CARBARYL

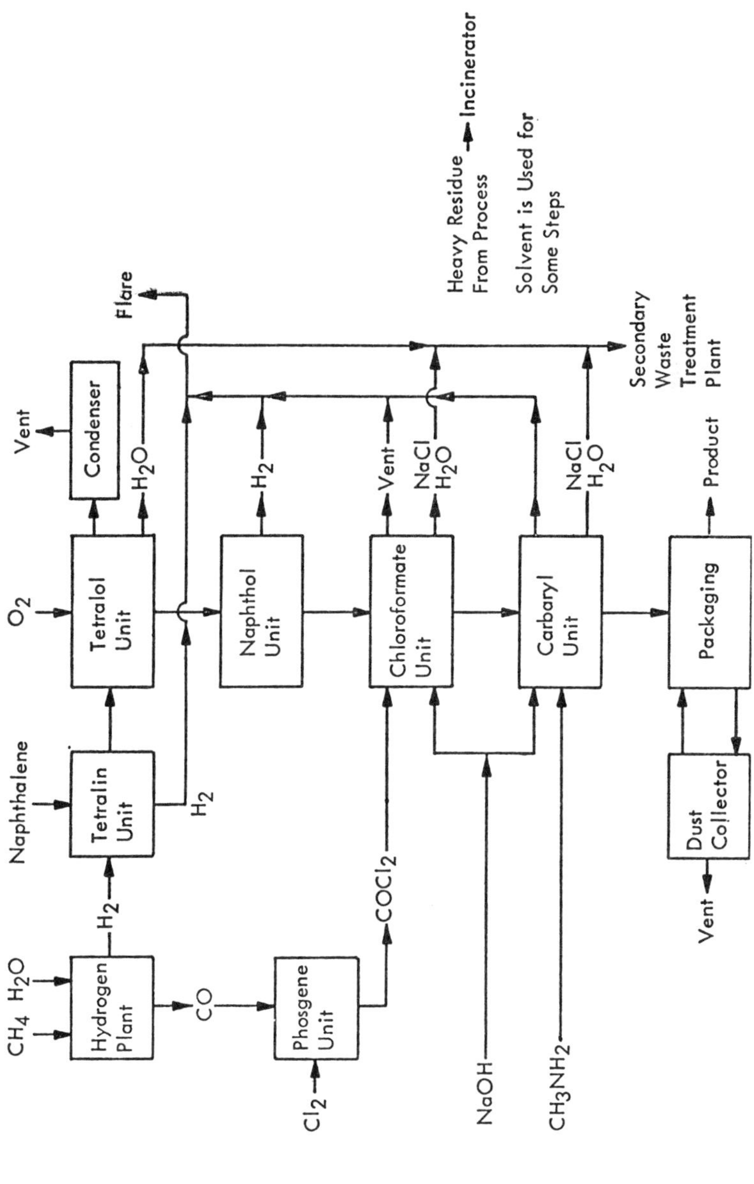

Source: Midwest Research Institute Report PB 213,782; 6/72

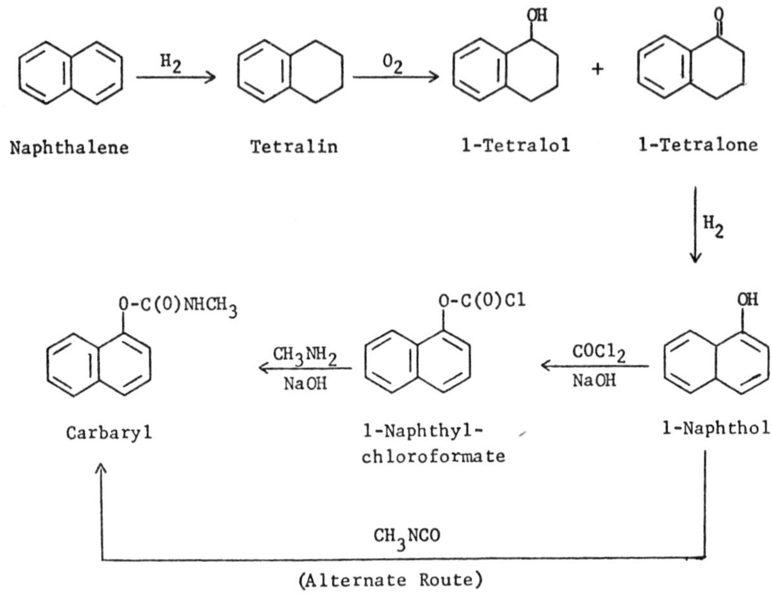

About 2 lb of chemicals (omitting NaOH and Cl_2) are required to produce 1 lb of carbaryl. By-product wastes are liquid streams, vents and some heavy residues. All process water goes into the plant secondary waste treatment systems. All toxic vents are flared or go to NaOH scrubbers. One nontoxic vent goes to a condenser and then is vented to the atmosphere. Standard hood systems with recycle of recovered material are used in the packaging process. No solid waste, as such, is generated by Carbaryl production. There is a heavy residue from the process that is burned in an incinerator.

There is a major shutdown per year in the carbaryl unit during the slack season. There are numerous maintenance cleanups of a smaller scale. Washings from these cleanups go into the process waste stream. Leaks have not been a major problem. Experience with leakers in the 5 gal plastic drums has been less than 0.1%.

CHLORDANE FROM CYCLOPENTADIENE

Chlordane is a broad spectrum insecticide of the chlorinated hydrocarbon variety. It is a solid material, insoluble in water and moderately toxic. Its chemical name and formula are shown on the following page. Technical chlordane is a mixture of 1,2,4,5,6,7,8,8-octachloro-4,7-methano-3a,4,7,7a-tetrahydroindane and related compounds.

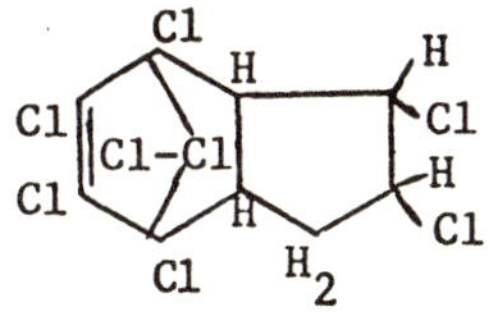

1,2,4,5,6,7,8,8-octachloro-4,7-methano-3a,4,7,7a-tetrahydroindane

It is produced by Velsicol Corporation at Marshall, Illinois using the following production chemistry.

$$\text{naphtha} \longrightarrow \text{cyclopentadiene } (C_5H_6)$$

$$Cl_2 + \text{NaOH (aqueous)} \longrightarrow \text{NaClO (aqueous)}$$

$$\text{NaClO (aqueous)} + C_5H_6 \longrightarrow C_5Cl_6 + \text{NaCl (alkaline solution)}$$

$$C_5Cl_6 + C_5H_6 \longrightarrow \text{chlordene } (C_{10}H_6Cl_6)$$

$$\text{chlordene} + Cl_2 \longrightarrow \text{technical chlordane } (C_{10}H_6Cl_8 + \text{related compounds})$$

The spent hypochlorite wastewater has about 2% NaOH and 400 ppm C_5Cl_6. The chlordane production and handling area is diked and all runoff and process wastes go to deep well disposal on the plant site.

The well is state approved and has been in use for several years. It goes 2,600 feet deep to a dolomite layer which is very porous and the discharge does not spread much, perhaps 5 acres in 6 years. The stratum already has salt in it. The shaft has three casings. A tube liner containing water at high pressure goes to 2,300 ft and the top 600 ft is further sealed to prevent surface water contamination.

Velsicol has a second well of 6,000 ft, completed in December 1971. It handles wastes from the resin part of the plant (these contain fluorides) which are being discharged. Those wastes have to be monitored for chlordane, for dissolved solids, BOD, etc. Velsicol monitors chlordane, but also has an outside lab check occasionally and the state also checks sometimes, levels are probably below 1 to 5 ppb (level which would be toxic to fish, LD_{50} = 1 ppb for chlordane; 25 to 30 ppb for heptachlor). Five companies in Illinois use deep-well disposal. Cost is $40 to $90 per foot.

Chlorine from the Cl_2 tanks is forced out under pressure, no losses, meets Chlorine Institute Specifications. Velsicol is installing automatic shut-offs. The tank cars are leased, used for chlordane only. Occasionally some are not needed and are returned to the railroad company following cleanup with detergent, the wash goes to deep well disposal. The tank trucks are received clean and dry from the trucking company, make a one-way trip and then are cleaned by the trucker before being sent to the next job. The trucks are self-draining and the trucker may use a bucket to catch drips, but he probably cleans the truck with detergent,

FIGURE 26: PRODUCTION AND WASTE SCHEMATIC FOR CHLORDANE

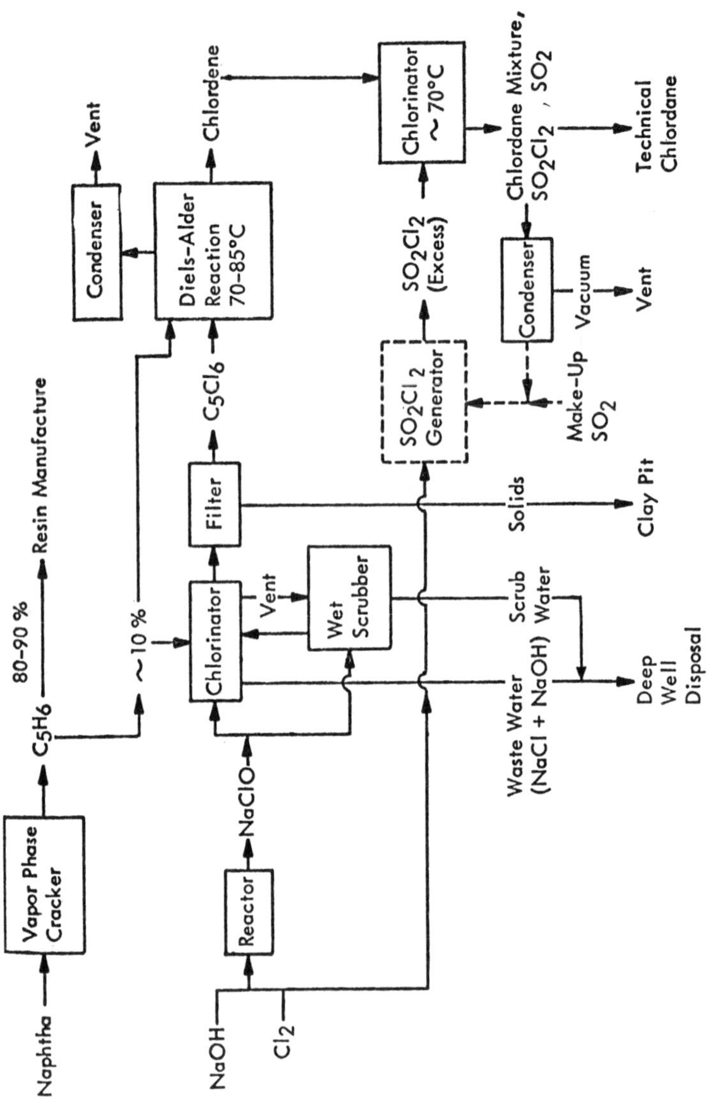

Source: Midwest Research Institute Report PB 213,782; 6/72

steam and air and dumps the washings in the sewer. Drums are lined, open top; filling, no ventilation problems, no controls; have had drum punctures, rail accidents, but no full truck spills; biggest spill in the last two years was three or four drums; drum (disposal) is the formulators' problem; recommended cleanup is kerosene wash and detergent wash but washings may be thrown on the ground.

Chlordane is very stable and does not degrade in detergent. The half-life in soil is about two years (although it does not migrate or biomagnify). The technical material is fairly viscous.

Emergency Procedure – send technical assistance as soon as word of spill is received. Recommend clay absorption or rag wipeup of spill with burial of material at least 12 inches below ground level. Do not recommend absorption on sawdust and burning.

CHLOROBENZENE FROM BENZENE

Where the end product is monochlorobenzene, benzene is chlorinated in the liquid phase reaction:

$$C_6H_6 + Cl_2 \rightarrow C_6H_5Cl + HCl$$

As an intermediate for phenol, the following oxychlorination reactions are used without the separation of the monochlorobenzene.

$$C_6H_6 + HCl + \tfrac{1}{2}O_2 \rightarrow C_6H_5Cl + H_2O$$

$$C_6H_5Cl + H_2O \rightarrow C_6H_5OH + HCl$$

The volume of the oxychlorination reaction is about twice that of the direct chlorination process. However, phenol production is outside the scope of this study. A projection indicates that, in 1980, the production will be approximately 700 million pounds. With the decline of DDT production, it is expected that the growth rate of monochlorobenzene will be small.

The raw materials are benzene, chlorine, iron turnings, and sodium hydroxide. The by-products are hydrogen chloride, dichlorobenzenes, heavies and spent caustic sludge. The main use of monochlorobenzene is as an intermediate, for example, DDT or nitrochlorobenzene.

Benzene and chlorine are fed to a liquid phase reactor containing iron turnings as a catalyst. The liquid phase is a boiling mixture of chlorobenzenes and benzene. The pressure is atmospheric and the temperature range is 80°to 100°C. The HCl from the reaction is scrubbed with recycled benzene for hydrocarbon removal and then absorbed in water forming 20°Bé hydrochloric acid.

The bottoms from the reactor are washed with sodium hydroxide to remove the last of the HCl and most of the dichlorobenzenes. After decanting from the sodium hydroxide solution, the product is fed to a stripping column, where the benzene is stripped off overhead and the benzene is recycled to the HCl scrubber.

The bottoms from the stripper are fed to a fractionating column where the pure monochlorobenzene is taken off overhead and where the bottoms are mostly dichlorobenzenes. Yields of monochlorobenzene are about 75%, while dichlorobenzene yields are 10 to 20%. There are two sources of air pollution, the vent on the tail gas absorber and the vent on the stripper. The vents are continuous. Benzene is the major air pollutant.

CHLOROMETHANE FROM VARIOUS RAW MATERIALS

In 1970, over 70% of the methyl chloride was made by chlorinating methane:

$$Cl_2 + CH_4 \rightarrow CH_3Cl + HCl$$

An older process, using methanol and hydrogen chloride, produced about 22% of the methyl chloride in 1970. The reaction is:

$$CH_3OH + HCl \rightarrow CH_3Cl + H_2O$$

For the past 10 years the annual growth rate has been 19%. For the future, the growth is expected to be at about half this rate. The main uses of methyl chloride are as an intermediate for producing silicones and tetramethyllead. Methylene chloride is made by chlorinating methane:

$$2Cl_2 + CH_4 \rightarrow CH_2Cl_2 + 2HCl$$

Its annual growth rate in the sixties was 14%. For the seventies, the growth rate is expected to decline to about half this rate, or about 7%. The main use of methylene chloride, about 40%, is as a paint remover. Chloroform is a coproduct from the chlorination of methane, along with methyl chloride, methylene chloride, and carbon tetrachloride. The reaction is:

$$CH_4 + 3Cl_2 \rightarrow CHCl_3 + 3HCl$$

The production growth rate for the past 10 years has been 12% per year. A production rate of 800 million pounds per year is expected in 1980. The main use (52%) of chloroform is as an intermediate for making Freon-22, monochlorodifluoromethane.

Carbon tetrachloride is made by three industrial processes. In 1970, approximately 40% of the carbon tetrachloride was made by the thermal chlorination of methane, namely:

$$4Cl_2 + CH_4 \xrightarrow{500^\circ C} CCl_4 + 4HCl$$

When methane is chlorinated, coproducts such as chloroform are produced. These coproducts show good growth rates. Chloroform, for example, has a 10% per year growth rate. Although methane chlorination does not have the best economics, continued growth of the process is forecast, because of the versatility of the process equipment and product distribution.

About 35% of the carbon tetrachloride was made by the thermal chlorination of propane, namely:

$$C_3H_8 + 8Cl_2 \longrightarrow CCl_4 + C_2Cl_4 + 8HCl$$

Recent developments in this process have been the use of partially chlorinated by-products, such as dichloropropane, as feedstocks. If the trend continues, this process may become dominant due to cheaper feedstocks and lower chlorine consumption. The coproduct, perchloroethylene, is a valuable product having a 1960 to 1970 annual growth of over 12%.

Approximately 25% of the carbon tetrachloride, or 225 million pounds, was made by chlorinating carbon disulfide, either directly or indirectly. This is an old and declining process. Most of the disulfide plants are located in the Northeast near the markets, but are being replaced by plants in the Gulf Coast area, which use cheaper feedstocks. Carbon tetrachloride is toxic, and is used mainly as an intermediate for freons.

Chlorination of Methane

The process sequence requires four main steps: reaction, HCl recovery, chlorides recovery and chlorides refining. High purity methane, chlorine and recycle methane are premixed and fed to the reactor. The reactor effluent, containing organic chlorides, HCl, excess methane, and only traces of chlorine, is cooled and fed to the HCl recovery system.

The first column in this system is an absorber designed for efficient HCl removal. The bulk of the absorbing liquor is HCl azeotrope (about 20% by weight HCl). The rich acid is thus above the azeotrope and allows stripping of anhydrous HCl. The second column distills off anhydrous HCl and produces the required azeotrope in the bottoms. For high yields of methyl chloride, the amount of recirculating methane is increased.

The HCl-free gases from the absorber are washed with caustic soda to remove final traces of HCl and are then ready for chlorides recovery. For the intermediate product distribution under discussion, compression, cooling, and drying with sulfuric acid are an economical combination.

The reactor is operated at 15 psig and 400° to 500°C. The yields are over 95%. There is one source of air pollution. It is a purge on the recycled methane to remove inerts. The vents from the product purification system are returned to the suction side of the recycle compressor. The raw materials are chlorine (dry gas), methane (99% plus), sulfuric acid, and caustic soda. The by-products are other chloromethanes, perchloroethylene, hydrogen chloride, heavy ends, and spent caustic sludge.

Hydrochlorination of Methanol

Methyl chloride is produced by the action of hydrogen chloride on methanol, with the aid of a catalyst, in the vapor phase. Vapors of methanol and hydrogen chloride are continuously mixed in approximately equimolecular ratios and passed through a preheater maintained at about 180°C.

The gas mixture is then passed at substantially atmospheric pressure through a converter at a temperature of 340° to 350°C. The converter is packed with previously ignited alumina gel of 8 to 12 mesh size or a similar acting catalyst, such as zinc chloride on pumice, cuprous chloride, or activated carbon. The converter is externally heated by electric coils or some other suitable means. Space velocities of about 275 cubic feet per hour per cubic foot of gross catalyst volume are generally used (based on gas volumes at STP).

The effluent gases from the reactor are scrubbed with water to remove excess HCl, and alkali wash, and a strong sulfuric acid wash (to dry the product). Crude methyl chloride is distilled under pressure and at -24°C to yield pure methyl chloride. The raw materials are 95% methanol, hydrogen chloride, and catalyst along with caustic and sulfuric acid. The by-products are heavies and excess HCl absorbed in water. There are no air pollution sources.

Thermal Chlorination of Propane

Fresh chlorine feed, together with recycled chlorine, and the propane feed are introduced into a vaporizer where they are mixed with recycled chlorides in the vapor space. The chlorine is in 10 to 25% excess, based on the propane. The recycle chloride diluent rate (controlled by heat input to the vaporizer) is designed to control adiabatic reactor temperature at 550° to 700°C and is usually about 75 mol percent of the total stream; its composition also serves to control CCl_4 to C_2Cl_4 ratio.

The mixed gases, at atmospheric pressure, are fed to a refractory lined reactor where they are rapidly mixed with the contents, thereby heated to ignition, and reacted adiabatically. The reaction is self-sustaining and supported by the considerable heat of reaction (after startup), but is tempered and controlled by the diluent action of the recycle chlorides.

The reactor effluent is essentially free of unreacted hydrocarbon and consists mainly of carbon tetrachloride, perchloroethylene, HCl and excess chlorine. It is rapidly quenched by intimate contact with a liquid which is largely perchloroethylene. The rapid quench serves to preserve the equilibrium ratio attained in the reactor and prevents formation of undesirable by-products.

It also serves to dissolve any quantities of hexachlorobenzene which may be present (these are small if reaction temperature is maintained below 650°C) and which might introduce difficulties due to deposition of solids on heat transfer surfaces and in vessels. Some heat economy is obtained from the hot reactor effluent, which serves to boil the contents of the quench tank and thus provide boilup for the CCl_4 column. Hexachlorobenzene is purged from the quench tank by allowing liquid to overflow to the recycle surge tank.

The CCl_4 column, operating on boilup from the quench tank, returns quench liquid rich in perchloroethylene as a bottoms stream. Fractionation results in an overhead which is largely free of perchloroethylene. The condenser yields carbon tetrachloride, the desired quantity of which is withdrawn as product, the remainder being used as reflux. Chlorides, free gases (HCl and chlorine) are also disengaged at the overhead. These gases are scrubbed with water to remove HCl, the effluent liquor forming the aqueous HCl by-product. They are subsequently dried with concentrated sulfuric acid and purged to maintain the balance of inert gases; the remaining chlorine (with some uncondensed chlorides) is recycled to feed. A perchloroethylene-rich stream is removed as a side stream from the CCl_4 column and processed to produce perchloroethylene product. Heavy ends are first removed by distillation and returned to the recycle surge tank. The overhead from the heavy ends column is fractionated in the C_2Cl_4 column where the desired quantity of perchloroethylene product is removed as the bottoms and the overhead (largely carbon tetrachloride) is sent to recycle.

The recycle surge tank serves as a reservoir for three recycle streams: (1) the quench tank overflow (largely perchloroethylene with some hexachlorobenzene); (2) heavy ends from the heavy ends column; and (3) the C_2Cl_4 column overhead (largely carbon tetrachloride). Proper control results in a recycle mixture of the desired composition, and this is fed to the vaporizer where the vapors are mixed with the feed materials.

To prevent buildup of hexachlorobenzene in the system, liquid is withdrawn from the vaporizer and sent to the heavies still. Here the more volatile components are stripped off and returned to the vaporizer, while heavy liquid consisting of hexachlorobenzene and some hexachloroethane is purged from the still pot.

The raw materials are propane, chlorine, catalyst ($CuCl_2$ and $BaCl_2$), water and sulfuric acid. The by-products are tetrachloroethylene (perchloroethylene), hexachlorobenzene, hydrogen chloride, hexachloroethane, and spent sulfuric acid.

There is one source of air pollution, the purge on the dry chlorine recycle, which removes inerts. This purge is continuous, with chlorine being the pollutant. The major cause of the pollution is impurities in the raw materials.

Carbon Disulfide Chlorination

Industrially, both direct and indirect chlorination of carbon disulfide are employed. In a direct chlorination process, a mixture of carbon tetrachloride, carbon disulfide and sulfur chlorides, recycled from a subsequent process step, is directly contacted with excess chlorine at about 30°C over a divided iron catalyst, converting over 99% of the carbon disulfide to carbon tetrachloride. Some fresh carbon disulfide feed may also be diverted to the primary chlorinator for purposes of controlling downstream loads.

It is essential that the reaction be complete but that overchlorination of sulfur to sulfur monochloride be avoided to prevent appearance of carbon disulfide and sulfur monochloride in the distilled crude product. Distillation of the reactor effluent yields a bottoms product and an overhead product of relatively pure carbon tetrachloride. This material may be treated with a base to destroy sulfur chlorides, and dried.

An indirect chlorination process is similar. Fresh carbon disulfide feed reacts with sulfur monochloride under conditions such that the carbon disulfide is all but completely reacted. However, the reactor product contains a fraction of a percent of carbon disulfide, as well as carbon tetrachloride and sulfur. A direct chlorination polishing reactor is used to convert this remaining carbon disulfide and to facilitate the subsequent distillation, where crude carbon tetrachloride is removed overhead and molten sulfur containing some sulfur monochloride is the bottoms product. The crude carbon tetrachloride may be purified as in the direct chlorination process, an alcoholic caustic treatment with redistillation is shown on the flow sheet.

Sulfur equivalent to the fresh carbon disulfide feed is separated from the bottoms product and may be returned to the CS_2 plant for processing. The residual sulfur stream is directly chlorinated to provide the sulfur chlorides chlorinating agent for the first reaction stage. The flow sheet shows the indirect process.

The raw materials are carbon disulfide, chlorine, iron catalyst, and caustic soda. The by-products are heavy ends, sulfur chlorides, sulfur, and spent caustic sludge. The quantities of air pollutants from the chlorination of carbon disulfide are very small. The neutralizer-dryer, using caustic soda, absorbs most of the noncondensables such as carbon dioxide or excess chlorine. The vacuum jets on the product still are vented back into the neutralizer-dryer.

COAL TAR DISTILLATION

Coal tar is a mixture of many chemical compounds (mostly aromatic) which vary widely in composition. The process of coal tar distillation separates these fractions into commercially valuable products.

In the plant visited by EPA investigators, crude coke oven tar is fractionally distilled into solvent, carbolic oil, road tar, creosote, and pitch fractions. These products are then purified or further fractionated into fine products. The processes of coal tar distillation, anthracene refining, pitch forming, and naphthalene refining, together with their associated wastewater sources, are briefly described in the following paragraphs; simplified process diagrams are presented in Figures 27 through 30.

Crude coke oven tar and dilute caustic solution are fed into a dehydration column. The vapor stream taken overhead from the column is condensed, and water is removed from the solvent and discharged to a sewer line. The liquid stream is sent then to a vacuum still and to a series of fractionators where crude carbolic oil, road tar, creosote, and pitch fraction are generated. There are two steam jets associated with the distillation columns; the condensates of these jets contain organic contaminants and are the major water pollution sources.

In the anthracene refining process, creosote is first washed with water in a crystallizer, and the creosote anthracene slurry is passed through filters and centrifuges to produce crude anthracene. The crude product is then sent to a crystallizer, where furfural is used to purify the product.

Refined solid anthracene is obtained after solid separation and drying steps. The liquid streams from the second-stage purification units are collected for furfural recovery. The aqueous stream discharged from the first-stage purification unit and the condensate of the steam jet associated with the furfural recovery unit are major wastewater sources. The liquid pitch from tar distillation is cooled by direct contact with water and then dried to form the final product. The contact cooling water is another major wastewater source.

FIGURE 27: COAL TAR DISTILLATION

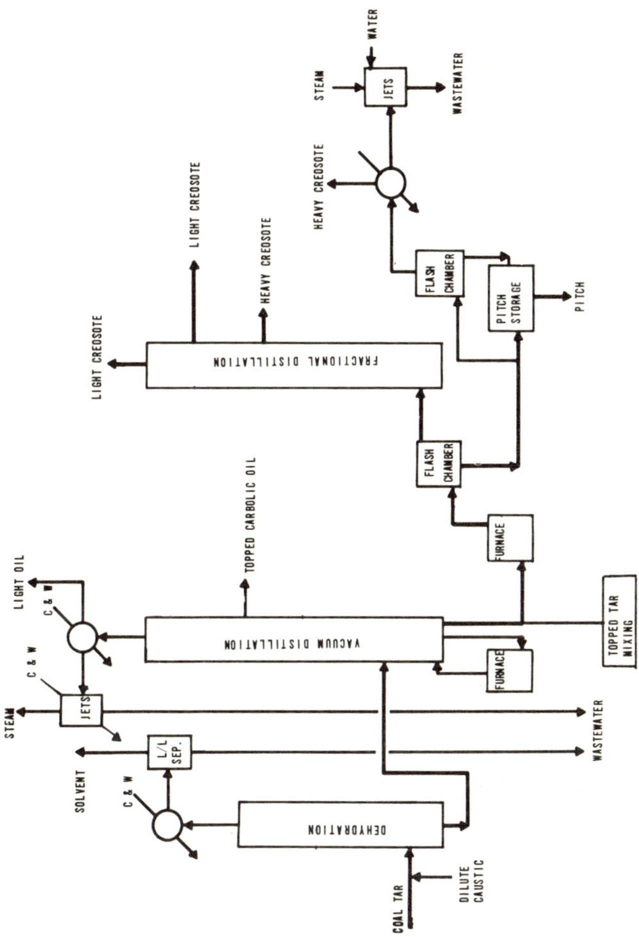

Source: Report EPA 440/1-73/009

FIGURE 28: ANTHRACENE REFINING

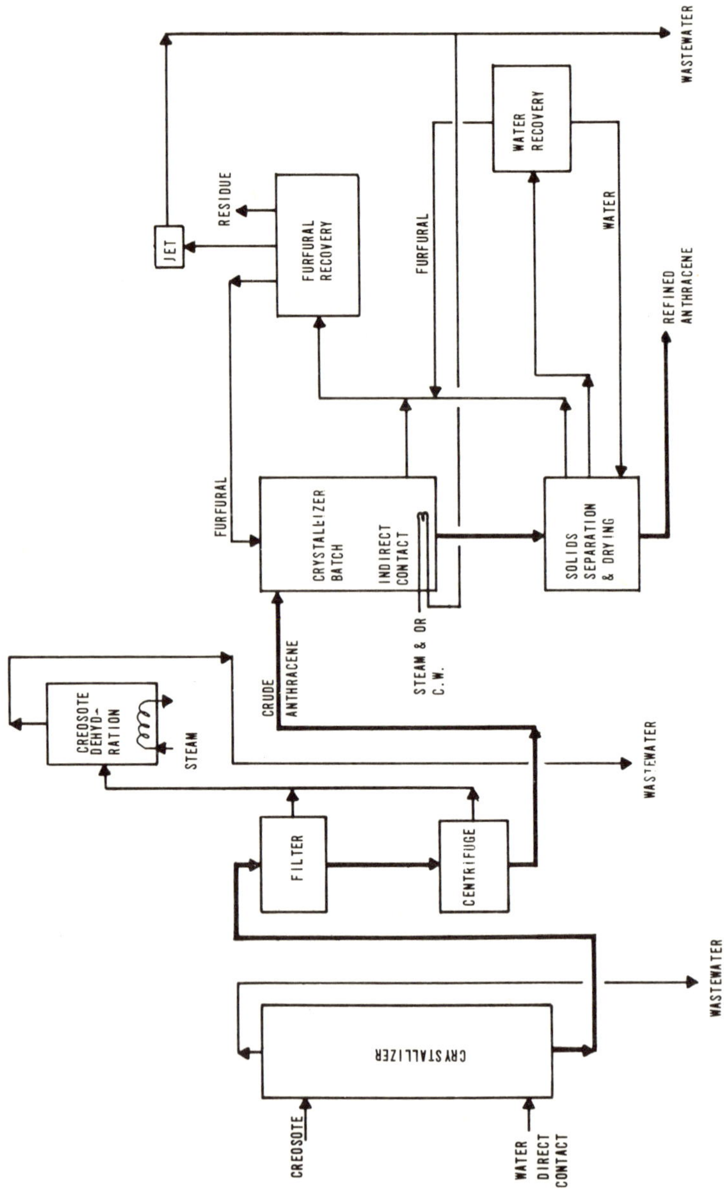

Source: Report EPA 440/1-73/009

FIGURE 29: EXTRACTION AND NAPHTHALENE REFINING

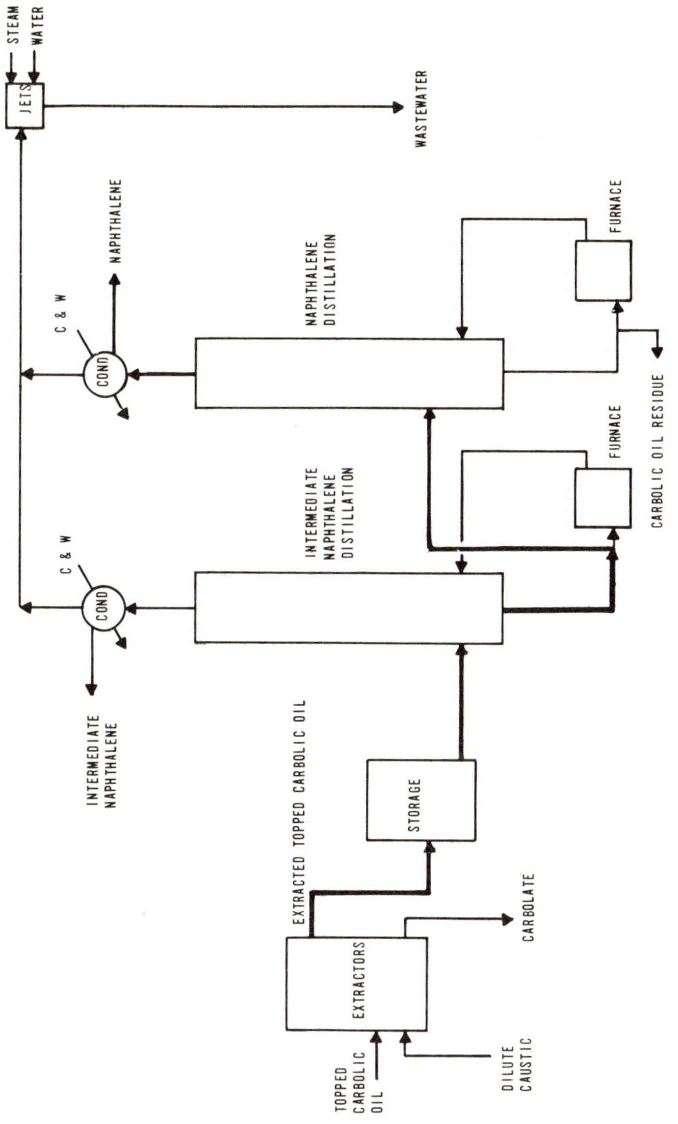

Source: Report EPA 440/1-73/009

FIGURE 30: PITCH FORMING

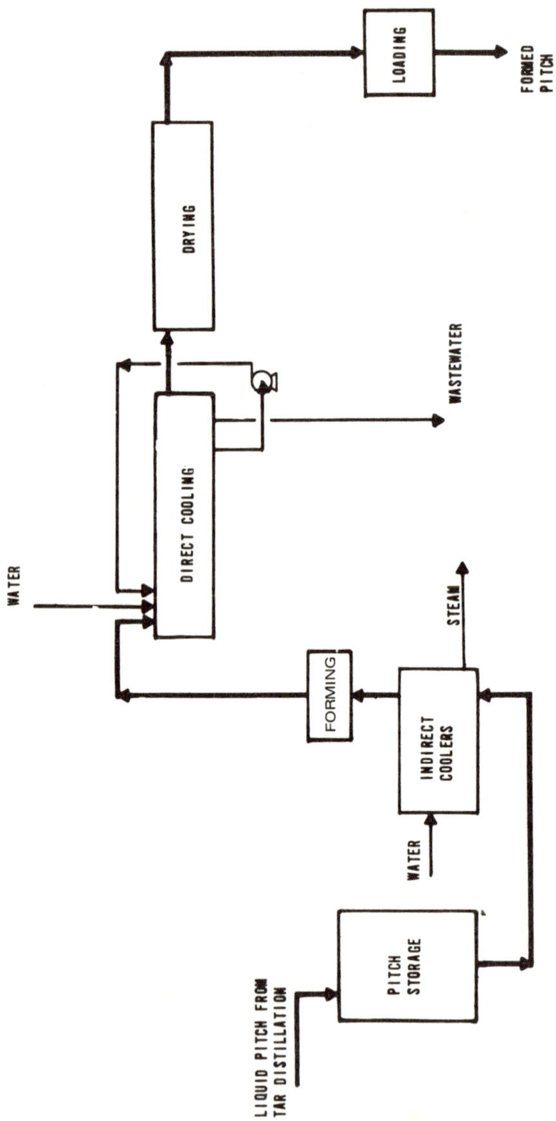

Source: Report EPA 440/1-73/009

The first step in naphthalene refining is extraction of topped carbolic oil with a caustic solution. The bottom layer in the extractors is the by-product of carbolate. The upper aqueous layer in the extractors is sent to a series of stills where naphthalene and intermediate products are generated. The only water pollution source is the condensate of the steam jets which are used to produce vacuum in the naphthalene stills.

End-of-pipe treatment and in-plant abatement have been achieved: segregation of clean water from process wastewater, replacement of barometric condensers with indirect condensers, installation of phenol recovery units, etc. These modifications have resulted in a low RWL. The characteristics of wastewater obtained from the plant survey are shown in the following tabulation.

	Coal Tar Distillation	Pitch Forming
Waste flow	405.3 gal/1,000 gal	126.1 gal/1,000 lb
COD	2,570 mg/l	61 mg/l
	8.68 lb/1,000 gal	0.064 lb/1,000 lb
BOD_5	833 mg/l	
	2.81 lb/1,000 gal	*
TOC	3,010 mg/l	
	10.16 lb/1,000 gal	*

*Not available

The historical data provided by the plant indicate that pitch forming has a waste flow of 200 gal/1,000 lb of product, with 0.13 lb of COD while the naphthalene refining has a waste flow of 408 gal/1,000 lb of product, with 0.86 lb of COD.

Although there is a variation between the survey and the historical data, the raw waste loads derived from the abovementioned abatements can be considered as representative of BPCTCA control technology of each individual process. However, standards for BATEA and BADCT should require that the remaining barometric condensers be converted to indirect condensers. Thus, the quantities of wastewater from the processes of coal tar distillation and naphthalene refining can be reduced, although RWL may not be correspondingly reduced.

CRESOL FROM TOLUENE

As in the case with other coal tar derivatives, the supply of coke oven by-product cresylics has failed to keep up with demand. p-Cresol was the first isomer to be synthesized commercially and is produced by sulfonation of toluene. The basic chemical equations are given below.

$$C_6H_5CH_3 + H_2SO_4 \longrightarrow (SO_3H)C_6H_4CH_3$$

$$(SO_3H)C_6H_4CH_3 + NaOH \longrightarrow (OH)C_6H_4CH_3 + Na_2SO_3$$

Toluene and a gas mixture of sulfur dioxide and sulfur trioxide are fed into a sulfonation reactor. The reactor effluent gas is passed through a caustic scrubber to remove unreacted sulfur dioxide. The liquid effluent from the reactor is

FIGURE 31: PARA-CRESOL

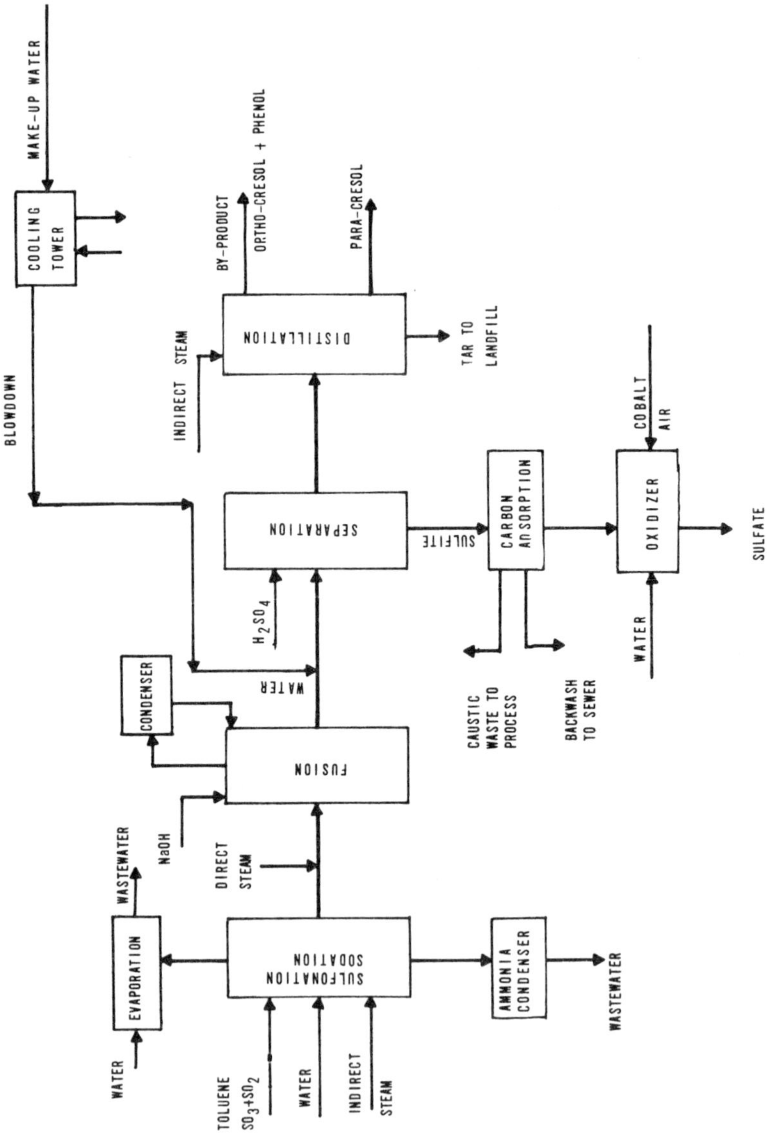

Source: Report EPA 440/1-73/009

first diluted with steam and then sent to a caustic fusion column, where crude p-cresol is produced. The crude product is then sent to a washing/separation column, where excess caustic solution is neutralized and two phases are formed. The aqueous phase is discharged from the system, and the organic phase is fractionated to obtain pure p-cresol.

Since the sulfonation reaction approaches 100% conversion of sulfur dioxide and trioxide, the vent gas scrubber water does not present a significant water pollution source. The major wastewater stream is the aqueous phase discharged from the sulfuric washing/separation column.

The average composition of this stream is 77% water, 15.2% sodium sulfite, 5.1% sodium sulfate, 0.4% cresylic compounds, and 1.7% other organic substances such as cresols, phenols, etc. The data obtained from Plant 1 are shown in the following tabulation.

Flow	1,291 gal/1,000 lb
COD	23,800 mg/l
	256 lb/1,000 lb
BOD_5	11,400 mg/l
	123 lb/1,000 lb
TOC	5,020 mg/l
	54 lb/1,000 lb

The sulfite and organic contaminants cause the high oxygen demand in the wastewater, while the cresol contaminant (10 mg/l) constitutes an odorous nuisance in the atmosphere.

According to the literature, the organic contaminants in the wastewater exhibit very strong antioxidant properties and present a difficulty to ordinary biological treatment processes. Several possible methods of controlling this wastewater discharge have been investigated.

The most promising scheme appears to be activated carbon adsorption of organic contaminants prior to oxidation, followed by chemical regeneration of cresylic compounds adsorbed on the carbon, to return a valuable product to the process, eliminate the odor problem, and reduce the discharge of pollutants.

A demonstration plant and its economics are briefly described. The system consists of two 4 ft diameter by 30 ft high columns of 304 L stainless steel. Each column is loaded to a height of 18.5 ft with approximately 6,000 lb of activated carbon.

The system was designed to have sufficient capacity for a one day operational cycle, requiring one column to be regenerated each 24 hours. Ten percent sodium hydroxide solution is used to regenerate spent activated carbon, and the desorbed cresylic compounds are recycled back to the process. During a seven month period, the columns were operated at an average superficial velocity of 3.2 gpm/ft. Influent concentrations during the period were 3,500 to 6,500 mg/l cresylic compounds, and effluent concentrations were between 0 and 700 mg/l cresylic compounds. During this time, 271,600 lb of p-cresol were returned to the process. This amount of p-cresol represents a value of $114,000. As demonstrated, the activated carbon system not only can recover p-cresol from the wastewater and turn it into profit, but also can decrease the RWL of the system. Furthermore, it improves the treatability of the wastewater. Consequently, to define BATEA and BADCT control technologies, an activated carbon system should be incorporated into the process.

Two other process routes for the manufacture of p-cresol are currently practiced: vapor phase methylation of phenol over alumina catalysts, and liquid phase oxidation of meta- and para-cumene.

CYCLOHEXANE FROM BENZENE

Cyclohexane can be obtained as a naturally occuring petroleum fraction or through the hydrogenation of benzene. The chemical reaction for the production of cyclohexane from benzene is given below.

$$C_6H_6 + 3H_2 \longrightarrow C_6H_{12}$$

The reaction is usually carried out in the liquid phase with a nickel-palladium or platinum catalyst at elevated temperature and pressure. Fresh feed (benzene) is combined with makeup and recycle hydrogen and preheated to reaction temperature by heat exchange first with reactor effluent and then with steam. The reaction effluent is cooled and flashed. Part of the vapor is used as recycle hydrogen, while the forward flow vent gas is chilled by refrigeration to minimize cyclohexane losses and is available as high pressure fuel gas. The separated liquid is sent to a column where the light end impurities are taken overhead.

The cyclohexane process surveyed utilized a C_6 hydrocarbon feedstock containing a high concentration of benzene. The only contact process wastewater associated with the process is a spent caustic wash containing 5 to 10 weight percent NaOH. The flow raw waste load for this stream is quite low and amounts to only 0.24 gallon per 1,000 lb of cyclohexane when expressed on a production basis.

The contact caustic wash was necessary in the operation of this process because of the high sulfur content of the feedstock. This sulfur would reduce the useful life of the precious metal catalyst if it were not removed prior to the hydrogenation reaction. It was not possible, based on this one survey visit, to determine if the sulfur content of the feed was abnormally high and whether or not other cyclohexane units would require this type of caustic treatment of the feed.

2,4-D FROM DICHLOROPHENOL

2,4-D is a herbicide of the chlorinated phenoxy type. It is a solid material. The salts are water-soluble; the acid and its esters are nearly insoluble. It is a moderately toxic, nonpersistent material. Its chemical name and formula are shown on the following page.

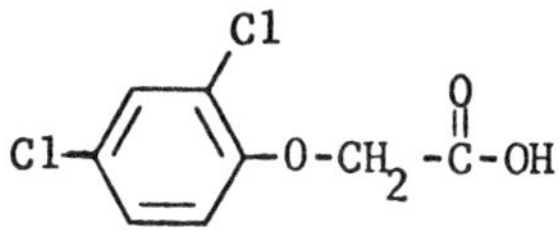

2,4-Dichlorophenoxyacetic acid
(usually as amine salts and esters)

It is produced by Dow at Midland, Michigan and by Chipman at Portland, Oregon according to the following production chemistry.

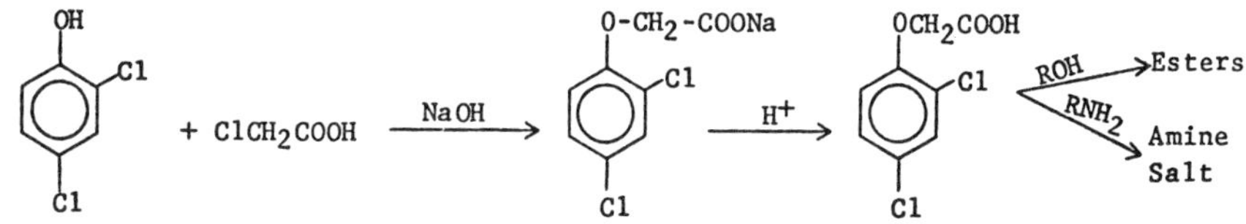

Dichloro-phenol	Chloroacetic acid	2,4-D Sodium salt	2,4-D

The Dow plant at Midland is on a very small river and had to face pollution and housekeeping problems much earlier than others. It has extensive wastewater treatment facilities and a huge incinerator equipped with a wet scrubber (for chlorinated tars especially). Wastewaters with special contaminants (especially from the tank car wash shed) go through special treatments such as a biologically active trickling filter before going to the general biological treatment plant.

Dichlorophenol imparts a taste to water at 0.0005 ppm and kills fish at 2 ppm. A 1968 spill of dichlorophenol plus some 2,4-D caused a fish kill and production-storage areas have since been diked to prevent recurrence. In emergency, a shot (holding) pond is automatically switched in.

For 2,4-D and 2,4,5-T manufacture nearly all the by-product chloride is recycled (charcoal recovery system) to electrolysis cells. By-product 2,6-D is difficult to degrade biologically and would interfere in waste treatment plant; therefore, it is removed (distillation or otherwise) and converted to pentachlorophenol which has value. (Dow says that a small producer could not afford to do this and also that the 2,6-D may run as high as 12% of product with poor technology.) Other by-products are converted by bleach-oxidation to CO_2, HCl, and removed.

Filter cakes, still bottoms, etc., from 2,4-D and 2,4,5-T production are accumulated and incinerated (1000°C) during the nongrowing season, especially not during April or May. Dryers are equipped with dust collection. Complete recycle is used. Dow furnishes workers outer clothing, which is washed in in-house

FIGURE 32: PRODUCTION AND WASTE SCHEMATIC FOR 2,4-D (DOW CHEMICAL)

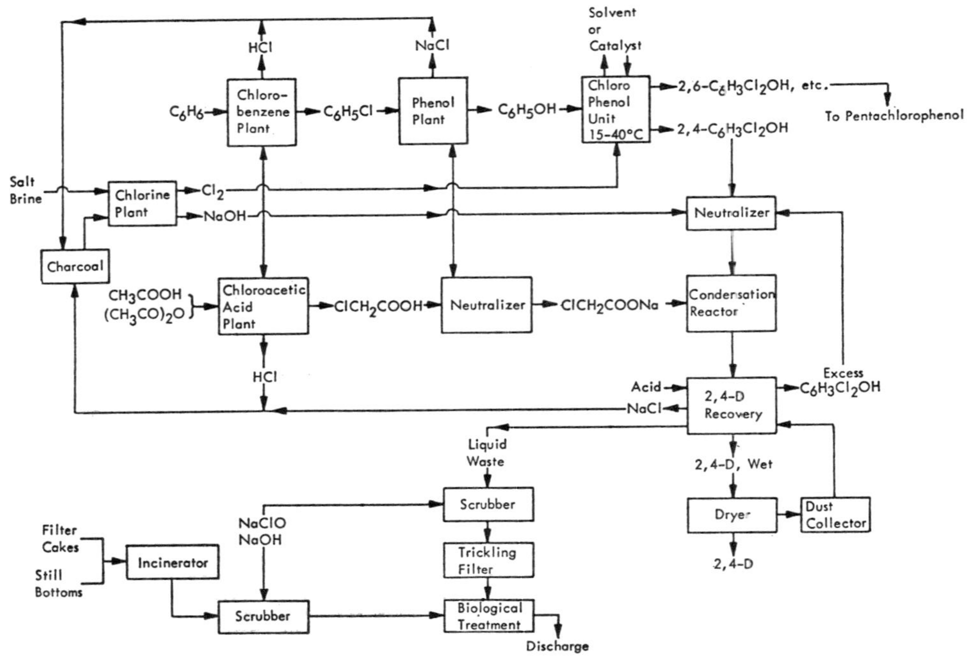

Source: Midwest Research Institute Report PB 213,782; 6/72

laundry with washwater going to waste disposal. Wipe cloths, etc., would also go to laundry or incinerator. Total loss of 2,4-D through air and shoes, etc., is now less than 2 lb/day, although it had previously been as high as 150 lb/day.

Drums used in shipments are one-way; any returns would be incinerated. Tank cars are owned by Dow and are used mostly for herbicides during rush season, but may be used for other Dow chemicals at other times. All are cleaned as needed in a special tank washing shed with detergent and high pressure.

Over $1 million has been spent for pollution control facilities at the Chipman production plant. The wastewater treatment facilities are centered in a continuous flow charcoal adsorption plant which is said to reduce phenolics below the 1.0 mg/l level, as described by T.B. Henshaw in *Chem. Eng.* 78, No. 12, 47-49 (May 31, 1971).

Pollution problems from pesticide production are said to be limited to occasional phenolic odor complaints. No area vegetative damage from 2,4-D and no corrosion damage or odor complaints from chlorine has occurred. No landfill of solid wastes is allowed in Oregon. Transportation problems have been apparently few: one incident of leakage by a defective tank car several years ago.

Filled drums receive spot check of closures and check in storage. Drum leakers are said to result from rough handling, not defects. Chipman has its own fire department and works with local department. Diking, etc., in the plant area is said to meet insurance requirements. No emergency holding pond appears to exist. Clothing of production personnel probably go to a commercial laundry.

DDT FROM CHLOROBENZENE

DDT is a broad spectrum insecticide of the chlorinated hydrocarbon type. Its chemical name and formula are as follows.

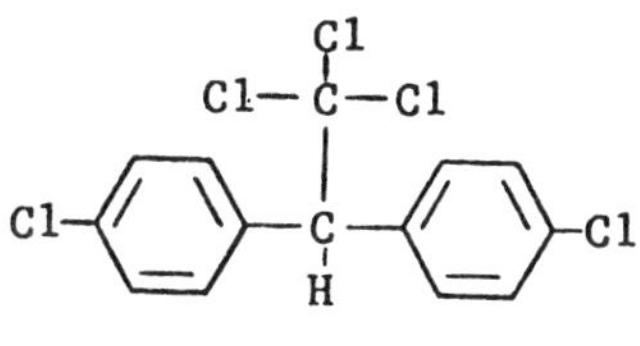

1,1,1-trichloro-2,2-bis(p-chlorophenyl)ethane

It is a solid, nearly insoluble in water. It is moderately toxic and persistent in the environment. It is produced by Montrose at Torrance, California using the following production chemistry.

$$C_2H_5OH + Cl_2 \rightarrow CCl_3CHO$$

$$C_6H_6 + Cl_2 \rightarrow C_6H_5Cl$$

$$CCl_3CHO + C_6H_5Cl \xrightarrow[\text{2. NaOH}]{\text{1. } H_2SO_4} CCl_3CH(C_6H_4Cl)_2 + H_2O$$

FIGURE 33: PRODUCTION AND WASTE SCHEMATIC FOR DDT

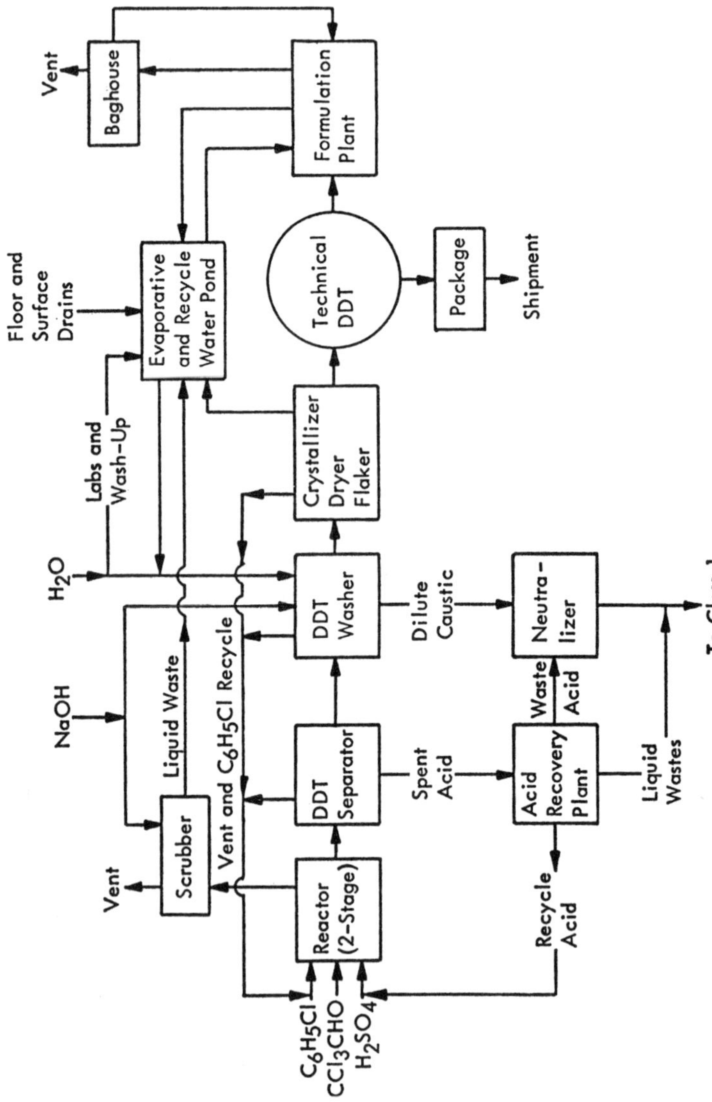

Source: Midwest Research Institute Report PB 213,782; 6/72

The process portion of the Montrose plant has no liquid waste outfall. Much of the water from the process goes to an open, concrete lined evaporative holding pond and is recycled. All drains and process sewers have been isolated from the city sewer system. Only sanitary waste and boiler blowdown water go to the city sewers.

The restroom lavatory basins, however, discharge to the holding pond system. Water consumption has been reduced from 20 million to 1.5 million gallons per month. Water from the recycle pond is also used for cooling water without filtration. This practice has caused no problem to date. The recycle water typically contains 10 to 15 ppm DDT.

From 20,000 to 25,000 gal/day of liquid waste are generated from the neutralization step and are removed by truck to a Class 1 dump. Some 10 to 15 cubic yards per day of solid waste bags, empty containers, etc., are also taken to this dump by a commercial disposal service. All vents from the second stage of the reactor are NaOH scrubbed and the liquid goes to the recycle water pond.

Equipment washdown is not a problem as this is normally done only during shutdowns. Washwater goes to the recycle pond. Spills and leakers have not been a major problem. One spill occurred when a truck carrying technical material had an accident and spilled DDT. The material was picked up along with the top 3 inches of soil and disposed of.

Prior to the modification of the waste treatment facilities DDT losses from this plant were apparently fairly high. DDT levels of 70 ppb were observed in the river below the plant as described by T.T. Schmidt, R.W. Risebrough and F. Gress in *Bull. Envir. Contam. Toxicol.* 6, 235 (1971).

Also, 0.5 million pounds of sediments from the river containing about 4,500 lb of DDT have been dredged up for treatment as described in *Air/Water Pollution Report,* page 285 (July 12, 1971).

DIELDRIN FROM ALDRIN

Dieldrin is a broad spectrum insecticide of the chlorinated hydrocarbon variety. Its chemical name and formula are as follows.

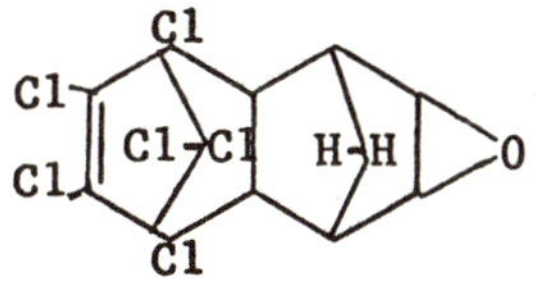

1,2,3,4,10,10-hexachloro-6,7-epoxy-1,4,4a,5,6,7,8,8a-octahydro-1,4-endo-exo-5,8-dimethanonaphthalene

It is a solid material, insoluble in water. It is very toxic and is persistent in the environment.

FIGURE 34: PRODUCTION AND WASTE SCHEMATIC FOR DIELDRIN

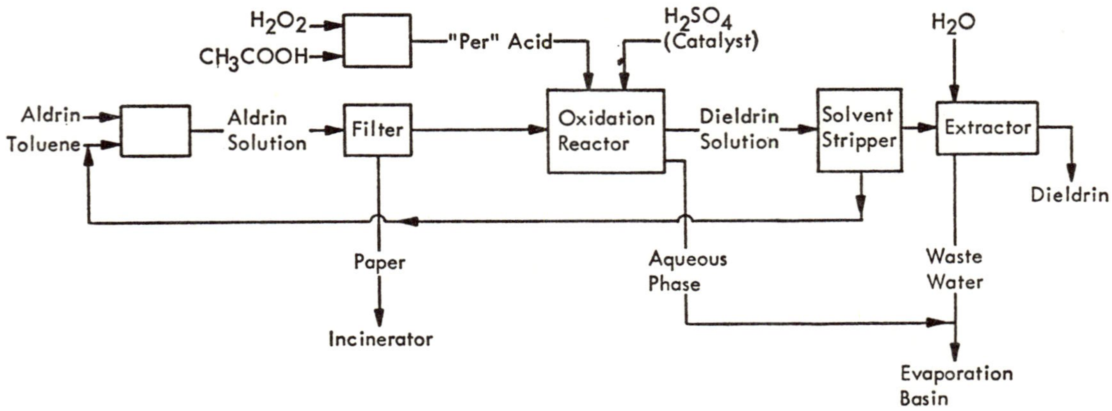

Source: Midwest Research Institute Report PB 213,782; 6/72

It is produced by Shell at Denver, Colorado according to the following production chemistry.

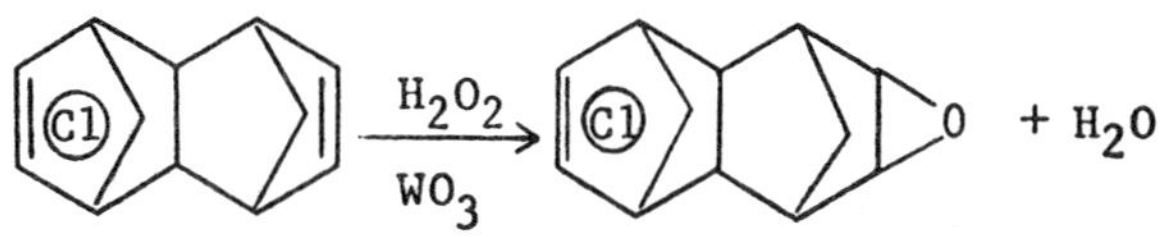

Aldrin Dieldrin

Liquid wastes go to evaporation basin (see aldrin). A small amount of solid waste, including filter paper, is burned in a smokeless-type incinerator. Equipment cleanup with toluene, which is apparently saved. See aldrin for further information.

DIMETHYL TEREPHTHALATE FROM TEREPHTHALIC ACID

The high purity monomer required for the development of polyester fibers and films is produced by converting terephthalic acid (TPA) to dimethyl terephthalate (DMT). However, with improved technology for the manufacture of fiber grade TPA, it is expected that most of the new fiber and film capacity installed will be based on purified TPA.

In the process for the esterification of TPA to DMT, preheated TPA and methanol are fed to a reactor in the presence of sulfuric acid as a catalyst. DMT in the reactor effluent is recovered and purified by conventional methods such as crystallization and distillation.

The water separated after condensation and the benzene used in the reactor to prevent the methanol from vaporizing too rapidly are the major water pollution sources. The wastewater may contain some alcohol, benzene, and pro-product or by-product losses.

Another water pollution source in the waste stream resulted from cleaning up scattered product resulting from leaks in various portions of the equipment. The characteristics of the wastewater obtained from plant surveys are shown in the following tabulation.

	Plant 1	Plant 2	Plant 3
Flow			
gal/1,000 lb	68.8	388	1,070
COD			
lb/1,000 lb	8.93	55.2	0.91
mg/l	15,000	17,000	102
BOD_5			
lb/1,000 lb	4.81	31.0	0.19
mg/l	8,400	9,580	21

(continued)

	Plant 1	Plant 2	Plant 3
TOC			
lb/1,000 lb	3.88	22.5	0.62
mg/l	6,800	6,950	69

Historical RWL data on process wastewater flow and COD were obtained at Plant 1. At this plant, there are actually two modules, with different production capacities, operating in parallel. The results of the analysis for probability of occurrence are summarized in the following tabulation.

	Flow RWL - - -(gal/1,000 lb)- - -		COD RWL - - -(lb/1,000 lb)- - - -	
	Unit A	Unit B	Unit A	Unit B
10% Occurrence	167	150	13.5	16.1
50% Occurrence	313	248	34	33.7
90% Occurrence	461	344	86.5	70.5
Ratio 90/50%	1.47	1.39	2.54	2.06

The analysis was based on consecutive 30 day average data collected by the manufacturer over a period of 24 months. The data show that there is only a slight variation between two units of different sizes at the same plant. However, the measured RWL is significantly lower than that from historical data. Again, the difference is due to the fact that historical data includes surface runoff caused by rainfall and housekeeping.

The survey data also reveal significant variations among plants. The high wastewater flow of Plant 3 is caused by steam jets with barometric condensers, while the low flow of Plant 1 is due to discharging steam jets directly into the atmosphere.

The variation in organic loadings between Plant 1 and Plant 3 is due mainly to different performance efficiencies of the solvent recovery units and to varying effectiveness of preventive measures for process leakages. The high RWL presented by Plant 2 is attributed to the low purity TPA manufactured by nitric acid oxidation. Plant 2 is scheduled to be phased out in the very near future.

To define BADCT and BATEA, it is certain that vacuum pumps with surface heat exchangers should be utilized in producing vacuum for process needs and that good performance of solvent recovery units should be required. Also, excellent preventive maintenance should be emphasized to reduce RWL. Process water usage and gross cooling usage are presented in the following tabulation.

	Process Water Usage lb/lb product	Cooling Water Usage lb/lb product
Plant 1	Not available	Not available
Plant 2	2	23,000
Plant 3	Not available	150

FIGURE 35: DIMETHYL TEREPHTHALATE BY ESTERIFICATION OF TEREPHTHALIC ACID

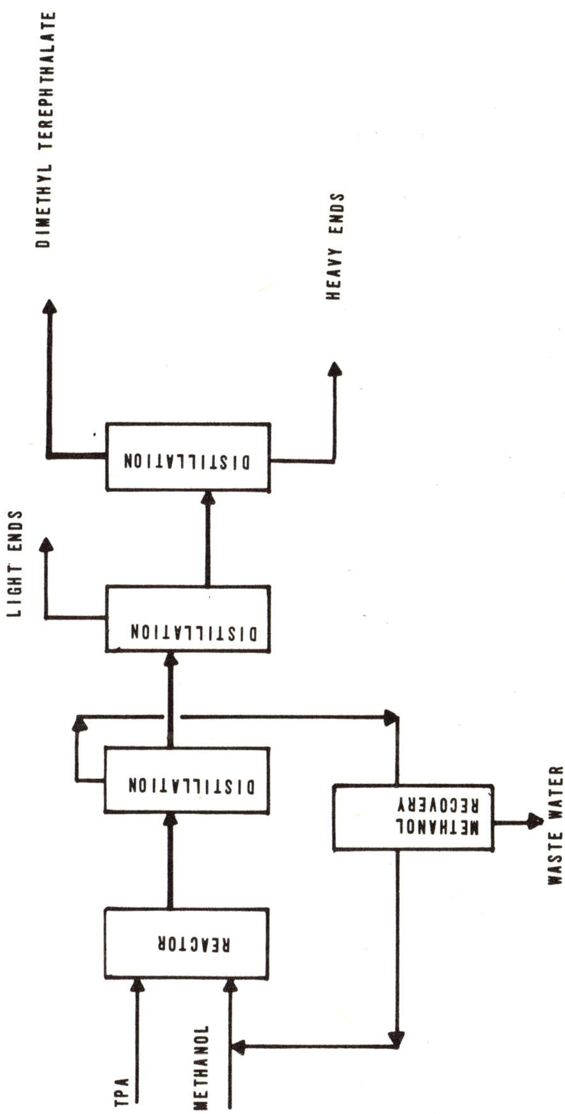

Source: Report EPA 440/1-73/009

An alternate route in the manufacture of DMT is the Hercules process. This synthesis involves liquid phase oxidation of p-xylene in acetic acid with a cobalt acetate or naphthenate as a catalyst to produce p-toluic acid. This is subsequently esterified with methanol to produce diethyl hydrogen terephthalate, which is finally esterified to form DMT.

DINITROTOLUENE FROM NITROTOLUENE

Dinitrotoluene is an explosive. It is mixed with NH_4NO_3 and acts as a promoter. Therefore, there is little production data on mono- or dinitrotoluene. However, in the Sixties, dinitrotoluene has become an important industrial intermediate for the production of polyurethanes. Dinitrotoluene is made by nitrating nitrotoluene,

$$C_6H_4CH_3NO_2 + HONO_2 \longrightarrow C_6H_3CH_3(NO_2)_2 + H_2O$$

An annual growth rate of about 20 to 25% is expected. Also, for nitrotoluene production, it should be noted that toluene nitrates easier than benzene. This process uses a mixed acid with an approximate composition of 72% H_2SO_4, 17% HNO_3 and 11% H_2O. The mixed acid is prepared by mixing 50 to 60% HNO_3 with 93% H_2SO_4. The nitrotoluene requires no feed preparation.

The process may be either batch or continuous. The same process steps are used for either process. For the continuous process, the reaction step consists of one or more reaction vessels in series, with all the feed going to the first vessel and with any subsequent vessels providing additional time for the reaction to go to completion. The reactors are provided with agitators and jackets or coils for cooling.

Mixed acid (as described under feed preparation), and either ortho- or para-nitrotoluene are fed separately and continuously as liquids to the reactor. A stoichiometric ratio of nitrotoluene to nitric acid in the mixed acid is used. The reaction is exothermic, and is controlled at 75° to 85°C. The liquid product from the reactor overflows to the decanter. The reactors vent to an absorber. The yield is 96%.

The liquid product from the reaction step is separated in the decanter to give a spent acid layer and a crude 2,4-dinitrotoluene layer. The spent acid layer is recycled (see recycle systems).

The crude 2,4-dinitrotoluene is washed and neutralized continuously in a series of steps. In the Meissner process these steps are as follows. The crude is washed, with water and with the wastewater from the second step washing, in a column in which both liquids flow upward concurrently. Air is fed to the bottom of the column, and flows upward with the liquids to provide agitation.

The liquids flow out the top of the column to a continuous decanter. The air vents to atmosphere from the top of the column. In the decanter the washwater separates out and is discarded. The organics layer is fed to a second stage column which operates in the same manner as the first column described above. The washwater consists of the wastewater from the third washing step and a dilute solution of sodium bicarbonate which neutralizes any acids present.

The liquids from this second column flow to a continuous decanter. The air is vented. In the decanter, a water layer separates from the organic layer and is fed to the first washing column. The organic layer is fed to a third stage of washing, which consists of a column and decanter similar to the other two stages. Water is used for washing. The organic stream from this stage is the product.

If ortho-nitrotoluene is used as the feed material, there will be some 2,6-dinitrotoluene present in the product. If a high purity 2,4-dinitrotoluene product is desired, it is obtained by distillation of the product from the third washing step.

Two recycle systems exist. The first system involves the spent acid from the reaction system. This acid is fed continuously to the spent acid washer, where it is mixed with a continuous flow of fresh nitrotoluene. Any residual nitric acid in the spent acid reacts with the nitrotoluene to form 2,4-dinitrotoluene and to produce a spent acid that does not require denitrification.

The spent acid and nitrotoluene are separated in a continuous decanter. The nitrotoluene layer becomes feed to the reaction step. The spent acid is fed to an H_2SO_4 concentrator, where the acid is concentrated to 93% H_2SO_4 for feed to the mixed acid system.

The second recycle system involves the absorber. The vents from the reactors go to the absorber, where any NO is oxidized and absorbed in water to produce nitric acid. The nitric acid is fed to the mixed acid system. The air is vented. There are two sources of air pollution. One is the vent from the absorber, which will consist of air and any unabsorbed oxides of nitrogen. The second is the vents from the washers, which will be air saturated with water and 2,4-dinitrotoluene.

DISULFOTON FROM ETHANOL

Disulfoton, also known as Di-Syston is a systemic insecticide of the organophosphate type. Its chemical name and formula are as follows.

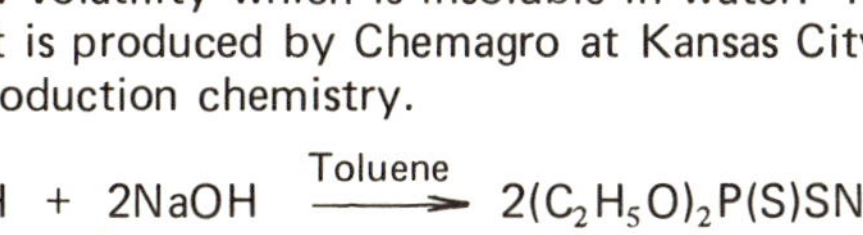

O,O-Diethyl-S-[2-(ethylthio)-ethyl] phosphorodithioate

It is a liquid of low volatility which is insoluble in water. It is highly toxic but is nonpersistent. It is produced by Chemagro at Kansas City, Missouri according to the following production chemistry.

$$P_2S_5 + 4C_2H_5OH + 2NaOH \xrightarrow{\text{Toluene}} \underset{\text{Diethyl Salt (DES)}}{2(C_2H_5O)_2P(S)SNa} + H_2S + 2H_2O$$

$$PCl_3 + \underset{\text{Thio Alcohol}}{3HOC_2H_4SC_2H_5} \longrightarrow \underset{\text{Chloro Thio Alcohol (CTA)}}{3ClC_2H_4SC_2H_5} + P(OH)_3$$

(continued)

$$(C_2H_5O)_2P(S)SNa + ClC_2H_4SC_2H_5 \longrightarrow \underset{\text{Disulfoton}}{(C_2H_5O)_2P(S)SC_2H_4SC_2H_5} + NaCl$$

A $1.9 million wastewater treatment plant recently came on stream. Treatment under caustic conditions (lime) removes some of the organophosphorus compounds during retention time of 36 hr. The pH is adjusted and a polyelectrolyte flocculent is used. Power plant water and sanitary wastes are handled and added. Final effluent is discharged to Blue River which probably has no more than parts per million range of any one pesticide now.

Solid waste from the treatment plant goes to commercial landfill. Chemagro is planning a biological treatment backup system. A process filter cake of diatomaceous earth is decontaminated [N_2 blow, toluene flush (recycle), steam, N_2 blow] then buried. The solids formulation line is well protected by a dust collection system which has a baghouse backed up by a Venturi-type wet scrubber (alkaline).

Formulation equipment is cleaned by solvent flush (which is used in next batch) and then by steam and water. Problems include trace emissions of odorants (more in production than in formulation), and disposal of containers. (Estimated 2 to 40 oz liquid left per drum.) H_2S flare and particulates emission from fuel combustion appear to be biggest pollution sources.

DYES AND PIGMENTS FROM AROMATICS

The manufacture and use of dyes and pigments constitute an important part of modern chemical technology. Because of the variety of products that require a particular material to give maximum coverage, economy, opacity, color, durability, and desired reflectance, manufacturers now offer many hundreds of distinctly different dyes and pigments.

Usually dyes are classified according to both the chemical makeup and the method of application. The manufacturers look at dyes from the chemical aspect, and arrange and manufacture them in groups, usually of like chemical conversions, while the users of dyes group them according to the methods of application.

The raw materials for the manufacture of dyes are mainly aromatic hydrocarbons, such as benzene, toluene, naphthalene, anthracene, pyrene, and others. These raw materials are almost never directly useful in dye synthesis. It is necessary to convert them to a variety of derivatives, which are in turn made into dyes. These derivatives are called intermediates. However, the industries which utilize either raw materials or intermediates to produce final product dyes are all subcategorized as the dye industry.

Because of the large number of compounds that are required, often in limited amounts, most dyes, if not all, are manufactured in batches. Since the purpose of the EPA study was to investigate process-related wastewater generation sources rather than to examine detailed unit processes/operations of manufacturing processes for each class of dyes/pigments, a typical manufacturing process for dyes is considered to illustrate the wastewater sources. Raw materials (which include aromatic hydrocarbons, intermediates, various acids and alkalies, and solvents)

FIGURE 36: MANUFACTURE OF DYES AND PIGMENTS

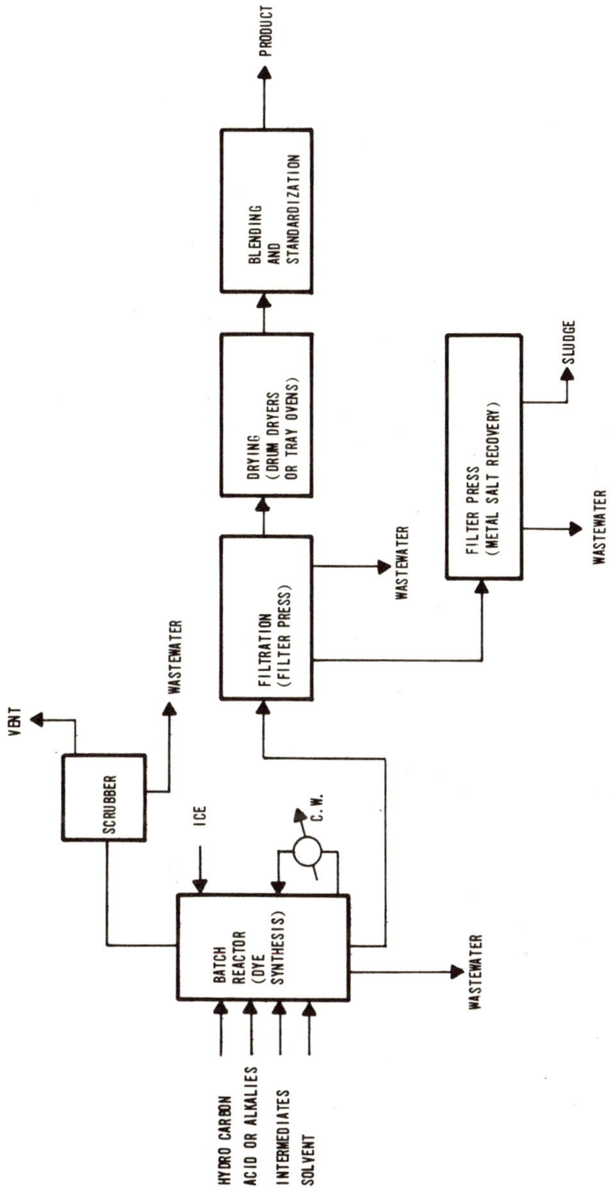

Source: Report EPA 440/1-73/009

are simultaneously or separately fed into the reactor, where the reaction is carried out ordinarily at atmospheric pressure. Because the reactions are exothermic, adequate temperature control is required to avoid side reactions. Temperature control is accomplished primarily by direct addition of ice to the reaction tank. When the reaction is complete, the dye particles salt out from the reaction mixture.

The vent gases taken as overheads from the reactor are continuously passed through a water scrubber before being discharged into the atmosphere. The liquid effluent from the reactor is then sent to a plate-and-frame filter press where the dye particles are separated from the mother liquor. The mother liquor is either directly discharged into sewers or sent to another filter press to recover some of the metal salts.

The filter cake is first washed with compressed air while still in the press. The moist cake is discharged into shallow trays which are placed in a circulating air dryer, wherein the moisture is removed at temperatures between 50° and 120°C. Vacuum dryers and drum dryers may also be used.

The dried dye is ground and mixed with a diluent, such as salt, to make it equal in color strength to a predetermined standard. Dilution is necessary because batches differ in their content of pure dye. Uniformity is assured by dilution to a standard strength.

The great majority of dyes and pigments are manufactured by the typical process described. However, the manufacture of some special dyes or pigments may require more or fewer processing steps. For example, in the manufacture of alkali-blue pigment, the process requires a steam ejector to produce vacuum for the batch reactors. The barometric condenser is then used to condense the exhaust steam. In the manufacture of Direct Blue 6 dye, the filter cake is not washed but merely freed from the adhering liquid by air drying.

The major water pollution sources of this process are the mother liquor from the filter press, the intermittent reactor cleanup waters, the draw-off from the vent gas scrubber, and the housekeeping cleaning waters. The data obtained from the plant survey are summarized in the following tabulation. Multiple data were collected at one of the plants, and these data were subjected to the analysis for probability of occurrence. The results of probability analysis are also shown in the tabulation.

Product	Sample I.D.	Flow	COD	BOD5	TOC
		gal/1,000 lb	lb/1,000 lb (mg/l)	lb/1,000 lb (mg/l)	lb/1,000 lb (mg/l)
1; Dye	Sample 1	13,700	1,075 (9,400)	220 (1,920)	450 (3,945)
Dye	Sample 2	13,700	652 (5,700)	126 (1,100)	269 (2,350)
2; Dye		21,050	175 (997)	59 (337)	60 (360)

(continued)

Product	Sample I.D.	Flow	COD	BOD5	TOC
		gal/1,000 lb	lb/1,000 lb (mg/l)	lb/1,000 lb (mg/l)	lb/1,000 lb (mg/l)
3; Dye	10% Occurrence	95,069	50 (63)	5 (6)	40 (51)
	50% Occurrence	95,069	1,850 (2,331)	79 (100)	790 (995)
	90% Occurrence	95,069	3,700 (4,662)	156 (197)	1,580 (1,991)
4; Pigment		124,000	4,925 (4,764)	1,470 (1,422)	819 (792)

Because of frequent changing of feed materials and desired products, dyemaking requires large amounts of water and of cleaning aids (such as detergent and bleach) to clean up reactors and filter presses on each reaction cycle. Chemical reactions involved are often exothermic and require strict temperature control. Due to the necessity of rapid cooling in order to avoid side reactions, direct cooling with ice, in addition to jacket cooling, is commonly practiced, and this also contributes a significant amount of wastewater.

While the high organic loading in the wastewater is primarily the result of incomplete crystallization and separation of dye products from the mother liquor, organic losses and cleaning aids from cleanup operations also contribute. In contrast to other organic chemical industries, jacket cooling water is required to be discharged into sewers to dilute the wastewater to be treated.

Reuse or recycle of wastewater from this type of process is deemed unfeasible, because the wastewaters are contaminated with many different salts, metal ions, and a high intensity of color, which will in turn contaminate the product.

EPICHLOROHYDRIN FROM ALLYL CHLORIDE

Epichlorohydrin was initially developed as an intermediate for synthetic glycerol, but more recently has become an epoxy resin component. Using a two-step process, it is made by chlorohydrinating allyl chloride:

$$ClCH_2{-}CH{=}CH_2 \; + \; HOCl \longrightarrow ClCH_2{-}CHCl{-}CH_2OH$$

followed by hydrolysis to epichlorohydrin:

$$ClCH_2{-}CHCl{-}CH_2OH \; + \; \tfrac{1}{2}Ca(OH)_2 \longrightarrow \tfrac{1}{2}CaCl_2 + H_2O + ClCH_2{-}\underset{\diagdown O \diagup}{CH{-}CH_2}$$

Epichlorohydrin is a compound that is a major raw material for epoxy and phenoxy resins and is used in the manufacture of glycerol. Although the hydrolysis reaction second step is not a chlorination process, in general the hydrolysis step immediately follows the chlorohydrination, and production figures are not available on the intermediate. The first reaction, the chlorohydrination of allyl chloride, can be carried out in a stirred tank at atmospheric pressure in the liquid phase, using water as the solvent. The temperature range is 30° to 80°C, with some heat evolution. Yields are over 90%. The reaction is complete when no more chlorine can be added.

The raw materials of the first step are allyl chloride, chlorine, and water. The by-products are trichloropropane, tetrachloropropyl ether and hydrogen chloride. Most of the air pollutants, such as HCl, are absorbed in the tail gas absorber.

Reactor products from the chlorohydrination of allyl chloride are treated with a lime slurry in a column-type reactor, using trichloropropane as a solvent. The temperature range is 70° to 100°C with some heat evolution. Atmospheric pressure is used. In a crude stripping column most of the water and trichloropropane are distilled off and recycled to the liming column.

The bottoms from the crude column are fed to the product distillation column, where the epichlorohydrin is withdrawn overhead. The bottoms of the product column are extracted with water, the water extraction being fed back to the reactor. The solids are settled out of the washed bottoms, and the liquor is returned to the solvent tank. The raw materials of the second step are dichlorohydrin and calcium hydroxide, while the by-products are calcium chloride and water. The reactor column is vented.

ETHYLBENZENE FROM BENZENE

Some ethylbenzene is recovered in refinery fractionation operations, but the majority is manufactured via the alkylation of benzene with ethylene. The alkylation reaction is as follows.

$$C_6H_6 + C_2H_4 \longrightarrow C_6H_5C_2H_5$$

Ethylene and feed benzene are combined with recycle benzene and polyaromatics, heated to reactor temperature, and introduced to the alkylation reactor. Off-gases from the reactor pass to the scrubbing system. The reactor effluent is passed to the separation section. Unreacted benzene is recycled, ethylbenzene is drawn off as the product, and polyethylbenzenes are recycled or drawn off as waste effluents.

If high purity benzene feedstock is used, the crude product is not required to be washed with caustic solution and water. However, the plant visited during the survey employs a feedstock containing some organic contaminants, and washing is necessary before the crude product is sent to the separation step. The washing step also removes any traces of the BF_3 promoter.

The major waste streams of this process are the spent caustic and washing streams used to wash the crude alkylate. Significant amounts of tars, benzene, ethylbenzene and other polymers will be found in these streams. Heavy aromatics fractions from the separation column are disposed of by incineration. The data obtained from the plant survey are summarized in the following tabulation.

Flow	37.7 gal/1,000 lb
COD	5,980 mg/l
	1.88 lb/1,000 lb
BOD_5	433 mg/l
	0.136 lb/1,000 lb
TOC	2,091 mg/l
	0.66 lb/1,000 lb

The alternate route in manufacture of ethylbenzene is a liquid phase reaction using aluminum chloride catalyst. The process requires much more extensive washing to remove highly acidic aluminum chloride catalyst. It is usually employed in combination with an ethylbenzene dehydrogenation step to produce styrene.

ETHYL CHLORIDE FROM VARIOUS RAW MATERIALS

In 1970, over 80% of the ethyl chloride was made by hydrochlorinating ethylene in this liquid phase reaction:

$$HCl + C_2H_4 \longrightarrow C_2H_5Cl$$

The remaining ethyl chloride was made by the following two reactions.

Thermal Chlorination of Ethane

$$Cl_2 + C_2H_6 \longrightarrow C_2H_5Cl + HCl$$

Hydrochlorination of Ethanol

$$HCl + C_2H_5OH \longrightarrow C_2H_5Cl + H_2O$$

The first of these reactions is handicapped by high cost chlorine and lower yields. The second is almost completely obsolete, due to the high cost ethanol, although the yields are very good.

The capacity and production of ethyl chloride is declining very slowly. The decline is probably due to the reduction in the use of tetraethyllead. However, the decline is offset by the increasing use of gasoline. It is estimated that the production will stay at 640 million pounds per year through 1980.

The main use of ethyl chloride is as an intermediate for making tetraethyllead. This use is in decline due to tetramethyllead growth and the environmentalists' fight against leaded gasolines.

Hydrochlorination of Ethylene

Ethyl chloride is produced by addition of hydrogen chloride to ethylene under anhydrous conditions in the presence of a catalyst such as aluminum chloride. Ethylene gas and substantially anhydrous hydrogen chloride are mixed in approximately equimolecular proportions and passed into a reactor partially filled with ethylene dichloride or a mixture of ethyl chloride and ethylene dichloride. In the presence of 0.2 to 0.3% aluminum chloride, at a temperature of 35° to 40°C, and a pressure of 40 psig, the exothermic hydrochlorination takes place. The vaporized products are fed into a column or flash drum, where the lower boiling ethyl chloride is separated from the heavier polymers. The crude ethyl chloride is refined by fractionation. Catalyst is continually withdrawn, and new makeup catalyst is added.

The overall yield based on ethylene is about 90%. The raw materials are ethylene, hydrogen chloride, and aluminum chloride catalyst. The main by-product is ethylene polymer oil. The only vent on the process is a purge on the recycled ethylene system, which keeps inerts (such as CO_2 and CH_4) from building up.

Chlorination of Ethane

Ethyl chloride is produced by the thermal chlorination of ethane:

$$Cl_2 + C_2H_6 \longrightarrow C_2H_5Cl + HCl$$

After preheating the ethane, equimolecular mixtures of ethane and chlorine are reacted in the vapor phase in a tank reactor at 375° to 475°C and 15 to 50 psig to yield about 75% ethyl chloride. The reacted mixture is quenched, scrubbed for HCl removal, dried, the excess ethane stripped off and the crude ethyl chloride fractionated. The excess ethane is recycled. The raw materials are ethane and chlorine. The main by-product is dichloroethane. The purge on the recycled ethane is the only source of air pollutants.

Hydrochlorination of Ethanol

The reaction of ethyl alcohol and hydrogen chloride in the presence of a catalyst yields ethyl chloride. Warm 95% ethyl alcohol is fed into the bottom of a jacketed glass lined reactor containing a 45% aqueous zinc chloride solution. Countercurrent to the alcohol, substantially anhydrous hydrogen chloride is fed into the aqueous zinc chloride solution, which is maintained at about 145°C.

The rates of feed are approximately 0.17 and 0.13 parts per hour per part of zinc chloride solution for the alcohol and hydrogen chloride, respectively. The reactor pressure is maintained at 30 psi. The catalyst concentration remains about constant, while vapors of ethyl chloride, water, and some hydrogen chloride are withdrawn from the reactor and passed countercurrent to a stream of water at 80°C in a glass lined scrubber.

The scrubber, operating at about 30 psi, removes the free acid from the incoming vapors. The scrubbed vapors are cooled to about 20°C; at the condenser pressure (30 psi) substantially complete condensation takes place. The liquid is passed into a decanter where ethyl chloride is separated from water. The process operates continuously to produce 99% ethyl chloride in a yield of 95 to 98% based on ethyl alcohol.

The raw materials are ethanol, zinc chloride catalyst, and HCl. The by-products are small and unimportant. The vent on the condenser is the only source of air pollutants.

ETHYLENE AND PROPYLENE FROM PARAFFINS

Ethylene and propylene are produced primarily by the pyrolysis of saturated hydrocarbons. In the United States, ethane and propane predominate as feedstock material. The chemical reactions for their pyrolysis are given below.

$$C_2H_6 \rightarrow C_2H_4 + H_2$$

$$2C_3H_8 \rightarrow C_3H_6 + H_2 + C_2H_4 + CH_4$$

The hydrocarbon feedstock is diluted with steam and passed through a pyrolysis furnace, where cracking takes place. Normal temperatures in the cracking section of the furnace are 1500° to 1600°F, and residence time is one second or less. The purpose of steam dilution is to depress any coking tendency within the furnace tubes.

In order that only the desired degree of cracking be obtained, the hot reactor effluent gases are cooled rapidly to a temperature which will quench the cracking reaction. Consequently, the cracked gases are cooled in a variety of ways, but usually at some point by direct contact with water in the quench tower.

After quenching, the cracked gases are compressed prior to treatment for removal of the contained acidic gases (CO_2 and H_2S). The acid gases are usually absorbed by some combination of systems using monoethanol amine (MEA), caustic, and water. The purified gas stream is then dried and further compressed before fractionation.

After compression, the dried, cracked gas is cooled to cryogenic temperatures, and hydrogen is flashed off and sent either to additional purification facilities or burned as fuel. The dehydrogenated stream flows to the demethanizer, where overhead methane is sent to fuel, and the C_2+ bottoms flow, under pressure, to the deethanizer.

At the deethanizer the C_3 and heavier materials are taken off as a bottoms stream and are sent to the depropanizer. The deethanizer overhead is selectively hydrogenated in the light acetylene converter, in order to remove trace amounts of acetylene; this stream then goes to the C_2 splitter, where the ethylene and ethane are separated. Ethane is recycled to the cracking furnace, and the overhead from the splitter is the product ethylene and is sent to storage.

The C_4+ depropanizer bottoms are sent to the debutanizer, and the overhead is selectively hydrogenated in the heavy acetylene splitter in order to remove trace amounts of methyl acetylene and propadiene. The depropanizer overhead goes to the C_3 splitter where the propane and propylene are separated.

The final tower in the fractionation train is the debutanizer, where various C_4 compounds are separated from the dripolene or pyrolysis gasoline fraction. The C_5 and heavier materials may be rejected as waste or may be used as a source of aromatics. The major areas of water usage in the cracking process relate to dilution steam requirements and the contact quench waters required in the cooling and primary separation of the cracked gas products. Pressure and hydrocarbon partial pressure are extremely important variables in the design and operation of

FIGURE 37: ETHYLENE AND PROPYLENE FROM PYROLYSIS OF HYDROCARBONS

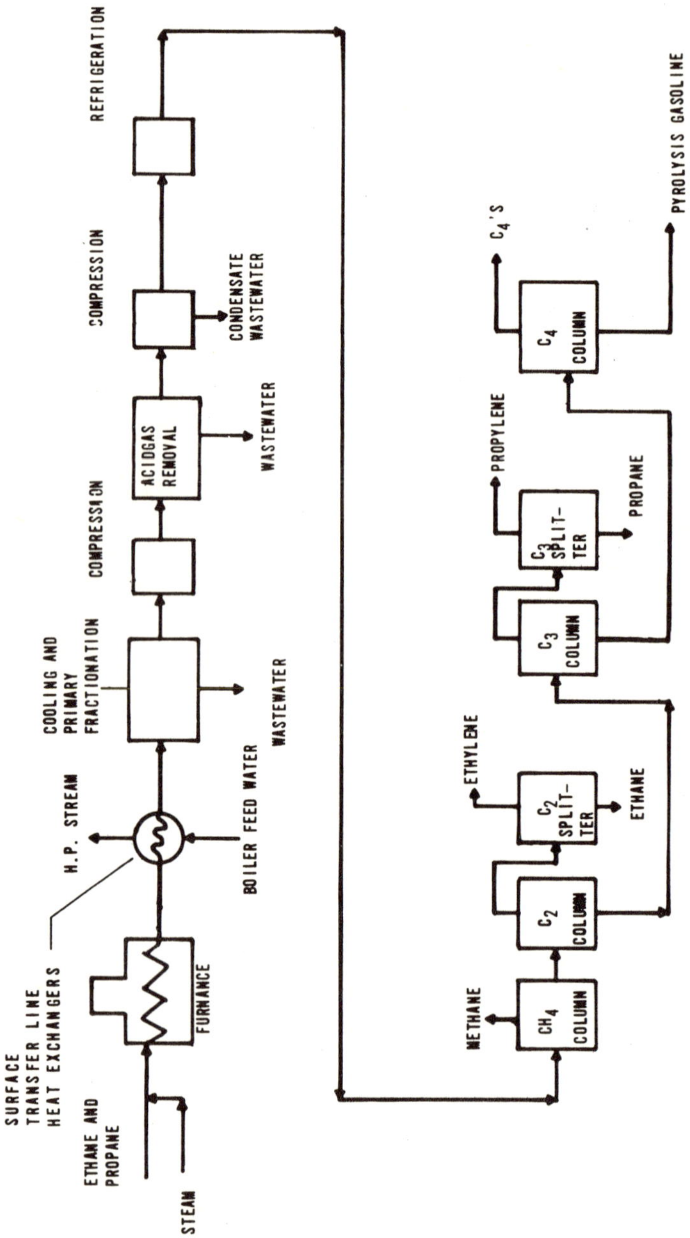

Source: Report EPA 440/1-73/009

ethylene plants. From an ethylene yield viewpoint, it is best to minimize pressure, or more specifically, hydrocarbon partial pressure. Low pressure is an economic problem since it increases the compression requirements following reaction. Instead of running the reaction at low pressure, steam is used as a diluent to reduce the hydrocarbon partial pressure. The steam also serves as a heating means for the cold feed.

For each feed there is an economic optimum of pressure and steam rate as they affect investments, operating costs, and product yields. Typical weight ratios of steam to hydrocarbon feed are as follows: gas feeds, 0.3; naphtha feeds, 0.5; and gas oil feeds, 0.7.

After cooling in the surface transfer line heat exchangers, the pyrolysis furnace effluent must be cooled further to a temperature suitable for economic compression. This cooling is generally carried out in a tray tower of more than one section. In addition to cooling the gases, heavy ends present in the furnace effluent must be scrubbed out. Because the quantity of heavy ends is very different when cracking gas oil, naphtha or ethane and propane, the tower design and function is different for each case.

In a plant where ethane or propane is cracked, the pyrolysis effluent contains very little hydrocarbon material that will condense at atmospheric conditions. Thus, when the gas is cooled to compressor suction conditions, only water and trace quantities of hydrocarbons condense. However, it is important that the hydrocarbon be removed from the gas since it is a tarry material that will foul the downstream processing equipment.

The hot gases enter the tower at the bottom and pass up through a baffled section passing through curtains of downflowing water. The gas is cooled to approximately 200°F (93.5°C) in the baffled section, and then passes to a tray section where cooling to approximately 105°F (40°C) takes place, the exact temperature being a function of the available cooling water temperature.

Fractionation between the heavy materials and the gasoline and lighter overhead also takes place in this trayed section. Heat is recovered in two stages so that maximum use is made of the heat in the gas. A larger settling drum is required to provide the separation between water, oil and tar.

In a plant where naphtha is being cracked, significant quantities of fuel oil are produced which can be separated in the quench tower, frequently termed the primary fractionator. There are a number of different designs for this area. One common design is a combination oil and water tower which eliminates overhead condensers and their attendant pressure drop ahead of the compressors. The lower section is a fractionator refluxed with cracked gasoline distillate which knocks down any fuel that might otherwise flow up the tower with the cracked gas.

The upper section is a water wash tower or spray condenser, where raw gasoline and dilution steam are condensed. The gasoline and water mixture is withdrawn from the bottom of this section as reflux while the net production goes to a distillate stripper before delivery to battery limits as raw product. The washwater can be used for certain low level heating services such as deethanizer and propylene splitter reboiling before returning to the top section of the tower. Gas oil cracking requires yet another type of design due to the very much larger quantity of gasoline and heavier material. The design of this tower begins to approach the design of crude oil distillation column. There are many possible designs for this, depending upon the products required.

The lower sections of the primary fractionator constitute distillation similar to a crude column, while the top sections are water wash sections. The bottoms from the fractionator are shown going to a vacuum flash tower to reproduce an additional vacuum distillate; this additional product would normally be required only when cracking heavy gas oils and when the fuel oil product is substantial.

The major wastewater sources in the cracking process are draw-offs from the water quench tower and the scrubber for removal of acid gases. Other possible sources are the water draw-offs from compressor interstages. The data obtained from the sampling program are summarized in the following tabulation.

Plant No.	Flow gal/1,000 lb	COD lb/1,000 lb (mg/l)	BOD5 lb/1,000 lb (mg/l)	TOC lb/1,000 lb (mg/l)
1	364	1.75 (533)	0.39 (130)	0.48 (259)
2	150	2.29 (1,827)	0.35 (279)	2.02 (1,617)
3	554	3.16 (684)	0.88 (189)	1.12 (242)
4	52.5	0.66 (1,502)	0.088 (200)	0.43 (980)
5		*	*	*
6	145	6.16 (5,110)	0.32 (265)	2.14 (1,770)
7	167	0.65 (467)	0.27 (192)	0.75 (538)

*Not available

Historical RWL data were also collected wherever they were available, and were subjected to analysis for probability of occurrence. The following tabulation presents the results of the analysis.

Plant	Occurrence	Flow gal/1,000 lb	COD lb/1,000 lb	TOC lb/1,000 lb
Plant 3	10%	*	1.6	0.6
	50%	*	4.0	1.05
	90%	*	6.4	1.51
Plant 5	10%	305	0.40	*
	50%	410	1.98	*
	90%	515	3.60	*

*Not available

The probability analysis for Plant 5 covered monthly average data for a period of 12 months; for Plant 3 it covered seven random daily samples for a period of 3 months. For the other plants there was not sufficient historical data for a full statistical analysis and comparison of the sampling data of all the plants.

Review of the available data reveals significant variations of RWL among the plants. The volume of wastewater per unit of product varies dependent primarily on the extent of scrubber water. At some plants, the use of steam strippers facilitates the reuse of quench water and minimizes the loss of hydrocarbons, thus generating a lower RWL; the organic loading in the wastewater is also affected by the performance of the quench towers and scrubbers. Higher RWL's in some cases are the result of contaminants in the feedstock.

The noncontact steam used in an ethylene plant is generated from extremely pure water because of the high pressure conditions. Most of this steam is recovered as condensate and returned to the boilers. A large quantity of steam, however, is used to contact dilution steam in the cracking reaction. When this steam is condensed from the process gases it is recovered as a fouled condensate and is not suitable for use as boiler feedwater or any other purpose. In some locations, even disposal of this steam is a problem.

Most of the boiler feedwater makeup in an ethylene unit is required to replace this condensed material. Since the boiler feedwater makeup must be suitable for high pressure steam generation, the net result is that the extremely pure water is degraded to fouled condensate. As plants have increased in size and steam pressures have increased, the differential cost in boiler feedwater treating has become significant and directed consideration to recovery of this water.

The obvious recovery method is to generate steam that is suitable for use as dilution in the cracking furnaces. This steam is required at 100 to 150 psig pressure and the water quality is not critical as in the case of high pressure steam. By removing the dilution steam requirement from the main steam system, the high pressure steam system becomes a closed loop and the only losses are to leaks and blowdown.

The principles of recovery of the condensed dilution steam for reuse are simple. In order for the condensed material to be used as feedwater for a vaporization system, it must be stripped of oils which would rapidly foul an exchanger used to vaporize this water. Then, since the water contains solids that must be purged from the system, heavy blowdown from the vaporizer is required to remove these solids.

A well-designed water system permits recovery of 90% of the steam used in hydrocarbon dilution and reduces the overall boiler feedwater makeup requirements to less than 20% of the requirements without a cleanup system. Unless boiler feedwater makeup is inexpensive because of existing high purity treating facilities, water cleanup is an economic addition to a new plant and is always included.

Since 1967, ethylene plants have incorporated the use of steam condensate strippers in order to reuse wastewater effluent and minimize hydrocarbon effluent in wastewaters. These facilities will generally require the use of a steam stripper and steam dilution. In addition, 2 pumps and a steam reboiler are required.

Investments for these facilities for a 5×10^8 lb/yr gas cracker are shown in Table 14 as $240,000.

TABLE 14: INVESTMENT FOR CONDENSATE STRIPPING*

I.	Process Water Stripper 4' & 8' I.D. x 39' High Including:	
	12 - Trays	
	1 - 240 GPM Pump	
	1 - 250 GPM Pump	
	1 - Filter	
	Instruments, Piping, Foundations, etc.	$160,000
II.	Dilution Steam Drum 7' I.D. x 20' High Including:	
	2 Heat Exchangers	
	Instruments, Piping, Foundations	$ 80,000
	TOTAL	$240,000

*For a 500 MM lb/yr ethylene plant using C_2/C_3 feed, totally installed, U.S. Gulf Coast location, 1973.

Source: Report EPA 440/1-73/009

Figures 38 and 39 are process flowsheets for quench tower loops without and with a condensate stripper. The increment operating costs are shown in Table 15. On ethylene product, it represents an increase of about $0.01 and on the wastewater reduction it represents $0.15/U.S. gallon of water saved.

The water draw-off from compressor interstages could be combined with the condensate stripping operation since it would only increase the quantity of water handled by about 20 to 25% without incurring any other handling problems. This water would probably be returned to the quench tower for reprocessing.

Table 16 presents the United States plants which are known to operate with condensate strippers.

FIGURE 38: WATER QUENCH WITHOUT CONDENSATE STRIPPER (500 MM LB/YR ETHYLENE PLANT WITH C_2 AND C_3 FEEDSTOCK)

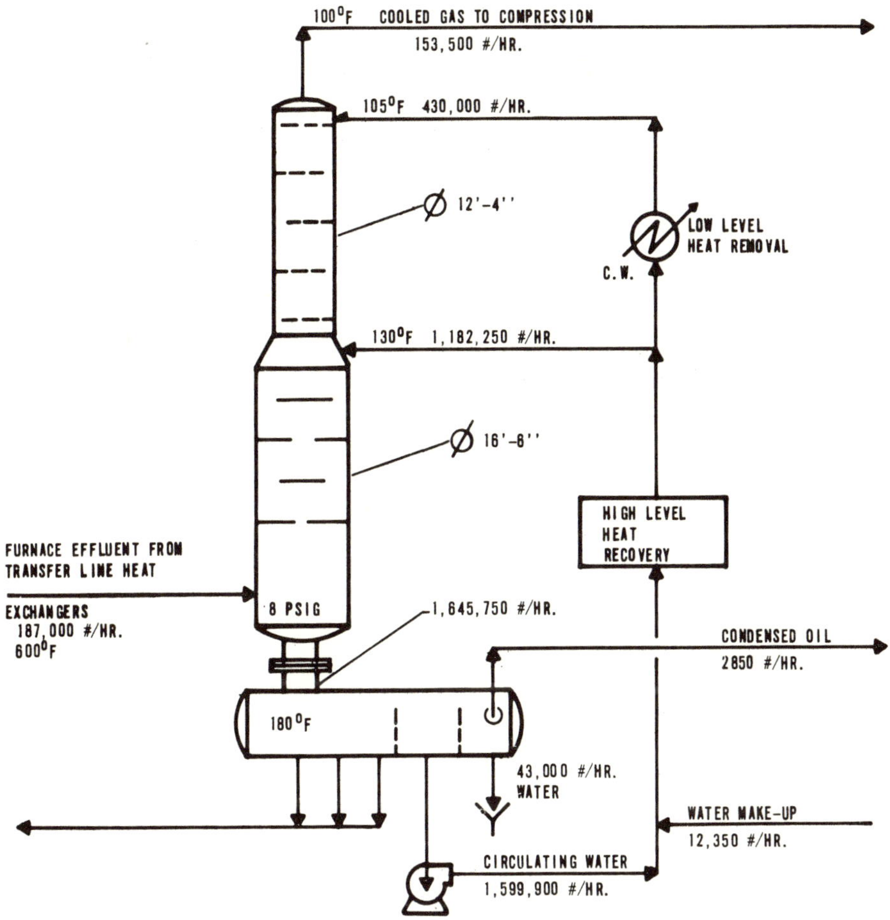

Source: Report EPA 440/1-73/009

FIGURE 39: WATER QUENCH WITH CONDENSATE STRIPPER (500 MM POUND PER YEAR ETHYLENE PLANT WITH C_2 AND C_3 FEEDSTOCK)

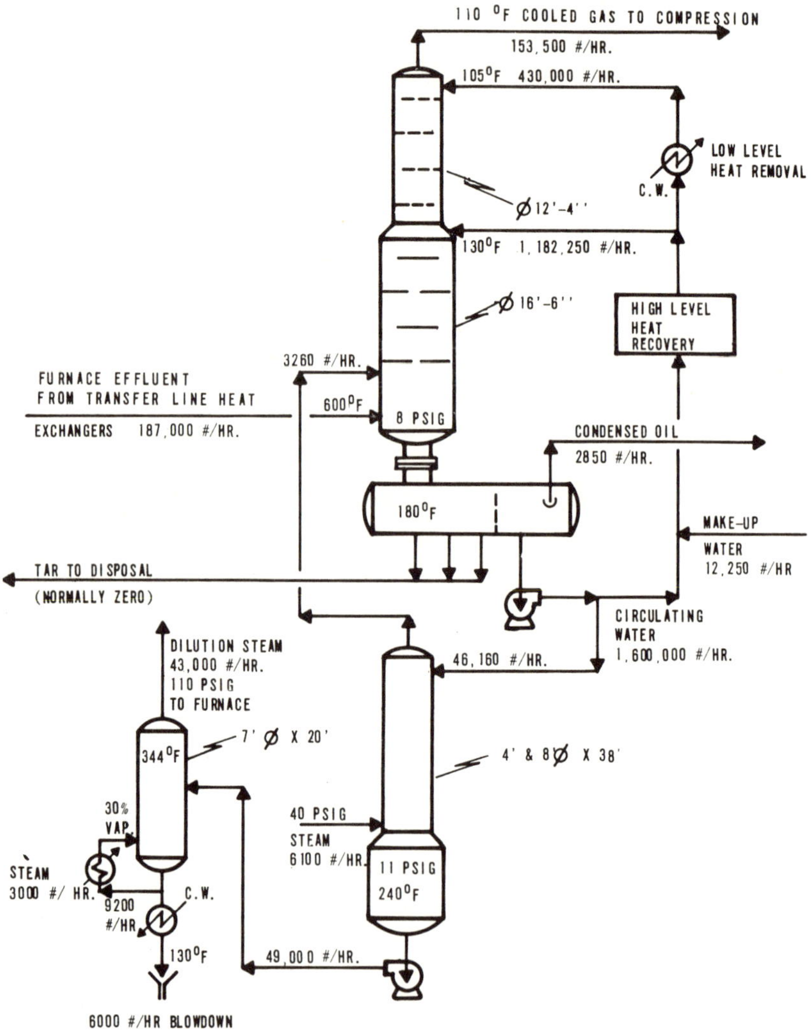

Source: Report EPA 440/1-73/009

TABLE 15: INCREMENTAL OPERATING COSTS FOR CONDENSATE STRIPPERS*

Incremental:

Steam - 6500 lb/hr (Stripping + Vaporization)
Electricity-Pumps - 40,000 Kwh
Saving of 74 GPM of wastewater or boiler feedwater

	Operating Cost/Year
Steam	$32,000
Power	400
	32,400
B.F.W. 74 GPM @ $.40/MUSG	(14,400) Credit
Net Utility Cost	$18,000

Investment Items:

Depreciation	24,000
Maintenance, Insurance and Other	11,000
	$35,000
TOTAL	$53,000/Year

Cost/lb of C_2 = .011¢

Cost/gallon of water saved = .15¢

*For a 500 MM lb/yr ethylene plant using C_2/C_3 feed.

Source: Report EPA 440/1-73/009

TABLE 16: U.S. ETHYLENE PLANTS USING CONDENSATE STRIPPERS

	Plant	Capacity
1.	Allied, Geismar	600 MM lb/yr Ethylene
2.	DuPont, Orange	750 MM lb/yr Ethylene
3.	Northern Petrochemical	800 MM lb/yr Ethylene
4.	Monsanto, Alvin	650 MM lb/yr Ethylene
5.	Union Carbide, Seadrift	1,400 MM lb/yr Ethylene
6.	Union Carbide, Texas City	1,250 MM lb/yr Ethylene
7.	Shell, Deer Park	1,000 MM lb/yr Ethylene
8.	Continental, West Lake	550 MM lb/yr Ethylene
9.	Cities Service, Lake Charles	440 MM lb/yr Ethylene
10.	Dow, Freeport	1,000 MM lb/yr Ethylene
11.	Union Carbide, Taft	500 MM lb/yr Ethylene
12.	Enjay, B.R.	1,200 MM lb/yr Ethylene
13.	Dow, Freeport	500 MM lb/yr Ethylene
14.	Amoco, Texas City	750 In construction
15.	Shell, Norco	1,000 Announced

Source: Report EPA 440/1-73/009

To define BADCT and BATEA technology, a steam stripper is required to reuse the wastewater from the quench tower. With the installation of steam stripper, contamination attributable to the quench water would be eliminated, and the resulting RWL's would be as shown below.

Plant No.	Flow gal/1,000 lb	COD lb/1,000 lb (mg/l)	BOD5 lb/1,000 lb (mg/l)	TOC lb/1,000 lb (mg/l)
1	364	1.57 (533)	0.39 (130)	0.48 (159)
2	50	1.83 (4,400)	0.19 (450)	1.77 (4,250)
3	11	0.77 (8,550)	0.43 (4,800)	0.13 (1,450)
4	52.5	0.66 (1,500)	0.09 (200)	0.43 (980)
7	10	0.48 (5,860)	0.12 (1,500)	0.63 (7,700)

The RWL's of Plants 1 and 4 are the same as in the previous tabulation because steam stripping has already been implemented at those plants. Plant 6 is missing from this tabulation because sampling at that plant (of the combined streams of quench water and scrubber water) precluded separate calculation of RWL without quench water.

The high wastewater flow of Plant 1 is attributed to high water usage in the scrubber; recycle of scrubber water could reduce the wastewater flow but not the organic loading. The calculated data for Plants 3, 4, and 7 can be considered as representative of the RWL for the process and can be used as criteria for BATEA and BADCT control technology.

A process developed by *J.R. Strausser and R.S. Kurland; U.S. Patent 3,507,782; April 21, 1970; assigned to Gulf Oil Corporation* provides for purification of quench water which is used to increase product yields by direct injection into cracking furnace effluent gases.

Water is highly suitable for quenching purposes since it is both effective and inexpensive. However, the employment of water in the quenching operation has one great attendant disadvantage. After treating the reaction effluent with the water, the quench medium contains significant amounts of dissolved and emulsified hydrocarbon oils, as well as heavy tar polymers and coke particles.

The oils comprise aromatic hydrocarbons and light polymers. These materials form extremely stable oil-in-water emulsions when the cracked gas stream is intimately mixed with the quench water. The resulting emulsions comprise from about 2,400 to more than about 6,500 parts of the oil per million parts of the emulsion. The stability of the emulsion is apparently due, at least in part, to a mutual affinity between the dispersed oil phase and the continuous water phase. Thus, the emulsion will resist efforts to resolve it into its component parts. These significant quantities of emulsified oils comprising aromatic hydrocarbons, and lesser amounts of olefinic hydrocarbons ordinarily require that the waste quench water be sewered, since it is undesirable to reuse this oil contaminated water in plant processing operations.

The oil content of this wastewater is sufficiently high that it would adversely affect process equipment if it were used in high temperature heat exchange equipment, e.g., shell and tube heaters, etc. On the other hand, the discharge of waste quench water which contains heavy polymer tars, finely divided coke particles, and the hydrocarbon emulsions, into local streams can create a pollution problem, since sufficient impurities are present in the wastewater to adversely affect fish and wildlife.

Accordingly, it would be desirable to purify this water to such an extent that it could be discharged into local streams without causing a pollution problem, or reused in process equipment. It would be further desirable to recover the dissolved and emulsified aromatic oils, since the aggregate of this material has many industrial and commercial applications. The aromatic and polymeric constituents might be individually recovered and employed as precursors for useful products, or the aggregate of the oil might be utilized, e.g., as a drying oil.

The use of chemical means for separating dissolved and emulsified hydrocarbons from aqueous media is undesirable, since it requires the employment of costly chemical reagents which may act as contaminants themselves either in the water or in the recovered oils.

The employment of simple gravity-type settling drums, commercial coalescers and filters for the separation of dissolved and emulsified hydrocarbons from aqueous media is undesirable and insufficiently effective in connection with emulsions of the presently described kind.

Thus, such means involve high operating and maintenance costs. Additionally, the conventional gravity-type coalescers yield rather crude and inconsistent separations, since they perform best for separating less stable emulsions. Likewise, filter coalescers of the commercial variety are expensive to operate.

Pollution control is carried out in this process by contacting the quench water containing dissolved and emulsified oils with an aromatic hydrocarbon solvent under conditions so as to demulsify and extract a majority of the dissolved and emulsified hydrocarbons thereby forming an extract phase comprising the aromatic hydrocarbons solvent and the demulsified hydrocarbons, and subjecting a resulting aqueous raffinate phase to a treatment with a finely divided, crystalline silica coalescing medium so as to coalesce and agglomerate the remainder of the dissolved and emulsified hydrocarbons.

The resulting aqueous phase is substantially hydrocarbon-free and may be reused in the various plant unit operations and processes or may be safely discharged into local waters. Likewise, the recovered hydrocarbon oils may be utilized as desired.

ETHYLENE DICHLORIDE FROM ETHYLENE

The direct chlorination of ethylene is carried out in the presence of a ferric chloride catalyst suspended in liquid ethylene dichloride.

$$C_2H_4 + Cl_2 \longrightarrow ClCH_2CH_2Cl$$

The raw materials are ethylene and acetylene-free chlorine. The temperature is about 90°C and the pressure about 7 pounds per square inch gauge. The reaction at these conditions is fast, complete, and exothermic, with some formation of 1,1,2-trichloroethane. The top of the tower-type reaction vessel is a fractionator, complete with reflux condenser.

The gas stream from the reactor is passed through a caustic scrubber, where the unreacted gases and a trace amount of hydrogen chloride are removed by a caustic solution. The liquid stream from the reactor is first sent to a distillation column to remove heavy ends and then to a wash tower, where a caustic solution is used to remove some impurities. The crude product is finally discharged to a distillation column for purification.

Liquid 1,2-dichloroethane, which is 99% pure, is withdrawn from near the top of the fractionator. Yields are about 95%. The vent on the reflux condenser, which runs continuously, is the only source of air pollutants.

There are two aqueous waste streams in this process. One is a liquid effluent from the scrubber and the other is the wastewater from the wash tower. The results of a survey at one plant are shown in the following tabulation.

Flow	96 gallons per 1,000 pounds
COD	6,050 milligrams per liter
	4.84 pounds per 1,000 pounds
BOD_5	Inhibitory
TOC	1,106 milligrams per liter
	0.89 pound per 1,000 pounds

A surface heat exchanger can be used to condense water vapor in the off-gas to the scrubber, while the remaining uncondensed gas from the reactor (which contains primarily unreacted ethylene and chlorine) can be totally recycled to the reactor. The scrubber and its wastewater can then be eliminated. With this modification, RWL of BADCT and BATEA for this process can be expected to have low values of 0.072 pound of COD and 0.106 pounds of TOC per 1,000 pounds of ethylene dichloride. Total process water usage of this process is 0.82 pound of water per pound of ethylene dichloride, and cooling water usage is 93 pounds of water per pound of product.

An alternate route in manufacturing ethylene dichloride is oxychlorination of ethylene with hydrochloric acid and air over a supported copper chloride catalyst. The characteristic wastewater stream from this process will contain most of the impurities found in the direct chlorination process.

In the oxychlorination of ethylene, pure ethylene, hydrogen chloride, and air are reacted in the vapor phase over a fixed bed catalyst of cupric chloride on alumina. Conditions are 250 to 315°C and 50 to 100 pounds per square inch

FIGURE 40: ETHYLENE DICHLORIDE (EDC) BY OXYCHLORINATION AND DIRECT CHLORINATION

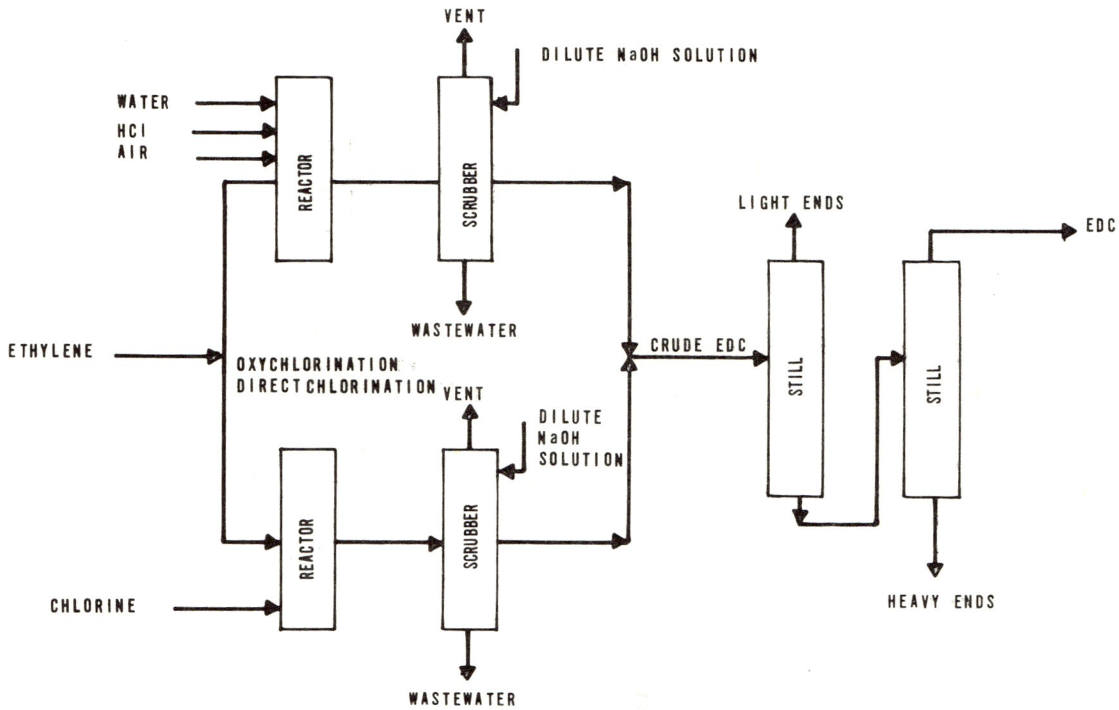

Source: Report EPA 440/1-73/009

gauge. The reaction is fast and exothermic. The resulting reaction products are passed through a caustic scrubbing tower where the DCE is condensed, the excess hydrogen chloride is absorbed and neutralized, and the inerts and non-condensable gases are vented. The bottoms from the caustic scrubbing tower are passed to a decanter tank, where the bottom layer is the crude product, DCE. The decanter bottoms are fractionated to yield a 99% pure product. The yields are over 95%.

The raw materials are ethylene, hydrogen chloride, air, caustic soda, and cupric chloride catalyst. The by-products are carbon dioxide, 1,1,2-trichloroethane, 1,2-dichloroethylene, water, and spent caustic soda. There is one source of air pollution, the vent from the caustic scrubbing tower.

ETHYLENE GLYCOL FROM ETHYLENE OXIDE

Ethylene glycol is produced from ethylene oxide by liquid-phase, acid-catalyzed hydration.

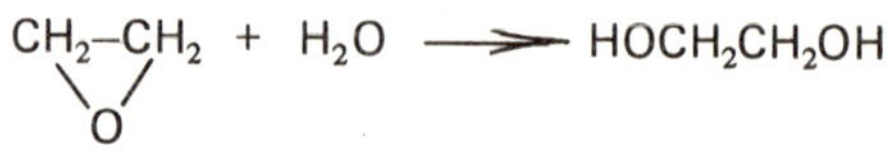

Ethylene oxide and water are reacted at about 300 psig and 180°C in the presence of sulfuric acid solution. By selection of the oxide-to-water ratio, it is possible to control the production of the mono-, di-, and higher glycols produced. Excess water is required for temperature control and to prevent the formation of undesirable by-products. Reactor effluent is dehydrated in a multiple-effect evaporator system. The effluent from the dehydration section is fed to a series of fractionators. The first tower removes water and traces of the light-ends, the second produces fiber-grade mono-ethylene glycol, and the subsequent towers produce diethylene and higher glycols.

The condensate from the dehydrator is partially recycled, and the remainder of this stream is the only source of water pollution in the process. The characteristics of this waste stream obtained from survey data is shown in the following tabulation.

Flow	584 gallons per 1,000 pounds
COD	1,800 milligrams per liter
	8.77 pounds per 1,000 pounds
BOD_5	69 milligrams per liter
	0.34 pound per 1,000 pounds
TOC	929 milligrams per liter
	4.53 pounds per 1,000 pounds

The high flow of the waste stream is caused by steam jets with barometric condensers which are utilized to produce vacuum for the multiple-effect evaporator system. If vacuum pumps with surface heat exchangers were to replace steam jets and barometric condensers, the flow of this waste stream could be significantly reduced. The condensate from the dehydrator could then be totally recycled back to the reactor. Consequently, BADCT and BATEA standards should require zero discharge from this process.

FIGURE 41: ETHYLENE GLYCOLS FROM ETHYLENE OXIDE

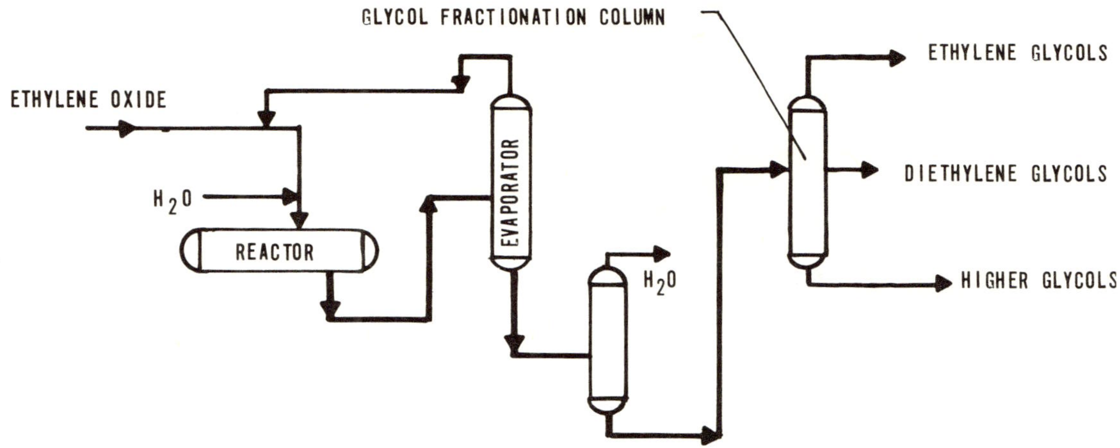

Source: Report EPA 440/1-73/009

ETHYLENE OXIDE FROM ETHYLENE

Most ethylene oxide manufacture is based on the direct vapor-phase oxidation of ethylene over a silver oxide catalyst.

$$C_2H_4 + \tfrac{1}{2}O_2 \longrightarrow \underset{\diagdown\, O\, \diagup}{CH_2{-}CH_2}$$

Oxidation takes place in the main reactor. Partial oxidation of ethylene to ethylene oxide and total oxidation to carbon dioxide and water are the two primary reactions. Ethylene oxide is recovered from the reactor effluent by absorption in a dilute aqueous solution.

In the process using high purity oxygen, the main absorber off-gas passes through a carbon dioxide removal system and is recycled to the reactor to reduce the ethylene partial pressure. When air is used as the oxidant, a secondary reactor system is employed to scavenge the remaining ethylene in the main absorber off-gas. Ethylene oxide is separated from water in the desorber, and the residual gases are discharged from the system. The combined liquid stream is fed to the ethylene oxide still, where the oxide product and residual water are separated.

The only waste stream generated in the direct oxidation process is the draw-off from the ethylene oxide separator bottoms. Process raw waste loads of this process obtained from plant visits are shown in the following tabulation.

	Plant 1
Flow	17.8 gallons per 1,000 pounds
COD	52,000 milligrams per liter
	7.7 pounds per 1,000 pounds
BOD_5	4,800 milligrams per liter
	0.71 pound per 1,000 pounds
TOC	19,650 milligrams per liter
	2.91 pounds per 1,000 pounds

	Plant 2
Flow	131.0 gallons per 1,000 pounds
COD	4,800 milligrams per liter
	5.26 pounds per 1,000 pounds
BOD_5	650 milligrams per liter
	0.71 pound per 1,000 pounds
TOC	2,699 milligrams per liter
	2.95 pounds per 1,000 pounds

The survey data show the same order of magnitude of raw waste loads in these two plants. However, the Plant 1 ethylene oxide separator operates more effectively and, consequently, generates a lesser volume of wastewater. Further reduction of RWL of Plant 1 is deemed unfeasible, and it should be considered as a representative RWL of this process in BPCTCA standards.

Ethylene oxide manufacture is usually accompanied by ethylene glycol manufacture. Since the wastewater from ethylene oxide contains 2% or more of ethylene

FIGURE 42: ETHYLENE OXIDE FROM ETHYLENE

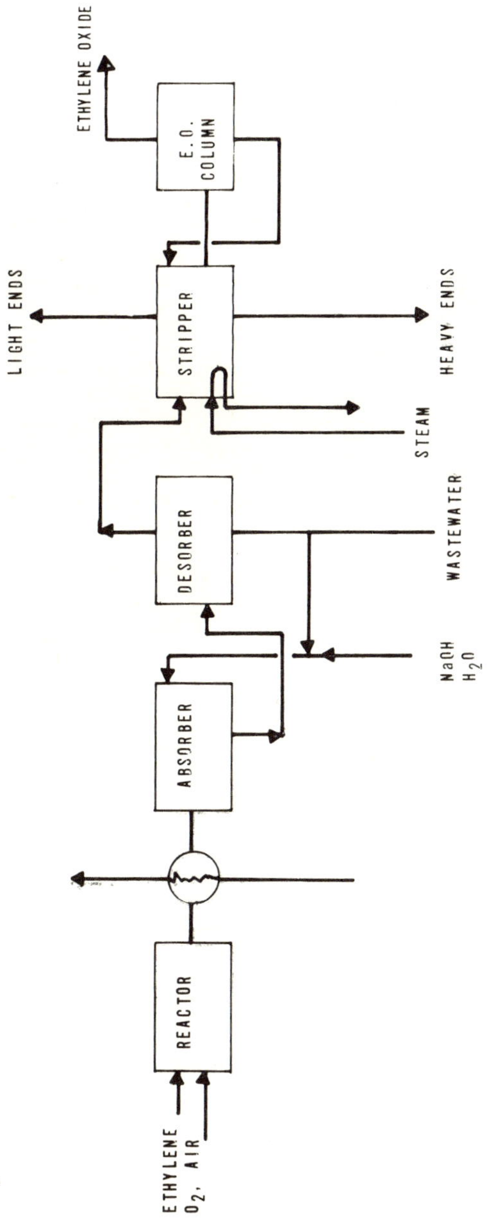

Source: Report EPA 440/1-73/009

glycol, this waste stream is usually sent to an ethylene glycol plant for further processing instead of being discharged into sewer lines. BADCT and BATEA standards, therefore, should require zero discharge from the direct oxidation process.

The high sulfate concentration in waste streams would disrupt the normal operation of biological treatment systems. Therefore, pretreatment or proper dilution with other waste streams is required.

Total process water usage (including steam directly supplied to the process) of this manufacturing process is approximately 0.25 pound of water per pound of ethylene oxide, while cooling water usage is 0.096 pound of water per pound of ethylene oxide.

An alternate route in the manufacture of ethylene oxide (used only by one chemical plant) is the chlorohydrin process. Ethylene, chlorine, and water are passed into a packed reactor, where they form ethylene chlorohydrin. The ethylene chlorohydrin is then reacted with hydrated lime to produce ethylene oxide. This process produces an aqueous lime slurry. The generation of this minimum-value by-product has led some producers to phase this process out.

Ethylene oxide production has grown nearly threefold in the last decade. Accompanying this growth has been a continuous increase in plant size, which has led to a corresponding decline in sales price.

FORMALDEHYDE FROM METHANOL

Formaldehyde is commonly manufactured by the oxidation of methanol. The process is a gas-phase reaction, operated with an iron-molybdenum oxide catalyst and a lean methanol-air mixture. The chemical reaction is given below.

$$CH_3OH + \tfrac{1}{2}O_2 \longrightarrow HCHO + H_2O$$

A mixture of methanol and water is vaporized by a closed steam loop, which circulates between the reactor and feed vaporizer. The reactants, mixed with air, flow through a thin layer of catalyst crystals in the reactor. The product gases are cooled by water, and product formaldehyde is recovered as a 50 to 52% aqueous solution by two-stage absorption. Product concentration is adjusted by controlling the amount of water supplied to the second stage absorber. The remaining unabsorbed gases from the abosrber are disposed of by incineration. A portion of the formaldehyde product may be passed through an anion exchanger to produce high purity formaldehyde by removing formic acid and sodium formate.

Wastewater streams generated in this process are intermittent. For example, wastewater from the washing of the absorber occurs at most twice per year. The contaminants in this stream are formic acid, methanol, formaldehyde, and ammonia. Wastewater created by regenerating the ion exchange units occurs three times per month at the plant visited. Another possible waste stream is withdrawn as an aqueous slip stream from the bottom of the feed vaporizer

FIGURE 43: FORMALDEHYDE BY OXIDATION OF METHANOL

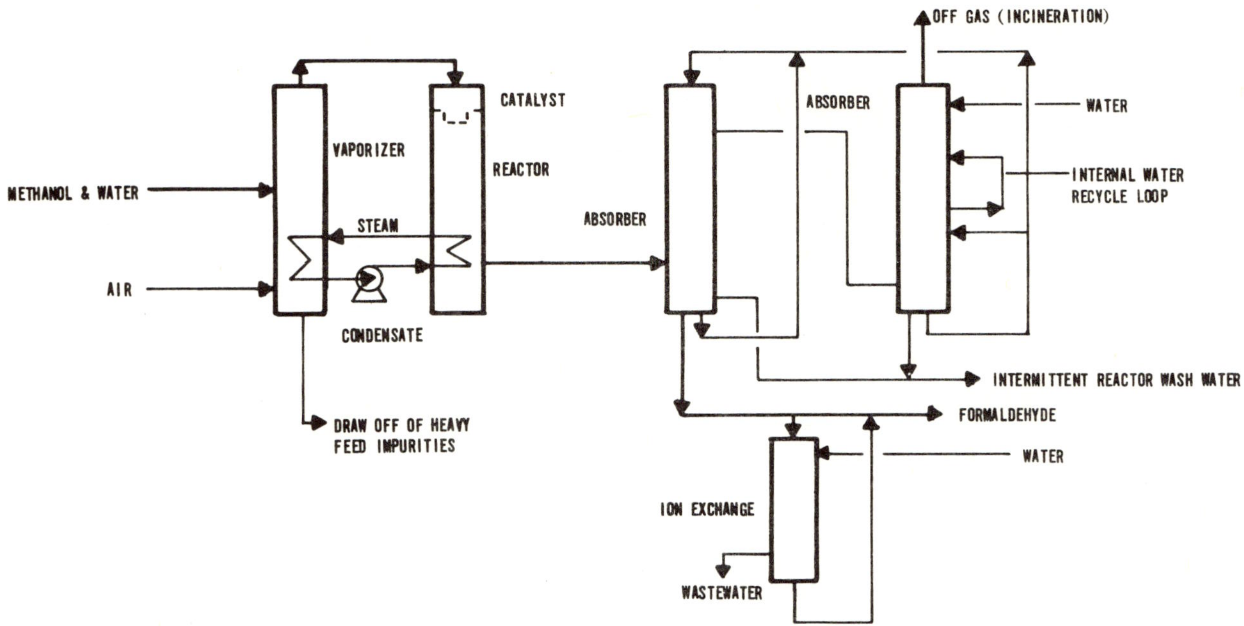

Source: Report EPA 440/1-73/009

whenever heavy impurities (such as acetone and oxygenated organics) occur in the methanol feed; the total flow of this waste stream, estimated by plant personnel, is about 131 gallons per 1,000 pounds of formaldehyde. A sample was not taken for analysis, since a continuous and representative sample is not available.

The alternate approach for formaldehyde manufacture from methanol involves a combined dehydrogenation and oxidation reaction over a silver or copper catalyst. This process operates with a rich methanol-air mixture.

About 90% of the formaldehyde produced in the United States is based on methanol as a raw material. The balance of the formaldehyde production is as a co-product of butane oxidation. The basic chemical reaction is summarized as follows.

$$2C_3H_8 + 2C_4H_{10} + 9O_2 \longrightarrow 14HCHO + 4H_2O$$

A process developed by *W. Riemenschneider and O. Probst; U.S. Patent 3,328,265; June 27, 1967; assigned to Farbwerke Hoechst AG, Germany* is a process for obtaining a sewage from a formaldehyde distillation concentration process that is practically free from formaldehyde. The process comprises introducing boiling water into the sump of the distilling zone, continuously feeding in the aqueous formaldehyde solution to be concentrated at a height of one quarter to four quarters of the total height of the distilling zone and discharging a product that is practically free from formaldehyde from the sump of the distilling zone. The procedure consists of a steam pressure distillation, in which case the discharged sump product can be drained off without additional intermediate treatment as sewage which is unobjectionable from a biological point of view.

LONG CHAIN ALCOHOLS FROM ETHYLENE OLIGOMERS

Long-chain alcohols are manufactured from ethylene in the presence of Ziegler catalysts. The process begins by reacting aluminum metal with ethylene and hydrogen to form triethyl aluminum (TEA). Ethylene is added to this compound at high pressures to give trialkyl aluminum compounds, which are then oxidized with dry air to aluminum trialkoxides. These are hydrolyzed by sulfuric acid to primary alcohols having an even number of carbon atoms. The basic chemical equations are summarized as follows.

$$3C_2H_4 + 1\tfrac{1}{2}H_2 + Al \longrightarrow (C_2H_5)_3Al$$

$$(C_2H_5)_3Al + nC_2H_4 \longrightarrow \begin{array}{l} R_1\diagdown \\ R_2\text{—}Al \\ R_3\diagup \end{array}$$

$$\begin{array}{l} R_1\diagdown \\ R_2\text{—}Al \\ R_3\diagup \end{array} + 1\tfrac{1}{2}O_2 \longrightarrow \begin{array}{l} R_1\text{—}O\diagdown \\ R_2\text{—}O\text{—}Al \\ R_3\text{—}O\diagup \end{array}$$

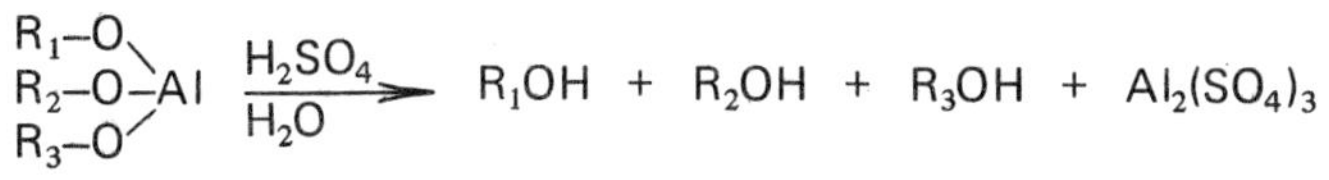

An atomized aluminum powder is first activated in a nonaqueous slurry media and next hydrogenated with dry hydrogen gas under pressure to give diethyl aluminum hydride. The hydride is then contacted with ethylene to produce TEA. Approximately 2 mols of TEA are recycled to the hydrogenator and 1 mol goes to the polymerization step. Recycle TEA solvent and aluminum are separated by means of a centrifuge.

In the polymerization section, TEA is reacted with ethylene under pressure to make trialkyl aluminum, which is then oxidized to produce alkoxides. A nonaqueous solvent such as toluene is circulated and recycled in this section. In the hydrolysis section, the alkoxides are hydrolyzed with sulfuric acid and water to yield alcohols and a solution of alum and water. The alum solution is separated from the alcohols in a decanter. The sulfuric acid residue is first neutralized with dilute caustic solution and next washed with hot water to remove sodium sulfate. In both the neutralization and wash steps, the alcohols are separated from the aqueous phase in decanters. The crude alcohols are then dehydrated and fractionated in a series of distillation columns to obtain pure alcohol products. Steam jets are used to produce vacuum in the stills.

The major water pollution sources in this process are the draw-offs from decanters and the condensate of the steam jets. Depending upon the desired concentration of the alum solution recovered, the cycle of decanter draw-off waters, and the modes of condensing ejected steam, the volume of wastewater per unit production will vary. Straight-chain alcohols are also obtained by the oxo reaction starting from straight-chain olefins and by direct oxidation of normal paraffins.

MALATHION FROM METHANOL

Malathion is a nonsystemic insecticide and acaricide of the organophosphate class. Its chemical name and formula are as follows.

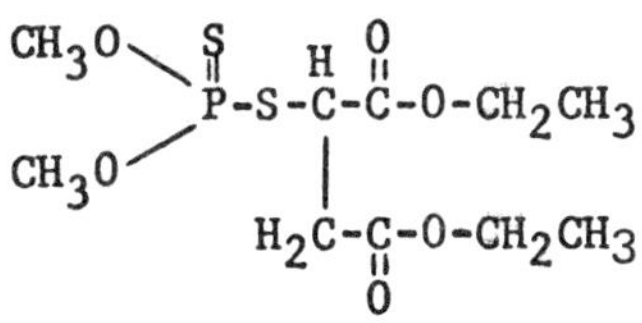

O,O-Dimethylphosphorodithioate of diethyl mercaptosuccinate

It is a liquid of low volatility which is slightly soluble in water. It has low toxicity and is nonpersistent in the environment. It is produced by American Cyanamid at Warners, New Jersey using the following production chemistry.

$$P_2S_5 + 4MeOH \longrightarrow 2\,(MeO)_2\overset{\overset{S}{\|}}{P}SH + H_2S$$

$$(MeO)_2\overset{\overset{S}{\|}}{P}SH + \begin{array}{l} HC\text{-}COOEt \\ \; \| \\ HC\text{-}COOEt \end{array} \xrightarrow{\sim pH\ 5.5} \begin{array}{l} (MeO)_2\overset{\overset{S}{\|}}{P}SCHCOOEt \\ \qquad\qquad\;\; | \\ \qquad\qquad\; CH_2COOEt \end{array}$$

The only by-product of malathion is H_2S, which is recovered as H_2SO_4 in a new sulfur recovery unit at the Warners, New Jersey plant of American Cyanamid. Emissions from this recovery unit meet New Jersey requirements of $<$5.5 lb NO_x per ton H_2SO_4 and $<$6.5 lb SO_2 per ton H_2SO_4. Some toluene is lost from the malathion distillation process, but emission factors are not known.

The acid production center, which was rebuilt two years ago and is used to produce both malathion and phorate, is generally considered to be the most hazardous area of the plant. This unit is completely diked, has separate fire walls, CO_2 blanket system, and no floor drains.

Washwater from the washing process as well as other liquid waste (spills, etc.) goes to a holding pond and is eventually barged to sea (150 to 200 miles out). Cleanup of equipment is limited to once or twice a year when changing products. Waste from this operation is barged to sea. Extensive precautions are taken against cross contamination at this time, and to date there have been no problems. Waste has been barged to sea for about four years.

The only solid wastes generated are filter cartridges. These are decontaminated with NaOH, and they are buried with lye in a landfill (on site). Dust in the formulation area is collected in a baghouse and recycled.

Containers are statistically spot checked before use. Defective lots of containers are rejected. Containers are not pressure checked. The storage of empty containers before they are used is very important in preventing leakers. Very few complaints about leaks have been received. They try to use the same carriers all the time (rail, truck, etc.), and this seems to help. However, LTL shipping has the potential for causing problems because of mixing shipments.

A process developed by *B.C. Wolverton; U.S. Patent 3,725,269; April 3, 1973; assigned to the U.S. Secretary of the Air Force* involves decontaminating and converting organophosphorus insecticides such as malathion into products which can be water washed and discharged into streams without the usual toxic effects to fish normally exhibited by unaltered insecticides.

The present decontamination process involves exposing the insecticide in a volume-to-volume ratio of 10 parts of a decontaminant to 1 part of the insecticide, the decontaminant consisting essentially of 12.5 to 25% by volume of monoethanolamine in dipropylene glycol monomethyl ether as a solvent and continuing the exposure for at least 30 minutes.

The resulting mixture of decontaminant and insecticide can be readily hosed

FIGURE 44: PRODUCTION AND WASTE SCHEMATIC FOR MALATHION

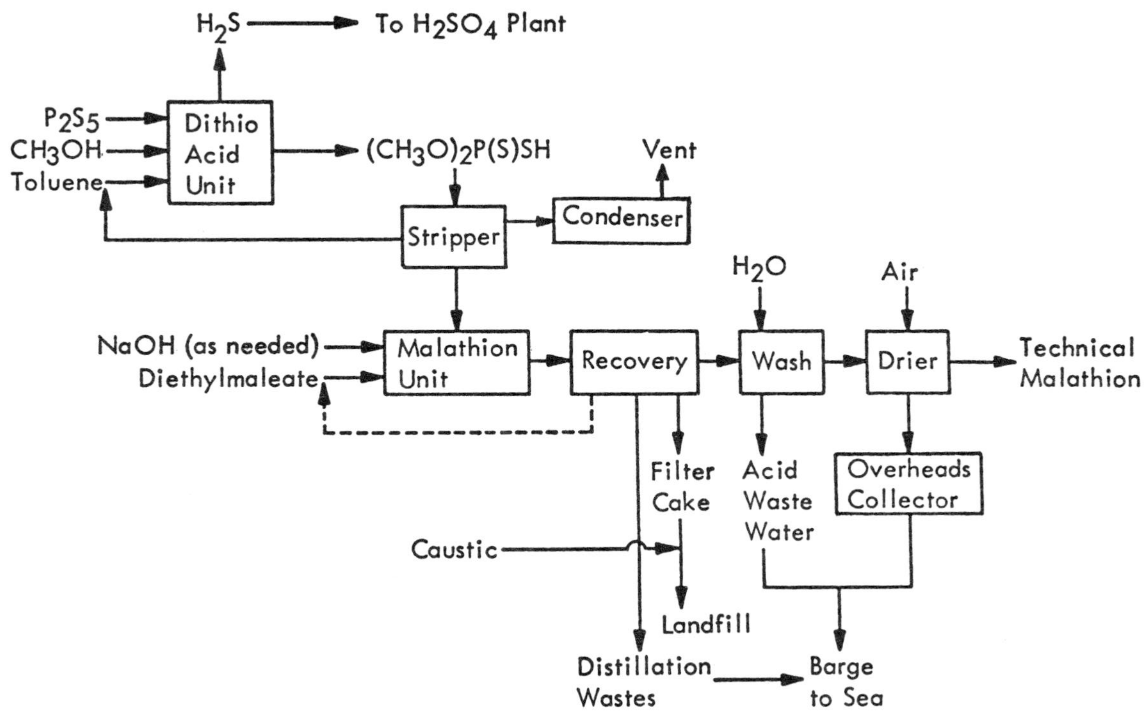

Source: Midwest Research Institute Report PB 213,782; 6/72

away with water without having the resulting run off deleteriously affecting any nearby streams and fish life therein.

METHANOL FROM NATURAL GAS

All of the processes for synthetic methanol involve the basic steps of steam reforming of natural gas plus addition of carbon dioxide to adjust the C/H ratio, compression, synthesis in a catalytic converter, and distillation for purification. The following reactions summarize the basic chemistry.

$$CH_4 + H_2O \longrightarrow \underset{\text{synthesis gas}}{CO + 3H_2}$$

$$CO + 2H_2 \longrightarrow CH_3OH$$

$$CO_2 + 3H_2 \longrightarrow CH_3OH + H_2O$$

The optimum atomic ratio for C/H in methanol synthesis is 1/4 as indicated above. However, carbon dioxide is added to take care of extra hydrogen resulting from steam reforming of natural gas.

The traditional conversion to methanol is carried out at high pressure (4,500 pounds per square inch gauge) in the presence of a chromium oxide-zinc oxide catalyst at about 650°F. However, a new process operates at only 750 psig and 500°F by using a new active copper catalyst. The much lower pressures allow the use of centrifugal compressors rather than reciprocating compressors, and also allow use of hydrogen rich synthesis gas without having to add carbon dioxide. Also, the conversion of natural gas to methanol is much higher in the low pressure process than that in the high pressure process.

The synthesis gas, after compression, is charged together with recycle gas to the reactor. The methanol-bearing gas leaving the reactor is cooled by heat exchange with air or water. The condensed crude methanol is separated from unreacted gas, which is recycled to the reactor. The flashed, gas-free crude methanol from the separator is purified by distillation.

The only wastewater stream from methanol plants using 100% natural gas feedstock is the aqueous stream from the final methanol distillation column. Processes which utilize off-gases from acetylene manufacture as feedstock introduce impurities into the system. These impurities must be removed before crude methanol can be purified. Usually, the impurities are first oxidized by a strong oxidizing agent; this is followed by sedimentation, filtration, and cation exchange. The results of survey date are shown in the following tabulation.

	Plant 1
Flow	59 gallons per 1,000 pounds
COD	320 milligrams per liter
	0.16 pound per 1,000 pounds
BOD_5	119 milligrams per liter
	0.059 pound per 1,000 pounds
TOC	107 milligrams per liter
	0.053 pound per 1,000 pounds

Plant 2

Flow	42.2 gallons per 1,000 pounds
COD	4,930 milligrams per liter
	1.74 pounds per 1,000 pounds
BOD_5	2,620 milligrams per liter
	0.92 pound per 1,000 pounds
TOC	583 milligrams per liter
	0.21 pound per 1,000 pounds

Plant 1 utilizes the low pressure system with natural gas as feedstock. Plant 2 uses off-gases from acetylene manufacturing as part of its feedstock. The RWL of Plant 1 is representative of the low pressure process, and can be considered as standard for BADCT and BATEA. The higher RWL of Plant 2 is due to impurities introduced with the off-gases. A preliminary fractionator can take the place of a series of treatments of crude methanol, and the impurities can be removed from the fractionator and disposed of by incineration. The RWL could then be expected to approach that of the low pressure process. There is no significant difference in wastewater characteristics between the high and low pressure processes. Only minimal carryover of metal catalyst is expected.

It is known that in the production of synthesis gas by the partial oxidation of a solid hydrocarbonaceous material there is obtained as a by-product a certain amount of finely divided carbon in the resulting product gases, carbon monoxide and hydrogen. It is also known that the presence of these solid carbon by-products in the gaseous products obtained from a synthesis gas generator interfere with the main reactions in subsequent processes for conversion of the synthesis gas products into hydrocarbons, ammonia, and oxygenated hydrocarbons, for example, in Fischer-Tropsch type synthesis, ammonia synthesis, or methanol synthesis.

Many methods have been proposed to effect separation of the free carbon particles from the synthesis gas products. These known carbon separation methods in general comprise treatment of the effluent products from the synthesis gas generator with water or water containing a low molecular weight oxygenated hydrocarbon in a scrubbing tower to effect removal of the entrained carbon particles therefrom and the removed carbon particles are then withdrawn from the scrubbing zone in the form of a carbon-water slurry. In general, the carbon content of the slurry varies from about 0.1 to 3% by weight dependent on the generator feedstocks and reaction conditions in the generator.

The carbon particles are thereafter separated from the carbon-water slurry in a conventional manner such as by filtration or by evaporation. Filtration is generally carried out using a rotary filter operated at subatmospheric pressure that is from about 5 to 10 psia. When the carbon is separated from the slurry by a filtration operation, the resulting water filtrate is commonly recycled to the scrubbing tower for reuse as scrubbing liquid by suitable piping, pumps, and the like.

A vacuum filtration type operation has many known disadvantages such as the cost of equipment, materials, handling facilities, and the like. One of the known disadvantages is that the carbon particles separated from the aqueous slurry are not suited for immediate use as fuel feedstock or in carbon black manufacture

FIGURE 45: METHANOL FROM NATURAL GAS

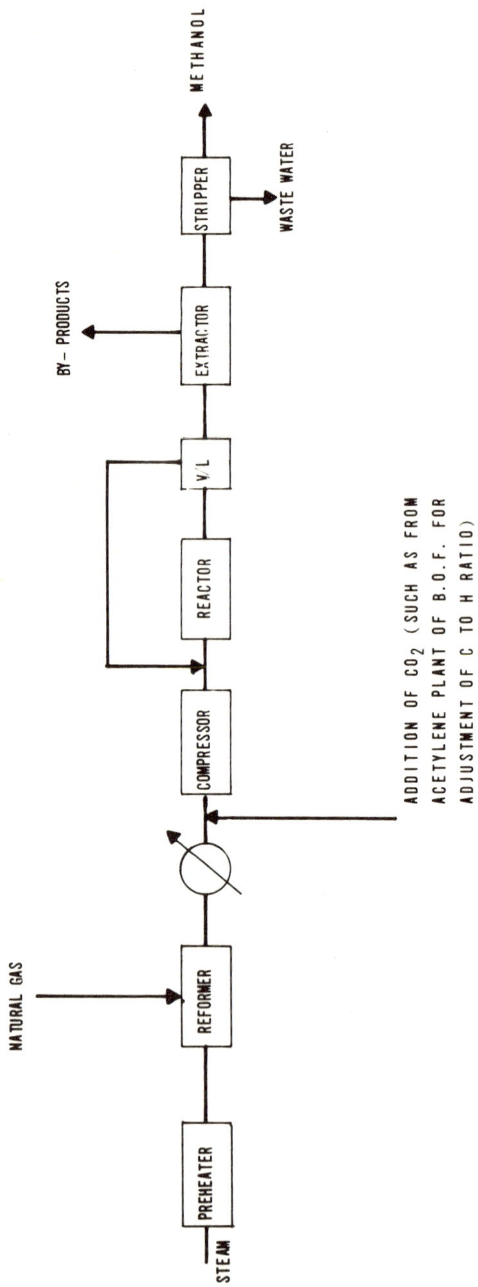

Source: Report EPA 440/1-73/009

because of the high water content of such particles. The recovered carbon particles in the form of filter cakes contain about 80 to 90% by weight water, the balance carbon.

In order to utilize the carbon particles recovered from a filtration operation in an efficient and economical manner, it is generally necessary to decrease the water content of the filter cake somewhat, particularly when the carbon is to be used as a fuel feedstock. One known method of decreasing the water content of the carbon filter cake is by an evaporation operation, but this method is not completely satisfactory in large scale industrial operations due to the considerable amount of heat required and also the equipment needed for such an operation. Solvent extraction has been employed heretofore in the separation of undesirable liquids including water from materials but such methods likewise have certain objectionable features such as solvent handling and solvent loss which make them unattractive to industry.

It is also known that the carbon particles dispersed in the carbon-water slurry can be recovered therefrom by permitting the slurry to remain in the quiescent state for a considerable period of time, that is, approximately about 2 or more days, depending on the concentration of the carbon in the slurry. On standing, the carbon particles gradually collect in the lower portion of the settling tank and they are recovered therefrom after withdrawal of the water layer such as by decantation.

The recovered carbon particles contain considerable amounts of water. This method of recovering the carbon particles is objectionable because of the need to provide large areas for settling tanks, ponds, etc., the length of time required for settling of the carbon and the relatively inefficient separation of the settled carbon particles from the aqueous liquid.

A process developed by *R.M. Dille; U.S. Patent 2,982,645; May 2, 1961; assigned to Texaco, Inc.* involves contacting the carbon-water slurry obtained in a synthesis gas generation process with a water-insoluble long carbon chain fatty acid or a glyceride thereof and thereafter recovering the separated carbon particles from the slurry.

Contacting the aqueous carbon slurry with the fatty acid or a glyceride thereof causes the carbon particles to collect in the upper portion of the aqueous liquid. The resulting separated carbon particles can be recovered from the upper layer by conventional flotation separation methods.

One method of recovering the separated carbon particles from the contacted slurry is by first drawing off the aqueous liquid at a low portion of the separation vessel and then collecting the separated carbon particles from a lower portion of the separator in a known manner. It is preferred, however, to effect recovery of the separated carbon particles from the contacted slurry by a skimming operation using a rotating rake. This process obviates the need for separating the carbon particles from the slurry by filtration. In addition, the water content of the recovered free carbon particles is materially reduced.

A process developed by *P. Visser and L.W. Ter Haar; U.S. Patent 3,694,355; September 26, 1972; assigned to Shell Oil Company* is one in which particles such as soot are removed from aqueous suspensions by a two-step process, the first step being to agglomerate the particles by contacting them with gentle agitation with a water-immiscible liquid, and then to contact the agglomerate-containing aqueous phase with a continuous phase that is water-immiscible under conditions such that the agglomerates enter the nonaqueous phase.

METHYLAMINES FROM METHANOL AND AMMONIA

Methylamines are synthesized by methanol and ammonia in the presence of catalyst to form a mixture of mono-, di-, and trimethylamine.

$$CH_3OH + NH_3 \longrightarrow CH_3NH_2 + H_2O$$

$$2CH_3OH + NH_3 \longrightarrow (CH_3)_2NH + H_2O$$

$$3CH_3OH + NH_3 \longrightarrow (CH_3)_3N + H_2O$$

Reactants are first preheated by the converter effluent, thereby recovering some of the exothermic reaction heat. The product stream is then flashed to remove the noncondensibles and is sent to the recovery system. First, ammonia is taken overhead and recycled, together with some trimethylamine. Next, water is added to break the TMA-ammonia azeotrope, and pure TMA is taken overhead from a distillation column. The mixture of mono- and dimethylamine is first dehydrated and then fractionated to separate DMA and MMA. The ratios of the three amines can be varied by changing reaction conditions.

This process uses water to scrub ammonia from all off-gases. The liquid effluent from the absorber is then flashed to recover ammonia. The major wastewater source, containing a significant amount of unrecoverable ammonia, is the bottoms from the flash column. The other two wastewater streams are the bottoms from the separation fractionators. The characteristics of the wastewater are summarized in the following tabulation.

	Sample No. 1
Flow	429 gallons per 1,000 pounds
COD	6,303 milligrams per liter
	22.56 pounds per 1,000 pounds
BOD_5	99 milligrams per liter
	0.351 pound per 1,000 pounds
TOC	11,634 milligrams per liter
	41.65 pounds per 1,000 pounds

	Sample No. 2
Flow	429 gallons per 1,000 pounds
COD	1,178 milligrams per liter
	4.21 pounds per 1,000 pounds
BOD_5	174 milligrams per liter
	0.62 pound per 1,000 pounds
TOC	3,808 milligrams per liter
	13.63 pounds per 1,000 pounds

FIGURE 46: METHYLAMINES FROM METHANOL AND AMMONIA

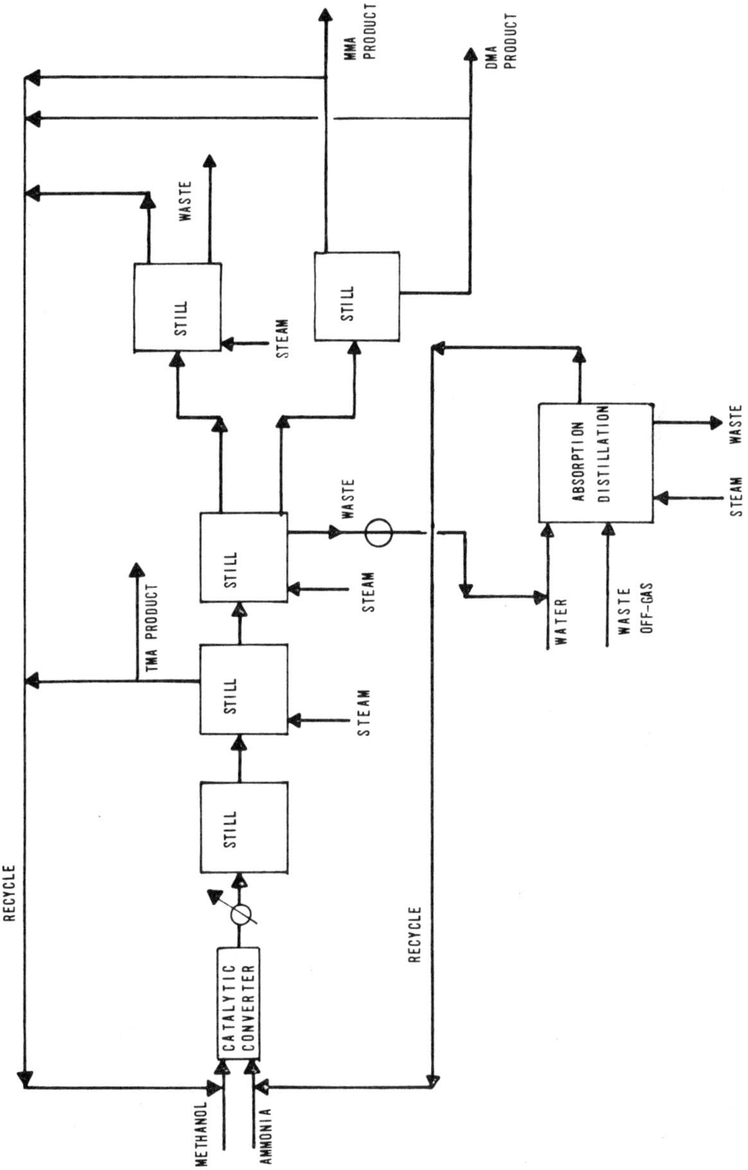

Source: Report EPA 440/1-73/009

The preceding data show significant variation. The extraordinarily high ratio of COD/BOD_5 is due to the ammonia contaminant which contributes to the measurement of COD but not to that of BOD_5. It is believed that Sample No. 1 was taken under the upset operating condition of the ammonia flash column.

Total process water usage, including steam directly supplied to the process, is 3.1 pounds water per pound of methylamines, while cooling water usage amounts to 16,700 pounds water per pound of product.

Minor process modifications such as reusing wastewaters from fractionators as ammonia absorption water can reduce the amount of wastewater. The ammonia content in the wastewater can be treated only by end-of-pipe treatment.

METHYL BROMIDE FROM METHANOL

Methyl bromide is a fumigant of the halocarbon class. Its alternate chemical name and formula are as follows.

CH_3Br
Bromomethane

Methyl bromide is a liquefied gas (BP 3.6°C) which is very toxic. It is probably not very persistent in the sunlight. It is produced by Dow Chemical at Midland, Michigan, Michigan Chemical, American Potash, Great Lakes Chemical and Vulcan according to the following production chemistry.

$$6CH_3OH + 3Br_2 + S \longrightarrow 6CH_3Br + H_2SO_4 + 2H_2O$$

As regards pollution control, methyl bromide is a colorless, odorless, poisonous gas which requires handling in a closed refrigerated system. By-product H_2SO_4 is recovered from the manufacturing process; some sold to fertilizer producers; some used to make bromates for photographic purposes. Dow gets approximately 90% recovery of CH_3Br on first pass (have some recycle streams). A possible by-product is $(CH_3)_2O$ and some competitors produce considerable amounts. The CH_3Br system is vented through a caustic scrubber and the product is dried with silica gel.

Dow has had almost no accidents with MeBr. They have had almost no leaks on cans or small cylinders; filled cans are checked in a water bath and with a halogen leak detector. Dow makes a 9 for 1 offer if a customer gets a leaker. They have recently put in an 18 month shelf life limitation on cans to guard against corrosion from outside in during storage under poor conditions. Dow uses a shielded type of valve on "pigs" which is drop-proof. Minor leaks from sticking valves have been handled easily. The biggest problem is the costly reconditioning given to all returned cylinders. Nearly all of these are Dow owned and are returnable. Chlorpicrin, a lacrymator, is added to some of the MeBr as a warning agent. See the discussion of 2,4-D for further details of Dow's liquid waste disposal system.

METHYL METHACRYLATE FROM ACETONE CYANOHYDRIN

Methyl methacrylate is produced by the acetone cyanohydrin process. The overall chemical reactions are given below.

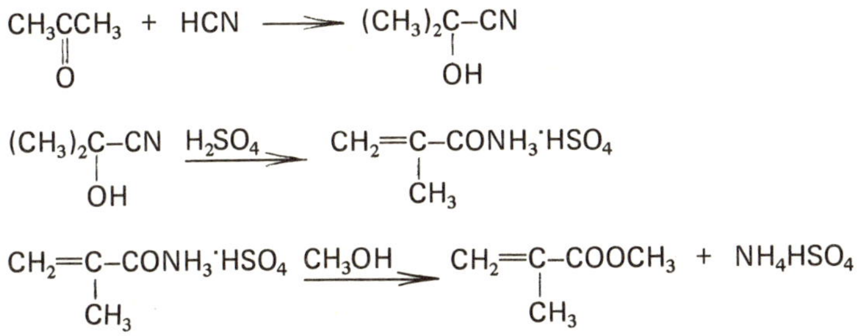

Acetone cyanohydrin is produced by the reaction of hydrogen cyanide and acetone with an alkaline catalyst in a cooled reaction kettle. The excess catalyst is neutralized, and crude acetone cyanohydrin passes to holding tanks. The salt formed by neutralization of the catalyst is removed in a filter press before the crude acetone cyanohydrin is fed to a two-stage distillation unit. Most of the water and acetone are removed and recycled overhead from the first column, and the remainder of the water is removed at high vacuum from the second column.

Acetone cyanohydrin and concentrated sulfuric acid are pumped into a cooled hydrolysis kettle to make the intermediate, methacrylamide sulfate, which is then sent to an esterification kettle to react with methanol continuously. To prevent polymerization, inhibitors are added at various points in the process. The esterified stream is pumped to the acid stripping column, from which the acid residue, made up of sulfuric acid (40% by weight), ammonium sulfate (28%), water (20%) and organic substances (10%) is sent to a Spent Acid Recovery unit (SAR). The recovered sulfuric acid is recycled back to the hydrolysis reactor.

The overhead stream from the acid stripping column is then distilled to remove methyl metacrylate and unreacted methanol, which is recycled. The last traces of methanol in the methyl methacrylate are removed by water extraction, after which the monomer is finally purified in a rerun tower.

The acid residue from the acid stripping column is the major waste stream generated in the process, and this waste stream is either sent to the SAR unit previously mentioned or is discharged into sewers. The waste streams generated as bottoms from various stills are combined with the acid residue for spent acid recovery. Water samples from streams leading to and leaving the SAR unit were taken for analysis, and the results are shown in the following tabulation.

	Into SAR	From SAR
Flow	260 gallons per 1,000 pounds	213 gallons per 1,000 pounds
COD	178,000 milligrams per liter	110 milligrams per liter
BOD_5	20,700 milligrams per liter	15 milligrams per liter
TOC	69,998 milligrams per liter	18 milligrams per liter

FIGURE 47: METHYL METHACRYLATE BY ACETONE CYANOHYDRIN PROCESS

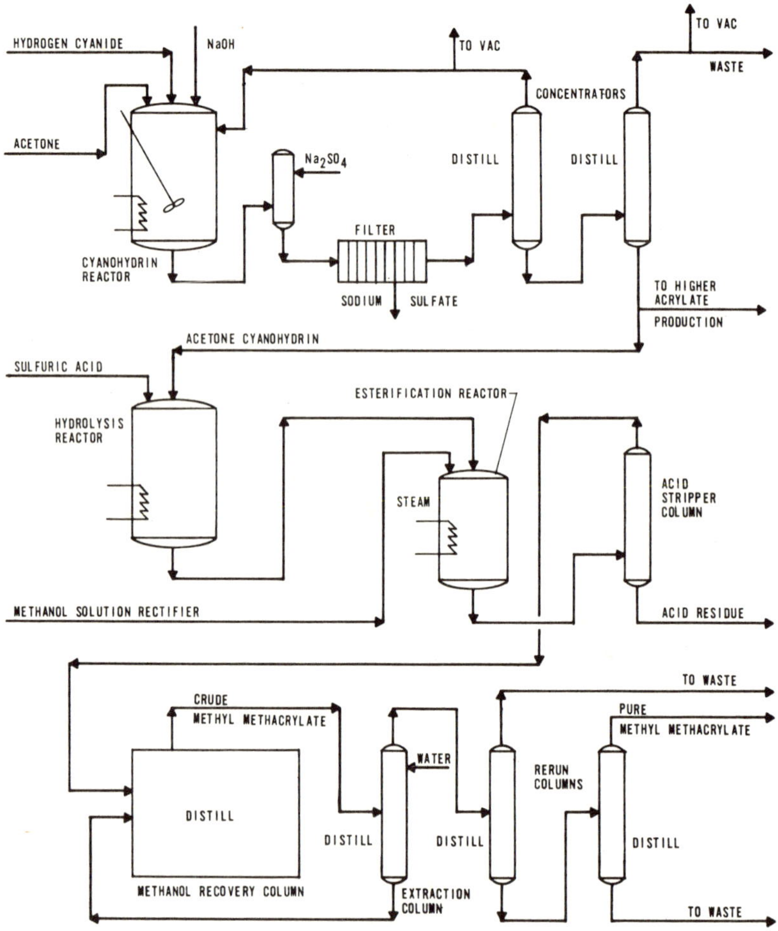

Source: Report EPA 440/1-73/009

A high concentration of floating solids was observed in the stream leading to the SAR, and it was impossible to obtain a well mixed sample. Therefore, samples from the stream were actually taken from the aqueous phase beneath the floating solids. The floating solids removed in the SAR were disposed of by incineration. High concentrations of metal contaminants such as copper and iron are indicated by the results of the analysis. Although a large portion of these metals are removed along with floating solids in the SAR unit, the metal concentrations in the streams discharged to sewers are still beyond the general discharge criteria for biological processes. Although sulfuric acid concentration had been reduced from 40% by weight in the influent to the SAR to 1% by weight in the effluent, the sulfate concentration in the discharge stream was still high enough to inhibit the normal functioning of the biological treatment process.

Because of the highly exothermic reactions involved, the process requires a large amount of cooling water. The survey data show that gross cooling water usage amounts to 366 pounds per pound of methyl methacrylate. Process water is introduced into the system in the form of direct steam stripping in the amount of 0.56 pound per pound of product.

To define BADCT and BATEA, this process should have a Spent Acid Recovery unit. Two types of SAR units have been devised, and descriptions of the equipment processing required, and estimated economics are presented in the following paragraphs.

Spent Acid Recovery by Neutralization: As shown in Figure 48, spent acid is neutralized with ammonia gas to form ammonium sulfate. The effluent from the neutralization tank is sent to crystallization and filtration units to separate ammonium sulfate from the aqueous solution. The economics of this unit are shown in Table 17.

Spent Acid Recovery by Complete Combustion: The spent acid solution (see Figure 48) is heated to such a high temperature (about 1000°C) that sulfuric acid decomposes into SO_2, O_2, and water vapor. Simultaneously, the organic substances are oxidized, and the contained ammonia converted to N_2 and water. The SO_2 gas stream is passed over a catalytic converter to oxidize the SO_2 to SO_3, which is then absorbed to form concentrated acid for recycle. The economics of this unit are shown in Table 18.

The economic analyses are based on the following flow rate and composition of spent acid.

H_2SO_4	=	245,000 pounds per hour
$(NH_4)_2SO_4$	=	16,500 pounds per hour
H_2O	=	13,500 pounds per hour
Organic Substances	=	6,150 pounds per hour

The acetone cyanohydrin process is the only methacrylate process used commercially in the United States. An alternate route used in Japan is nitric acid oxidation of isobutylene to methacrylic acid, followed by esterification with methanol.

FIGURE 48: SPENT ACID RECOVERY UNITS

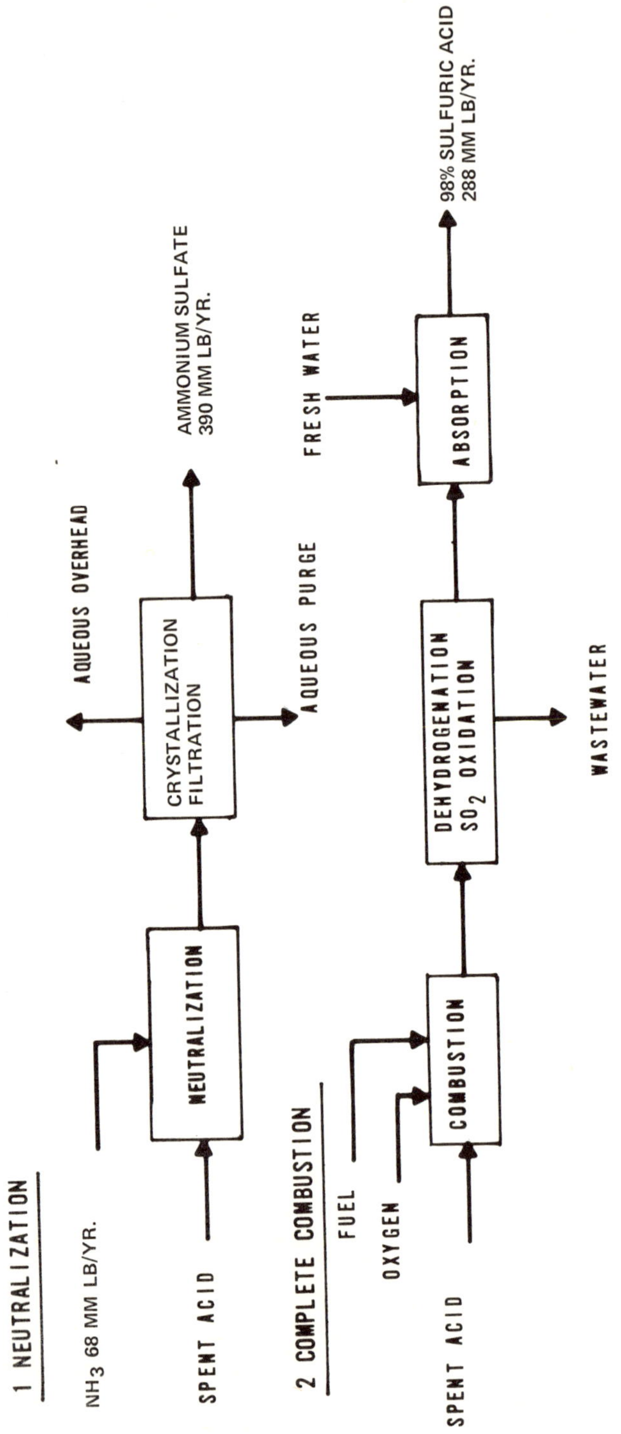

Source: Report EPA 440/1-73/009

TABLE 17: ECONOMICS OF SPENT ACID RECOVERY BY NEUTRALIZATION*

Investment	
Battery Limits	= $2,200,000
Off-site	= 800,000
Total Investment	$3,000,000
Operating Costs	
Utilities	$/yr
Steam: 720,000 M lb @ 60¢/M lb	= $ 430,000
Power: 10,000,000 Kwh @ 0.8¢/Kwh	= 80,000
Cooling Water: 2,000,000 M gal @ 3¢/M gal	= 60,000
	$ 570,000
Chemicals	
NH_3: 68,000,000 lb @ 2¢/lb	= $1,360,000
Amortization	= 440,000
Labor	= 200,000
Sub-total	$2,000,000
Return on Total Investment @ 20%	= $ 600,000
Total Annual Cost	= $3,170,000
Net Revenue from Recovered Ammonium Sulfate 390 MM lbs/yr @ 0.70¢/lb	= $2,730,000

*Based on 485,000,000-lbs/yr Spent Acid Recovery plant.

Source: Report EPA 440/1-73/009

TABLE 18: ECONOMICS OF SPENT ACID RECOVERY BY COMPLETE COMBUSTION*

Investment	
Battery Limits	= $3,000,000
Off-site	= 1,000,000
Total Investment	$4,000,000
Operating Costs	
Utilities	$/yr
Fuel: 800,000 MM BTU/yr @ 50¢/MM BTU	= $ 400,000
Power: 3,000,000 Kwh @ 0.8¢/Kwh	= 24,000
Cooling Water: 750,000 M gal @ 3¢/M gal	= 22,500
	$ 446,500
Amortization	= $ 600,000
Labor	= 100,000
	$ 700,000
Return on Total Investment @ 20%	= $ 800,000
Total Annual Cost	= $1,946,500
Net Revenue on Recovered H_2SO_4 144,000 tons/yr @ $20/ton	= $2,880,000

*Based on 485,000,000-lbs/yr Spent Acid Recovery plant.

Source: Report EPA 440/1-73/009

NITROBENZENE FROM BENZENE

Nitrobenzene is produced by reacting benzene and nitric acid in the liquid phase according to the following reaction.

$$C_6H_6 + HONO_2 \longrightarrow C_6H_5NO_2 + H_2O$$

In the Sixties, the annual growth rate was 12%. For the Seventies, an annual growth rate of 10% is expected. About 97% of the nitrobenzene production is used to make aniline.

A mixed acid is used in the reactor. This acid has a composition of 53% H_2SO_4, 39% HNO_3, and 8% H_2O. The mixed acid is prepared by mixing 93% H_2SO_4, 97% HNO_3 and 50 to 60% HNO_3. The benzene requires no feed preparation.

The reaction is carried out either in batches or continuously. For batch operation, the kettles are sized for 1,000 to 1,500 pound quantities of benzene and are equipped with agitators. They are jacketed and also have internal coils for cooling.

The reaction cycle requires 2 to 4 hours. The benzene is charged to the reactor along with a heel of spent acid. The mixed acid is added slowly below the surface of the benzene. The temperature is maintained at 50° to 55°C, but is allowed to rise to 90°C toward the end of the reaction to promote complete reaction. The reactor vents to an absorber.

An emergency drown tank is provided, located below the reactor. In case of a runaway reaction in the reactor, the batch is dumped to the drown tank, which contains water to kill the reaction. When the reaction is complete, the reactor batch is transferred to a separator tank. The overall process yield from benzene is 95 to 98%.

In the separator tank, the crude nitrobenzene separates from the spent acid by gravity. The spent acid either goes to provide a heel for the next reactor batch or to acid concentration. There is a loss of 0.5% yield of nitrobenzene in the spent acid.

If the nitrobenzene is to be used to make aniline (most of the nitrobenzene which is made is so used), the crude nitrobenzene from the separator is transferred to the aniline process. If a nitrobenzene of high purity is desired, the crude nitrobenzene is transferred to a washer where it is washed with water and sodium carbonate solution to remove residual spent acid. The crude is then distilled to obtain the desired pure nitrobenzene product. The residuals from distillation are wastes.

The vents from the reactor go to an absorber, where any NO_x is oxidized with air and absorbed in water to produce nitric acid. The unreacted air is vented. The acid stream goes to acid concentration. In the acid concentration system, spent acid and acid from the reactor vent absorber first go to the top of a denitrating tower, which is heated with steam. As the liquids flow down through the tower, rising vapors strip all HNO_3 and oxides of nitrogen. The water in the feed remains with the sulfuric acid, which is drawn off the bottom of the tower. The overhead gases go to the bleacher. The dilute sulfuric acid goes to a sulfuric acid concentrator.

In the bleacher the vent gases are oxidized with air and pass through a reflux condenser where any HNO_3 present condenses and drops back to the bleacher. Nitric acid with a concentration of 97% is drawn off from the bottom of the bleacher and goes to the mixed acid system. The off-gases from the reflux condenser go to an absorber column, where additional oxidation takes place and the oxides of nitrogen are absorbed in nitric acid and water to produce nitric acid. The nitric acid from the bottom of the column is recycled to the denitrating column. The air from the absorber is vented. The sulfuric acid from the bottom of the denitrating column goes to a sulfuric acid concentrator where the acid is concentrated to 93% sulfuric acid. The sulfuric acid produced goes to the mixed acid system. The water removal is a waste. There are two sources of air pollution. These are the absorber on the reactor vents and the absorber on the acid concentration system. Both of these vent air and unabsorbed oxides of nitrogen.

NITROCHLOROBENZENE FROM CHLOROBENZENE

Nitrochlorobenzene is produced by nitrating chlorobenzene.

$$C_6H_5Cl + HONO_2 \longrightarrow C_6H_4ClNO_2 + H_2O$$

The chloro group increases the reactivity of benzene, making the reaction easy to carry out. The alternate route to nitrochlorobenzene, namely, nitration of benzene followed by chlorination, does not proceed smoothly, and has lower yields and more side reactions. There is no industrial process based on these reactions. About 80% of production is paranitrochlorobenzene. Due to fluctuations in the pesticide market, no sustained growth in nitrochlorobenzene production is expected. Nitrochlorobenzene is used as an intermediate.

Mixed acid containing 53% H_2SO_4, 35% HNO_3 and 12% H_2O is prepared by mixing 50 to 60% HNO_3 with 93% H_2SO_4 and 97% HNO_3. The latter two acids are produced in the acid concentration system.

The reaction is usually a batch process. A typical process uses a 2,640-gallon reactor equipped with an agitator and with a jacket and internal coils for cooling. The reactor is charged with 5,500 pounds of spent acid and 10,000 pounds of monochlorobenzene. The agitator is started, and 15,600 pounds of mixed acid is fed over a period of 9 hours. The temperature is permitted to rise from 20° to 50°C during the 9 hours. When the reaction is complete, the batch is allowed to settle for 3 hours. The spent acid layer is drawn off to the spent acid washer and the crude chloronitrobenzene is transferred to the crude washer.

The spent acid is washed with a fresh charge of monochlorobenzene, to be used in a subsequent nitration batch to remove nitrobodies from the acid. The acid is then sent to acid concentration. The crude chloronitrobenzene is washed 3 times with water, followed by washing with dilute caustic and water. The charge is then dried at 90° to 100°C under vacuum. The crude chloronitrobenzene contains about 65% of the para isomer, about 34% of the ortho isomer and 1% of the meta isomer. These are separated and purified by alternate crystallization and distillation steps.

The crude is first cooled to 16°C in a crystallizer. About 50% of the para isomer crystallizes out as a pure component and is separated from the mother liquor. The mother liquor is then distilled under vacuum. The first cut is primarily unreacted monochlorobenzene, with traces of meta-chloronitrobenzene. This is redistilled to recover the chlorobenzene for recycle and a meta crude product containing 60% meta-chloronitrobenzene. The second cut is predominantly the para isomer. This is crystallized out to obtain the para isomer as a product, while the mother liquor is recycled to the mother liquor tank. The third distillation cut is a slop cut and is returned to the mother liquor tank. The still residues are distilled without the column. The overheads go to a crystallizer, where the ortho isomer product crystallizes out. The mother liquor from this crystallization is returned to the mother liquor tank. The residues from the final distillation are a waste material. Product purities are not available.

There are several recycle streams which result during the product purification steps. These were discussed under product purification above. Another recycle stream is the spent acid stream from the reaction. This acid is fed to the denitrating column. In this column, the oxides of nitrogen and any HNO_3 are driven overhead as a gas while the water is held back by the sulfuric acid. The sulfuric acid leaves the bottom of the column at about 60 to 70% H_2SO_4 and is fed to a sulfuric acid concentrator where the water is driven off and a 92 to 93% sulfuric acid is produced for feed to the mixed acid tank.

The overheads from the denitrating tower go to a bleacher, where they contact air and the NO is oxidized. The gases then pass through a reflux condenser where 97% HNO_3 is condensed and returns to the bleacher and is transferred to the mixed acid tank. The gases from the reflux condenser go to the absorber where the oxides of nitrogen are absorbed in water and some fresh nitric acid. The bottoms from this absorber are recycled to the denitration tower.

There are two usual sources of air pollution. The first is the vent from the reactor, which contains oxides of nitrogen. The second vent is from the absorber in the acid concentration system, which contains air and unabsorbed oxides of nitrogen. If the plant has a drum type sulfuric acid concentrator instead of a distillation type concentrator, a vent exists which will contain flue gases and entrained sulfuric acid vapor and organics vapor.

NITROPARAFFINS FROM PARAFFINS

Nitroparaffins are made by nitrating light aliphatics in the vapor phase.

$$C_3H_8 + HONO_3 \longrightarrow C_3H_7NO_2 + H_2O$$

Commercial Solvents has pioneered this family of compounds, having their main production plant in Louisiana with a semiworks in Peoria. The nitroparaffins are used as intermediates and solvents. The annual growth rate is estimated at less than 5%.

Two major raw materials are used. These are propane and 50 to 60% nitric acid. Minor raw materials are the unidentified chemicals used in the chemical treatment step, and hydroxylamine hydrochloride required for makeup for losses in the absorption and stripping steps.

The reaction is carried out continuously. Propane is vaporized and heated to 370° to 450°C, and is fed to the reactor at 125 to 175 psig. In the reactor, 50 to 60% nitric acid is sprayed into the propane gas. About 5 mols of propane per mol of HNO_3 are fed. The heat of reaction vaporizes the nitric acid. The reactor operates adiabatically. The desired reaction products are nitromethane, nitroethane, 1-nitropropane and 2-nitropropane. A typical distribution is 10% nitromethane, 25% nitroethane, 25% 1-nitropropane, and 40% 2-nitropropane. By varying the reaction conditions, the relative amounts of nitroparaffin may be controlled. Normal ranges, with propane as a raw material, are nitromethane, 10 to 30%; nitroethane, 20 to 25%; nitropropanes, 55 to 65%.

The purities of the finished nitroparaffin products are all 99% plus nitroparaffins. The minimum purity of the individual products is nitromethane, 95%; nitroethane, 90%; 1-nitropropane, 94% and 2-nitropropane, 94%. With recycling, the nitroparaffin yield, based on nitric acid, approaches the theoretical yield. Some 90 plus mols of nitroparaffins are produced per 100 mols of HNO_3 used up. No information is available on yields based on propane fed.

The product gases from the reactor pass through a cooler and condenser in series. The nitroparaffins condense and the mixture is brought down to atmospheric pressure. The remaining vapor contains nitric oxides, some inerts, unreacted propane, and a small quantity of aldehydes, ketones, etc.

The cooled mixture is fed to an absorber where all components except nitric oxides, inerts, and unreacted propane are absorbed in a solvent such as hydroxylamine hydrochloride. The off-gases go to a recovery process which is discussed under recycle systems. The main reason for the absorption step is to insure that the aldehydes, ketones, etc., are not recycled with the propane.

The bottoms from the absorption column go to the stripper where steam is used to strip the reaction products from the solvent. The solvent is removed from the bottom of the stripper and is recycled to the absorber. The overheads from the stripper are condensed and fed to a decanter. The bottom layer, which is water, is returned to the stripper. The light layer is fed to the heads column, where aldehydes, ketones and other light boilers are separated as overheads and become a by-product. The bottoms from the heads column are treated with chemicals (unidentified) and then washed with water to remove chemical products that would cause corrosion in the subsequent purification steps.

The reaction products, after the washing step, consist of nitroparaffins and higher boiling compounds saturated with water. This mixture is fed to a drying column where the water is removed overhead and returned to the decanter following the washing step. The bottoms from the drying column are fed to four distillation columns in series, where the nitroparaffins are separated into four products; nitromethane, nitroethane, 2-nitropropane, and 1-nitropropane. The bottoms from the last column are high boilers (heavies) and are a waste product.

The off-gases from the absorber consisting of nitric oxides, unreacted propane, and CO_2 are fed to a separation system. Details of this system are not available. However, it is assumed that it consists of an adsorption step which separates the unreacted propane from the nitric oxides and CO_2. The propane is recycled to the reactor. The nitric oxides and CO_2 pass through an oxidizer where NO is oxidized. This is followed by an absorber where the oxides of nitrogen are absorbed in water to produce nitric acid. The nitric acid is recycled to the reactor. The off-gases, consisting of air, N_2, CO_2, and unabsorbed oxides of nitrogen, are vented. This process produces one by-product, the overheads from the heads column, containing aldehydes and ketones. This is either a waste product or undergoes further processing to produce salable products. The only pollution source is the vent from the nitric acid absorber. This vent will contain air, CO_2 and unabsorbed oxides of nitrogen.

OXO PRODUCTS FROM OLEFINS

The oxo process is a broadly applicable technology which is used to produce aldehydes which are usually converted to the corresponding alcohols. The process is used on a number of feedstocks, the two most important being propylene and alpha olefins, to produce linear alcohols for plasticizers and surfactant usage. 2-Ethylhexanol, produced primarily from propylene via n-butyraldehyde, is the most important oxo chemical in terms of volume. The basic chemical reactions are given below.

$$C_3H_6 + CO + H_2 \longrightarrow CH_3CH_2CH_2CHO + (CH_3)_2CHCHO$$

$$2CH_3CH_2CH_2CHO \longrightarrow CH_3CH_2CH_2CH{=}\underset{\displaystyle C_2H_5}{\underset{|}{C}}{-}CHO + H_2O$$

$$CH_3CH_2CH_2CH{=}\underset{\displaystyle C_2H_5}{\underset{|}{C}}{-}CHO + H_2 \longrightarrow CH_3CH_2CH_2CH_2{-}\overset{\displaystyle H}{\overset{|}{\underset{\displaystyle C_2H_5}{\underset{|}{C}}}}{-}CH_2OH$$

Carbon dioxide, natural gas, and steam are passed into a synthesis gas reactor to produce water gas (1:1 ratio of H_2O and CO) which is then mixed with propylene in a liquid-phase reactor in the presence of a cobalt solution. The reaction is carried out under pressure and the reactor is maintained approximately isothermal. A liquid-gas mixture of aldehydes and unreacted materials is taken overhead from the reactor, cooled, and then separated in successive high and low pressure flashing stages, whence unreacted synthesis gas is recycled to the oxo reactor. The catalyst cobalt is then removed continuously from the liquid phase. The liquid product, containing n-butyraldehyde, iso-butyraldehyde, and solvent, is separated in two distillation columns.

n-Butyraldehyde is then sent to a condensation reactor, where the subsequent reaction is carried out at moderate temperature and atmospheric pressure in the presence of strong base such as sodium or potassium hydroxide. Continuous removal of the water produced during reaction drives the aldol condensation to completion. The unreacted C_4 aldehyde is separated from the product 2-ethylhexenal by distillation and is recycled to the condensation reactor.

The 2-ethylhexenal produced is then hydrogenated to 2-ethylhexanol in the presence of a solid nickel catalyst in a pressurized reactor at 50 to 100 atmospheres. After being washed with caustic solution and water, the reactor effluent is sent to a fractionator to recover the product 2-ethylhexanol.

FIGURE 49: OXO CHEMICALS MANUFACTURE

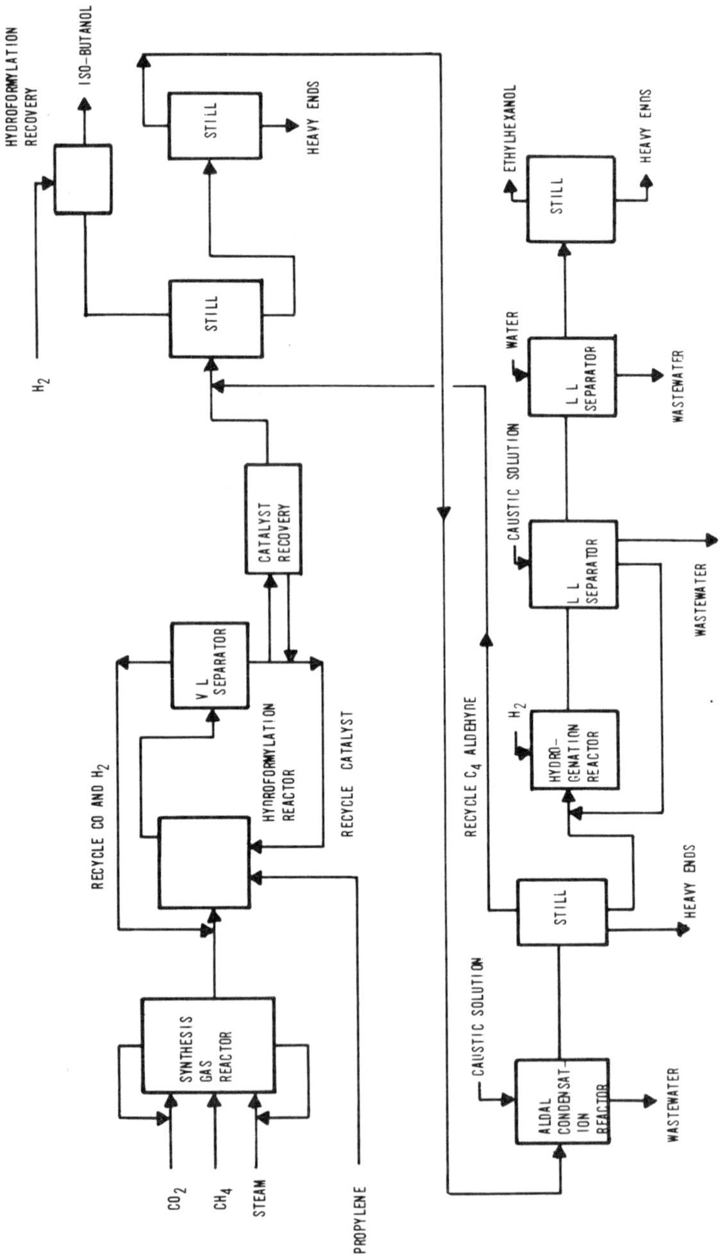

Source: Report EPA 440/1-73/009

The major wastewater streams in oxo chemical manufacturing are the water removed from the aldol condensation and the water used in washing the crude product before fractionation into final product. The wastewater may contain some intermediates, product, and by-product losses. No significant catalyst loss from the reactor is expected. Heavy ends from various stills are disposed of by incineration.

The characteristics of the wastewater obtained from the plant survey are summarized in the following tabulation. It should also be noted from the results of analyses that the oil concentration in the waste stream is beyond the limits of the general discharge criterion for biological treatment processes.

Flow	420 gallons per 1,000 pounds
COD	1,212 milligrams per liter
	4.25 pounds per 1,000 pounds
BOD_5	900 milligrams per liter
	3.15 pounds per 1,000 pounds
TOC	549 milligrams per liter
	1.92 pounds per 1,000 pounds

Other than reusing the aldol condensation water as washwater, it is deemed unfeasible to further reduce RWL of the process by any in-process modification. Consequently, RWL presented can be considered as standards for BADCT and BATEA control technology for this manufacturing process.

An alternate route in oxo chemical manufacturing is based on a new catalyst system. By carrying out the hydroformation reaction in an alkaline medium using phosphine-promoted cobalt carbonyl processes, 2-ethylhexanol and butanol can be produced directly in one step. Olefin feed and the recycled catalyst stream are charged to the first of a series of packed reactors at control rates. Synthesis gas (H_2/CO molar ratio = 2/1) is fed separately to each reactor. The stream taken overhead from the final reactor is directly sent to the product recovery column. The bottoms from the product recovery column will contain the catalyst complex dissolved in a mixture of alcohols and heavy ends. This stream is recycled to the first reactor with periodical purging to remove the built-up heavy ends.

PARATHION FROM p-NITROPHENOL

Parathion is a broad-spectrum nonsystemic insecticide of the organophosphate class. Its chemical name and formula are as follows.

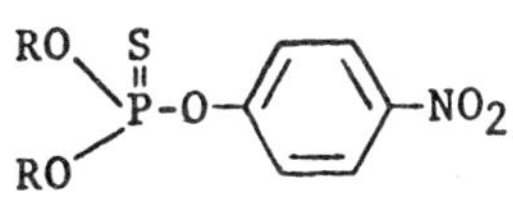

O,O-Diethyl O-p-Nitrophenyl Phosphorothioate
and
O,O-Dimethyl O-p-Nitrophenyl Phosphorothioate
parathion, R = C_2H_5
methyl parathion, R = CH_3

Parathion is a liquid of low volatility which is insoluble in water. It is very toxic but is nonpersistent in the environment. Methyl parathion is a low melting solid (MP 36.5°C) which has low solubility in water. It is very toxic but is nonpersistent in the environment. Producers include Monsanto at Anniston, Alabama, American Potash, Stauffer and Shell using the following production chemistry.

$$P_2S_5 + 4ROH \longrightarrow 2(RO)_2\overset{\overset{S}{\|}}{P}SH + H_2S$$

$$(RO)_2\overset{\overset{S}{\|}}{P}SH + Cl_2 \longrightarrow (RO)_2\overset{\overset{S}{\|}}{P}Cl + HCl + S$$

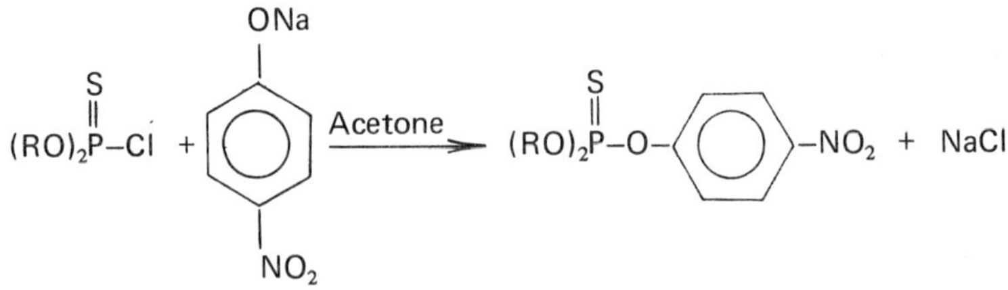

Liquid wastes, including by-product HCl–NaCl, go to biological treatment facilities, and then into the Anniston sewer system in the case of the Monsanto plant. City specs are for <1 ppm parathion. They have had no problem complying on all effluent components. By-product H_2S is burned in a flare, and by-product S is burned. Waste solvent is also burned. No solid wastes are produced except sludge from the biological oxidation system, which is recycled and discharged at a slow rate into the sewer.

Spills within the plant are washed down, and go into the waste treatment system. Spills outside are handled by Monsanto personnel. The frequency of outside spills has been about 8 to 10 per year, including customer problems. Soda ash breaks down the parathions rather quickly, and is used for decontamination. Equipment cleanup is required only 1 or 2 times per year and presents no major problem.

PERCHLOROETHYLENE FROM PROPANE

In 1970, 80% of the tetrachloroethylene production was made by the thermal chlorination of propane:

$$C_3H_8 + 8Cl_2 \longrightarrow CCl_4 + C_2Cl_4 + 8HCl$$

The remaining 20% was made from acetylene, which is being phased out of DuPont's Niagara Falls Plant (60 million pounds per year) and Detrex's Ashtabula Plant (25 million pounds per year).

$$C_2HCl_3 + Cl_2 \longrightarrow C_2HCl_5$$

$$C_2HCl_5 \longrightarrow C_2Cl_4 + HCl$$

For the future, all tetrachloroethylene is expected to come from propane. In the sixties, the annual growth rate was over 12%. For the seventies, a 7% growth rate is expected. About 70% of the tetrachloroethylene is used as a dry cleaning solvent. The second use is as a chemical intermediate.

In the process for thermal chlorination of propane, fresh chlorine feed, together with recycled chlorine, and the propane feed are introduced into a vaporizer where they are mixed with recycled chlorides in the vapor space. The chlorine is in 10 to 25% excess, based on the propane. The recycle chloride diluent rate (controlled by heat input to the vaporizer) is designed to control adiabatic reactor temperature at 550° to 700°C and is usually about 75 mol percent of the total stream, its composition also serves to control CCl_4–C_2Cl_4 ratio.

The mixed gases, at atmospheric pressure, are fed to a refractory lined reactor where they are rapidly mixed with the contents, thereby heated to ignition, and reacted adiabatically. The reaction is self-sustaining and supported by the considerable heat of reaction (after start-up), but is tempered and controlled by the diluent action of the recycle chlorides.

The reactor effluent is essentially free of unreacted hydrocarbon and consists mainly of carbon tetrachloride, perchloroethylene, HCl and excess chlorine. It is rapidly quenched by intimate contact with a liquid which is largely perchloroethylene. The rapid quench serves to preserve the equilibrium ratio attained in the reactor and prevents formation of undesirable by-products. It also serves to dissolve any quantities of hexachlorobenzene which may be present (these are small if reaction temperature is maintained below 650°C) and which might introduce difficulties due to deposition of solids on heat transfer surfaces and in vessels.

Some heat economy is obtained from the hot reactor effluent, which serves to boil the contents of the quench tank and thus provide boil-up for the CCl_4 column. Hexachlorobenzene is purged from the quench tank by allowing liquid to overflow to the recycle surge tank.

The CCl_4 column, operating on boil-up from the quench tank, returns quench liquid rich in perchloroethylene as a bottoms stream. Fractionation results in an overhead which is largely free of perchloroethylene. The condenser yields carbon tetrachloride, the desired quantity of which is withdrawn as product, the remainder being used as reflux. Chloride-free gases (HCl and chlorine) are also disengaged at the overhead. These gases are scrubbed with water to remove HCl, the effluent liquor forming the aqueous HCl by-product. They are subsequently dried with concentrated sulfuric acid and purged to maintain the balance of inert gases; the remaining chlorine (with some uncondensed chlorides) is recycled to feed.

A perchloroethylene-rich stream is removed as a side stream from the CCl_4 column and processed to produce perchloroethylene product. Heavy ends are first removed by distillation and returned to the recycle surge tank. The overhead from the heavy ends column is fractionated in the C_2Cl_4 column where the desired quantity of perchloroethylene product is removed as the bottoms and the overhead (largely carbon tetrachloride) is sent to recycle.

The recycle surge tank serves as a reservoir for three recycle streams: (1) the quench tank overflow (largely perchloroethylene with some hexachlorobenzene), (2) heavy ends from the heavy ends column, and (3) the C_2Cl_4 column overhead (largely carbon tetrachloride). Proper control results in a recycle mixture of the desired composition, and this is fed to the vaporizer where the vapors are mixed with the feed materials.

To prevent buildup of hexachlorobenzene in the system, liquid is withdrawn from the vaporizer and sent to the heavies still. Here the more volatile components are stripped off and returned to the vaporizer, while heavy liquid consisting of hexachlorobenzene and some hexachloroethane is purged from the still pot.

The raw materials are propane, chlorine, catalyst ($CuCl_2$ and $BaCl_2$), water and sulfuric acid. The by-products are tetrachloroethylene (perchloroethylene), hexachlorobenzene, hydrogen chloride, hexachloroethane, and spent sulfuric acid. There is one source of air pollution, the purge on the dry chlorine recycle, which removes inerts.

PHENOL FROM CHLOROBENZENE

$$C_6H_5Cl + 2NaOH\ (Excess) \longrightarrow C_6H_5ONa + NaCl + H_2O$$

$$C_6H_5ONa + HCl \longrightarrow C_6H_5OH + NaCl$$

The feed materials (chlorobenzene and excess caustic solution) are fed into a liquid-phase reactor, and the effluent is discharged into a decanter. The upper layer of unreacted chlorobenzene is recycled back to the reactor. The bottom layer of sodium phenate is neutralized to produce a mixture of phenol and brine; this mixture is then decanted. The upper layer is sent to a fractionator, where pure phenol is obtained, and the bottom brine stream is passed through an activated carbon bed to remove the remaining phenol, which is eventually recycled back to the reactor.

The chlorobenzene process is used by only one company in the United States. The major wastewater source in this process is the brine solution from the second decanter, which is contaminated with phenol and acetic acid. However, an activated carbon system and chlorination reactor, both being considered as parts of an integral system of the process, are used to remove phenol by adsorption and to destroy the acetic acid component. The effluent from the system is totally recycled for chlorine production. The adsorbed phenol is desorbed with caustic solution to form sodium phenate, which is recycled back to the reactor. Therefore, the process is free of discharge and can be considered as a standard for BADCT and BATEA.

The activated carbon system has been claimed to be effective in reducing phenol concentration from about 100 milligrams per liter down to 1 milligram per liter.

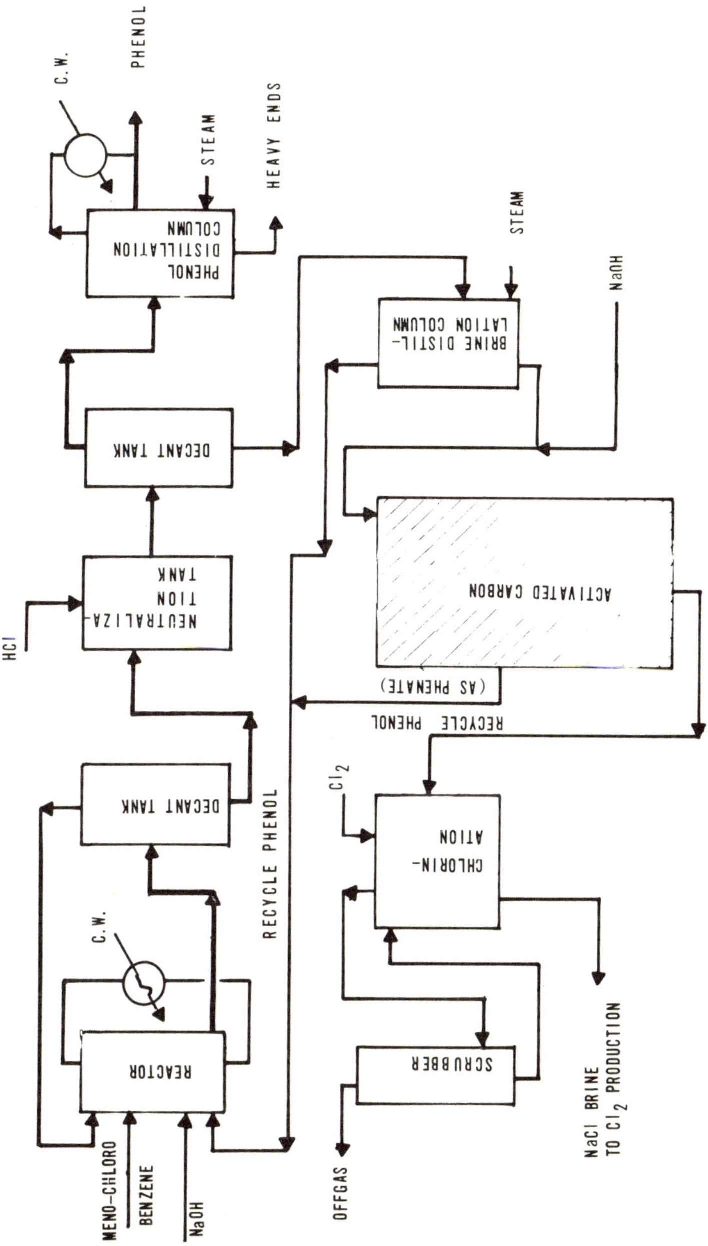

FIGURE 50: PHENOL FROM MONOCHLOROBENZENE

Source: Report EPA 440/1-73/009

The saturated activated carbon beds can be regenerated with caustic solution by desorbing phenol into phenate salt. The salt is then recycled to the oxidation reactor. With this system, phenol is recovered for reuse, and the RWL of the process is reduced as well. Consequently, BADCT and BATEA should require a steam stripper/dephenolizer with an activated carbon system to achieve a low RWL standard.

Several other process routes in manufacturing phenol are currently practiced. These include the Hooker-Raschig process, toluene oxidation, and sulfonation. However, the manufacture from cumene is by far the most important, and it is predicted that all phenol capacity installed over the next ten years will be based on this process.

Gross cooling water usages for the processes differ greatly: 3.85 and 463 pounds of water per pound of phenol, respectively, for the chlorobenzene and cumene processes. The Hooker-Raschig and sulfonation processes are briefly described in the following paragraphs.

The Hooker-Raschig process is a two-step, vapor-phase reaction. A benzene chlorination reaction is carried out at 400°F with air, over a copper and iron chloride catalyst. The copper-iron catalyst oxidizes the hydrogen chloride to chlorine and water. The chlorine attacks the benzene ring to yield chlorobenzene and additional hydrogen chloride. The chlorobenzene is then hydrolyzed over silica at 900°F to yield phenol and hydrogen chloride. There is no net production of hydrogen chloride since it is continually converted to usable chlorine. The net products are, therefore, phenol and water.

The sulfonation process is a liquid-phase reaction. Benzene is first reacted with sulfuric acid to produce benzenesulfonic acid, which is then converted to phenol by caustic fusion. The sulfuric acid employed in this process is totally lost.

PHENOL FROM CUMENE

The cumene process is currently the most popular route and the one upon which most expansions will be based. The manufacture of phenol from cumene is carried out by a process involving the following basic steps:

(a) Oxidation of cumene with air to form cumene hydroperoxide

$$C_6H_5C(CH_3)_2 + 2O_2 \longrightarrow C_6H_5(CH_3)_2OOH$$

(b) Cleavage of cumene hydroperoxide to form phenol and acetone

$$C_6H_5C(CH_3)_2OOH \longrightarrow C_6H_5OH + CH_3COCH_3$$

Cumene and air are fed to a liquid-phase reactor, operating at 25 to 50 psig and 130° to 140°C, in the presence of a small amount of alkali, to produce the hydroperoxide intermediate. Reactor liquid effluent is fed to a fractionating tower, where unreacted cumene is recovered and recycled to the reactor. Cumene hydroperoxide from the fractionator is fed to a hydrolysis reactor where the cumene hydroperoxide is cleaved to phenol and acetone with the aid of a sulfuric acid catalyst. Typical operating conditions are 5 psig and 150° to 200°F, and conversion is essentially complete, with minimal formation of undesired by-products. The crude phenol-acetone mixture is passed through an ion exchange system and then fed to a series of tower fractionation trains, where pure phenol and co-produced acetone are separated from light and heavy ends and other by-products.

The cumene oxidation process recycles the water present in the hydroperoxide reactor. Water from the dilute sulfuric acid in the cleavage reactor is also recycled. The only significant wastewater stream is generated by water scrubbing the vapor effluent from the cleavage reactor; this stream contains dissolved sulfuric acid, sulfates, and oxygenated organic compounds.

The major parameters of surveyed RWL data from two cumene oxidation plants are summarized in the following tabulation. The results of the analyses also show that phenol and oil contaminants in wastewaters from both plants are in excess of general discharge criteria for biological treatment processes and would interfere with the normal functioning of such processes.

	Plant 1
Flow	279.6 gallons per 1,000 pounds
COD	4,770 milligrams per liter
	11.1 pounds per 1,000 pounds
BOD_5	2,410 milligrams per liter
	5.6 pounds per 1,000 pounds
TOC	194 milligrams per liter
	0.45 pound per 1,000 pounds

	Plant 2
Flow	164 gallons per 1,000 pounds
COD	84,304 milligrams per liter
	11.5 pounds per 1,000 pounds
BOD_5	17,575 milligrams per liter
	24 pounds per 1,000 pounds
TOC	77,406 milligrams per liter
	105.6 pounds per 1,000 pounds

The survey data show a significant difference in RWL between the two plants. The lower RWL of Plant 1 is attributed to the installation of dephenolizer facilities (steam stripper). These facilities are considered as part of the process rather than end-of-pipe treatment, since phenol is recovered at this unit and recycled back to the oxidation reactor. The higher RWL of Plant 2 is attributed mainly to the disposal of concentrated light ends and heavy ends from acetone and phenol fractionators into the sewers, instead of by incineration as commonly practiced. RWL represented by Plant 1 can be logically considered as standard for BPCTA control technology.

FIGURE 51: PHENOL FROM CUMENE

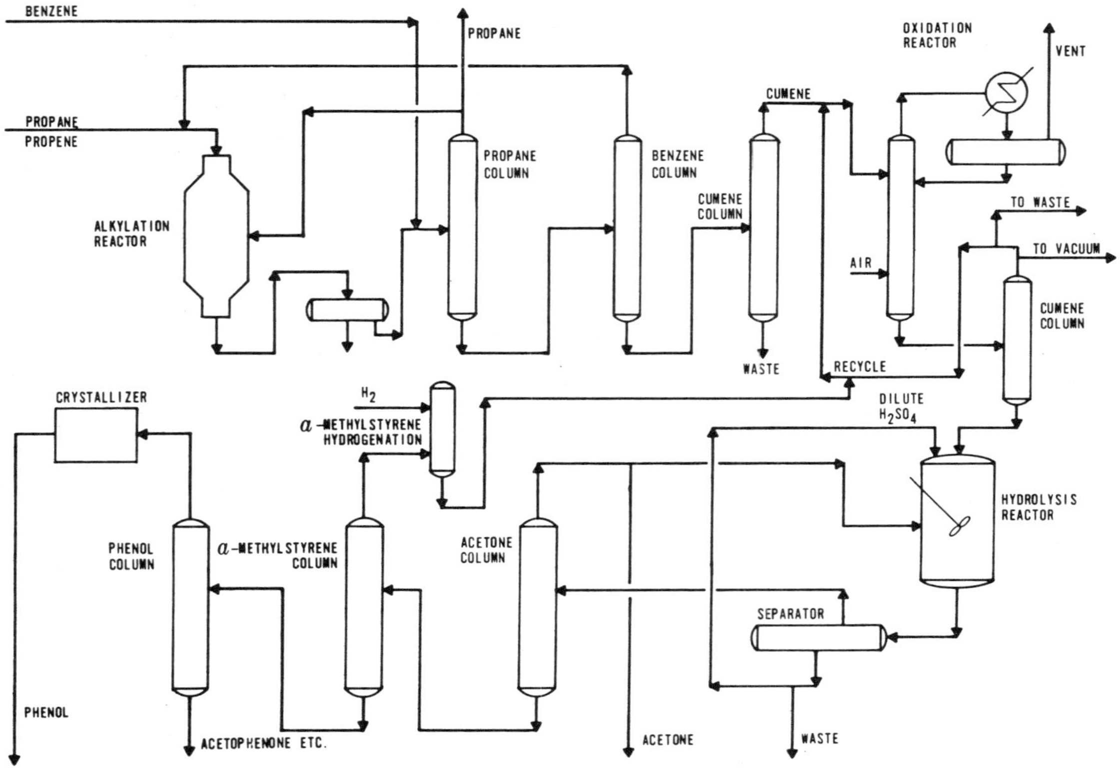

Source: Report EPA 440/1-73/009

PHORATE FROM ETHANOL

Phorate is a systemic insecticide of the organophosphate class It has the following chemical name and formula.

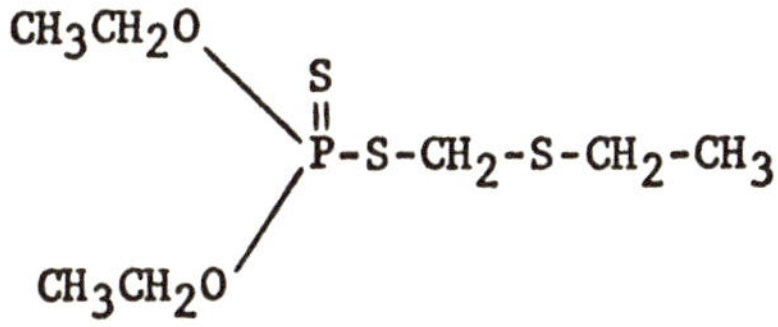

O,O-diethyl S-(ethylthio)-methyl phosphorodithioate

It is a liquid of low volatility which is insoluble in water. It is highly toxic but is nonpersistent in the environment. It is produced by American Cyanamid at Warners, New Jersey according to the following production chemistry.

$$P_2S_5 + 4EtOH \longrightarrow 2(EtO)_2\overset{\overset{S}{\|}}{P}SH + H_2S$$

$$(EtO)_2\overset{\overset{S}{\|}}{P}SH + H_2C{=}O \longrightarrow (EtO)_2\overset{\overset{S}{\|}}{P}S\text{-}CH_2OH$$

$$(EtO)_2\overset{\overset{S}{\|}}{P}\text{-}S\text{-}CH_2OH + EtSH \longrightarrow (EtO)_2\overset{\overset{S}{\|}}{P}\text{-}SCH_2SEt$$

The major by-product is H_2S which is recovered (see malathion). Some ethyl mercaptan is burned and discharged as SO_2 to the atmosphere. Other wastes are handled as those for malathion. Equipment is generally cleaned once per year at the end of the 6 months or so production run and wastes barged to sea.

PHOSGENE FROM CO AND CHLORINE

Phosgene is made by reacting chlorine and carbon monoxide.

$$Cl_2 + CO \longrightarrow COCl_2$$

Phosgene is very toxic, and was used as a poisonous gas in World War I. However, it is an excellent intermediate, especially for isocyanate production. In the sixties it had an annual growth rate of 26%. For the seventies an annual growth rate of 15% is forecast. Its main use (about 60%) is as an intermediate for making tolylene diisocyanate.

Carbon monoxide and chlorine gases are dried, filtered, metered, and mixed, and then fed to a reactor with a very slight excess of CO. The reactor consists of water-cooled iron tubes filled with activated charcoal. Reactor conditions are 200°C and 2 to 4 pounds per square inch gauge. The reaction is highly exothermic. The hot gases from the reactor are cooled to about 0°C, and liquid phosgene condenses out. The noncondensables are scrubbed with benzene or toluene to remove the last traces of phosgene, and are vented. Yields are 99%.

The raw materials are carbon monoxide (dry), chlorine (dry), and activated charcoal. There are no by-products. There is only one source of air pollutants, the noncondensables from the solvent absorber.

PHTHALIC ANHYDRIDE FROM NAPHTHALENE OR XYLENE

Phthalate esters, particularly DOP (dioctyl phthalate), DIOP (di-isoctyl phthalate), DDP (didecyl phthalate), and DTDP (di-tridecyl phthalate) are the backbone of the plasticizer industry. They contribute low-temperature flexibility, resilience, high impact strength and good electrical properties to vinyl resins. Hence, the manufacture of phthalic anhydride contributes an important staple product for the chemical industry.

Phthalic anhydride has traditionally been made from naphthalene, a by-product of the coking of coal. The increased availability of ortho xylene as a product of the catalytic reforming of petroleum fractions has made it a desirable alternative raw material however, both from the standpoint of price and of availability, even in times of national emergency. The reactions involved, using these two raw materials, are as follows.

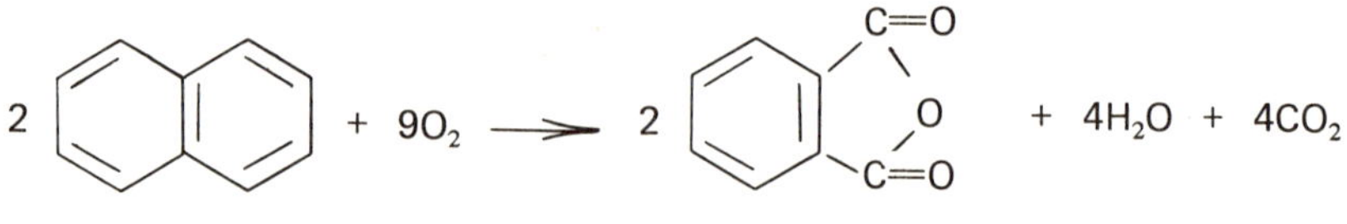

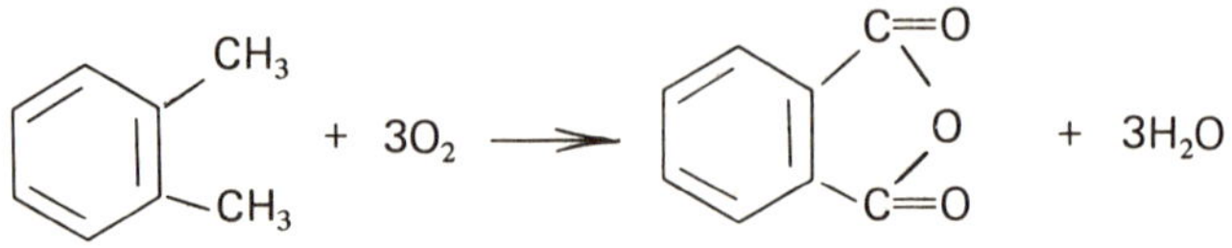

In the processes for the preparation of phthalic anhydride by the gas phase oxidation of naphthalene and mixtures of naphthalene and o-xylene, it is necessary for reasons of safety, to carry out the oxidation with a large excess of air. After the phthalic acid anhydride has been separated from the reaction gas, a considerable amount of waste gas remains. The waste gas contains in addition to about 81 to 82 volume percent of nitrogen, 16 to 17 volume percent of oxygen, 1 to 2 volume percent of carbon dioxide, 0 to 0.5 volume percent of carbon monoxide, as well as small amounts of sulfur dioxide (about 100 to 150 mg/cubic meter of gas at normal conditions), and sulfur trioxide (about 15 to 30 mg/cubic meter of gas at normal conditions), along with varying amounts of organic substances and entrained amounts of finely divided reaction products.

The organic and entrained substances present in the waste gas are mainly small quantities of phthalic anhydride, maleic anhydride, benzoic acid, and 1,4-naphthoquinone. Because of the strong corrosive properties of these substances, it is not possible to release into the atmosphere the waste gases of the naphthalene oxidation without previously carefully purifying the same.

According to the prior art, various processes have been recommended for purifying the waste gases. For example, the waste gases have been burned catalytically or thermally directly after they leave the apparatus used for condensing the phthalic anhydride. During such an oxidation process, all organic substances present in the waste gases are eliminated. Such methods are uneconomical and require a high energy expenditure for the necessary heating up of the waste gases.

Depending upon the type of heating means employed in the direct oxidation process, a considerable additional charging of the atmosphere with sulfur dioxide often cannot be avoided. A further disadvantage of burning up the waste gases is that all the potentially valuable reaction products still present in the waste gases are destroyed. This makes it impossible to recover part of the considerable expenditure for this purification process by recovering these reaction products.

A water washing step is often employed for the waste gas purification. Suitable processes and apparatus are disclosed in Kirk & Othmer's *Encyclopedia of Chemical Technology,* volume 7 (1951), pages 97 through 101. Power washers, for example, ventilator washers, (suction washers) are preferred. With respect to the purity of the waste gases, such washing systems are quite satisfactory. However, it is obvious that the water washing method only prevents the contamination of the atmosphere, while the drain water of the washing step is loaded with impurities. As a result, an indispensable second step of this method is the purification of the drain water.

A particular and inherent disadvantage of purifying the waste gases of the naphthalene oxidation by water washing resides in the poor water solubility of the 1,4-naphthoquinone present in the waste gases in varying amounts. When the water is sprayed into the waste gas, the 1,4-naphthoquinone is eliminated therefrom in the form of a solid and not in dissolved form. As a result, not only extensive clogging of pipelines and apparatus occurs, but an imbalance is created in the apparatus and there is danger of destroying or damaging washing devices having rapidly moving parts incorporated therein.

Substances which under normal circumstances are quite soluble in water, such as, for example, the dicarboxylic acids, are partially prevented from being dissolved by the presence of solid 1,4-naphthoquinone. Because of this phenomenon which is probably the incrustation of the organic acids by the 1,4-naphthoquinone, the quantity of solids obtained is always considerably larger than the theoretical amount of 1,4-naphthoquinone contained in the waste gas. The elimination of such precipitates is necessary for avoiding pressure increases in the overall apparatus and pipelines. The capacity of the process is decreased by the necessity to interrupt at regular and relatively long intervals the otherwise continuously conducted naphthalene oxidation.

The cleaning operations conducted during the down period of the plant represent an annoying and also very expensive working step since only manual operations are involved. The toxicity of the 1,4-naphthoquinone makes it additionally necessary to maintain special protective measures for the personnel since it is a strong skin irritant.

A need exists for a simple method permitting the purification of the waste gases obtained during the above mentioned oxidation process, preferably by means of a water washing step even though 1,4-naphthoquinone has poor water solubility.

A process developed by *G. Hoffmann and A. Striebeck; U.S. Patent 3,370,400; February 27, 1968; assigned to Chemische Werke Huls AG, Germany* is based on the discovery that the waste gases of the gas phase oxidation of naphthalene and mixtures of naphthalene with o-xylene can be treated with concomitant purification by watering the waste gases with an aqueous maleic acid solution which dissolves naphthoquinone, extracting the used washing solution with an organic solvent which is water immiscible, and subsequently concentrating the residual washing solution by evaporation.

A process developed by *D.C. Ferrari and C.G. Bertram; U.S. Patent 3,624,984; December 7, 1971; assigned to The Badger Company Inc.* is one in which effluent waste gases from phthalic anhydride and maleic anhydride plants are effectively water washed of residual organic matter (98 to 99% removal) in a wet scrubber using recycled water to concentrate the organic pollutants in the scrubber liquor. A concentrated liquid purge (blowdown) from the scrubber recycle circulating loop is directed to a thermal incinerator where the purge is vaporized and the organic pollutants are oxidized to nonpollutant products.

PROPYLENE GLYCOL FROM PROPYLENE OXIDE

Propylene glycol is produced by the hydration of propylene oxide according to the equation:

$$CH_3\overset{}{C}H{-}CH_2\ (\text{with O bridging, epoxide}) + H_2O \longrightarrow CH_3CHOHCH_2OH$$

A typical propylene glycol plant produces about 5.4 gallons of wastewater for each pound of propylene glycol produced. The large amount of the wastewater and its high salinity make this a particularly difficult waste to handle by conventional techniques. Thus the overall objective of an EPA sponsored program as described by M.A. Zeitoon and W.F. McIlhenny of Dow Chemical Company in EPA Report 12020 EEQ 10/71 has been to systematically characterize the physical, chemical and biological nature of the brine waste and to develop a method of treatment. More specifically, the objective was to examine several alternative treatment methods and to determine the technical, engineering and economic feasibility of each. The processes examined were:

Biological Treatment – The object was to develop a biological oxidation process that would require a minimum of dilution of the brine waste or that exhibited a maximum tolerance of salinity at high purification rates. The microbial metabolism of ethylene and propylene glycols in the presence of sodium chloride was the subject of a subcontract supported by EPA and carried out by the Texas A. and M. Research Foundation. Organisms were to be sought in areas exposed to glycol and glycerin wastes, from hyper-saline soils, from known cultures, and from other saline sources including seawater and naturally saline lakes and enclosed bays.

Solvent Extraction with Secondary and Tertiary Amines – The solvent extraction process as developed for the reclamation of saline water, if applicable to the glycol wastewater, could enable recovery of the glycol in a relatively pure water product. The raffinate would be a saturated salt solution low in organics that could be used as feed to chlorine cells for the production of the caustic required in the hydrolysis. This would then represent a completely closed cycle with no waste to be discharged.

Adsorption and Recovery of the Glycol – The adsorptive capacity of activated charcoal for ethylene and propylene glycols was known to be low, but investigations of several regeneration methods that would result in the recovery of the glycol as a product were part of this program. The use of activated carbon as a pretreatment for an activated sludge process, or as a tertiary treatment to polish the biologically treated waste were also to be investigated.

Membrane Separation – Many potential chemical and physical processes for the removal and recovery of glycols from the hydrolyzer effluent are inoperable because of the presence of the large amounts of salt. Membranes that would allow selective separation of the salt and the glycol would allow these other processes to be employed. The screening of potential membranes highly permeable to NaCl but rejecting propylene glycol was done by the Gulf South Research Institute in New Orleans, Louisiana, as a part of a separate R and D grant (Project No. 12020DQC) awarded to the State of Louisiana Department of Commerce and Industry by the Environmental Protection Agency.

The final objective was to develop sufficient pilot plant information, based on the most economically attractive of the processes examined, for the conceptual design of a waste treatment facility. Hopefully, the process developed and engineered would then be applicable to other organically contaminated waste brines.

PROPYLENE OXIDE FROM PROPYLENE

Starting in the early sixties, there was a rapid growth in the capacity for propylene oxide (PO) as the chlorohydrin plants for ethylene oxide were modified for PO production. The production growth rate exceeds 10% per year. A projection indicates a 1980 production of over 3,000 million pounds. The two industrial reaction steps for making PO are shown on the following page.

Chlorohydrination:

$$C_3H_6 + HOCl \longrightarrow C_3H_6(OH)Cl$$

Hydrolysis:

$$C_3H_6(OH)Cl + \tfrac{1}{2}Ca(OH)_2 \longrightarrow \tfrac{1}{2}CaCl_2 + H_2O + C_3H_6O$$

Though a direct oxidation plant (known as the Oxirane process) has been built in Texas for the production of 170 million pounds of PO per year, great expansion of this process in the seventies is not expected. However, after 1980, the Oxirane process will probably dominate. The reaction is:

$$C_3H_6 + \tfrac{1}{2}O_2 \longrightarrow C_3H_6O$$

The main use of PO is as an intermediate for making polyols.

Propylene, chlorine, and water are introduced into the bottom of a packed tower where they react under controlled conditions to form propylene chlorohydrin. The chlorine/propylene/water ratio of the feed to the tower is so chosen that the liquid effluent leaving the tower contains about 5% propylene chlorohydrin. The temperature of the effluent from the tower is about 50°C, and the pressure is atmospheric. Yields of chlorohydrin are about 90%. Unreacted propylene, which is in excess in the tower feed, leaves the top of the tower and is scrubbed with a dilute caustic solution to remove hydrochloric acid and any residual chlorine, and then is recycled. A portion of the recycled propylene is vented to control inert gases.

The liquid stream leaving the chlorohydrin tower is pumped to the flash hydrolyzer, where the chlorohydrin solution is mixed with an excess (10 to 20%) of 10% solution of milk of lime. By heating the mixture with live steam at about 100°C and in a vacuum of 1.5 pounds per square inch absolute, the PO is removed from the hydrolysis zone as it forms. The vapors are partially cooled to remove part of the water, and then sent to dual distillation columns to separate the PO out overhead. The bottoms are mostly water and dichloropropane.

Yields of PO, based on the entering propylene, are about 80% with dichloropropane the main by-product. The other by-products are sodium chloride, calcium chloride, chloropropenes, and heavies. The raw materials are propylene, chlorine, water, lime, and caustic soda. The hydrolyzer is vented back into the lime feed system.

The caustic hydrolysis of the chlorohydrin may also take place using sodium hydroxide in the aqueous phase. One molecule of sodium chloride is produced for each molecule of propylene oxide formed.

$$CH_3CHCl{-}CH_2OH + NaOH \longrightarrow CH_3\overbrace{CH{-}CH_2}^{O} + NaCl + H_2O$$

The hydrolyzer is a separation column in which the overhead product is the oxide and the bottoms contain the unreacted chlorohydrin, any excess caustic, the NaCl produced and several organic materials at low concentrations. The largest single such component in concentration is glycol which results from the overlap in the rates of the several reactions.

STYRENE FROM ETHYLBENZENE

Styrene is produced by vapor-phase dehydrogenation of ethylbenzene over supported zinc oxide, magnesium oxide, and iron oxide catalysts. Steam is used as the diluent.

$$C_6H_5C_2H_5 \longrightarrow C_6H_5C_2H_3 + H_2$$

Feedstock ethyl benzene and superheated steam are mixed in a dehydrogenation reactor. After being condensed, the reactor effluent goes to a separator, where three phases are formed. The uncondensed gases are passed through a scrubber where organic vapors are removed by the scrubbing water. The water phase is removed from the separator and discharged from the system, and the organic dehydrogenated mixture passes to the distillation section.

Since the dehydrogenation reaction operates at about 60% ethylbenzene conversion, it is necessary to fractionate the process unreacted ethylbenzene for recycle. Styrene will polymerize at temperatures approaching its normal boiling point; therefore, it is necessary to operate the styrene-ethylbenzene distillation under vacuum to prevent styrene loss due to polymerization.

The draw-offs from separator and scrubber are two of the three major wastewater pollution sources in the process. The other source is a steam-ejector system used to produce vacuums for distillation columns. The survey data derived from plant visits are summarized as follows.

Plant 1

Flow	2,810 gallons per 1,000 pounds
COD	219 milligrams per liter
	5.13 pounds per 1,000 pounds
BOD_5	69 milligrams per liter
	1.62 pounds per 1,000 pounds
TOC	22 milligrams per liter
	0.53 pound per 1,000 pounds

Plant 2

Flow	657 gallons per 1,000 pounds
COD	426 milligrams per liter
	2.34 pounds per 1,000 pounds
BOD_5	70 milligrams per liter
	0.381 pound per 1,000 pounds
TOC	22 milligrams per liter
	0.12 pound per 1,000 pounds

The smaller amount of wastewater in Plant 2 is attributed to its use of steam jets with surface heat exchangers in contrast to the steam jets with barometric condensers used in Plant 1, and also to effective operation of the scrubber system. Use of untreated river water as quenching water for the barometric condensers at Plant 1 introduces some contaminants into the wastewater stream. Plant 2 discharges uncondensible vapors (consisting of some organic contaminants) from surface heat exchangers into the atmosphere.

To achieve BADCT and BATEA control technology, the steam jets (with either surface or barometric condensers) should be replaced by vacuum pumps. RWL for BADCT and BATEA can then be expected to be lower than that represented by Plant 2.

An example based on a 5×10^8 pound per year styrene plant has been devised for illustrating the advantages of vacuum pumps over steam jets. A description is given in the following paragraphs.

A two-stage steam ejector system is currently used to obtain the vacuum in the distillation section. The ejector system illustrated uses surface exchangers for both inter and after condensers. A schematic flow sheet, depicting steam and effluent flow rates and effluent composition, is presented in Figure 52. The effluent steam from the ejectors contains a fair amount of organics and represents a source of pollution. The cost of operating the two-stage ejector system is presented in Table 19. Some producers reportedly fractionate the ejector effluent stream and recycle the organics back to the process. However, it is not known if this technique is widespread or successful. Note that the use of barometric condensers will result in an excessively large effluent stream.

FIGURE 52: STYRENE-ETHYLBENZENE DISTILLATION

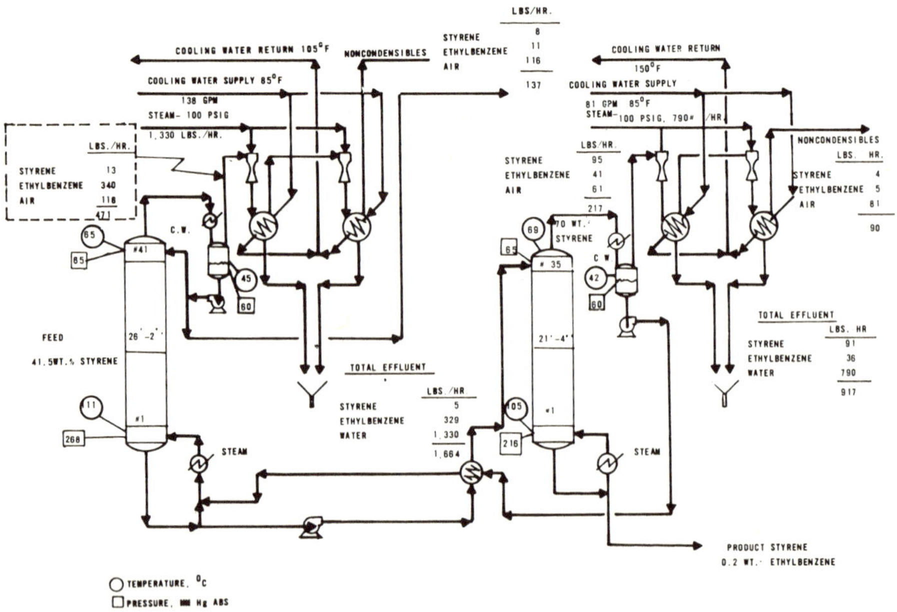

Source: Report EPA 440/ 1-73/009

TABLE 19: OPERATING COST OF TWO-STAGE STEAM EJECTORS*

	Two-Tower System			
	Tower No. 1		Tower No. 2	
Investment, $ (for ejectors etc.)	10,000		7,400	
Utilities	Lb/Hr	$/Yr	Lb/Hr	$/Yr
Steam, x 55¢/M lb	1,330	6,000	790	3,600
Cooling Water, 2.5¢/M gal	GPM	$/Yr	GPM	$/Yr
ΔT=20°F	150	1,800	89	1,100
Total Utilities, $/Yr		7,800		4,700
Investment Related	$/Yr		$/Yr	
Maintenance Material and Labor, 2% of Investment	200		150	
Plant Overhead, 65% of Maintenance	130		100	
Insurance and Taxes, 1.5% of Investment	150		110	
Depreciation, 10% of Investment	1,000		740	
Total Investment Related Expenses, $/Yr.	1,480		1,100	
Product Losses	Lb/Hr	$/Yr	Lb/Hr	$/Yr
Styrene, 7.0 ¢/lb	13	7,500	95	54,500
Ethylbenzene, 3.5 ¢/lb	340	97,600	41	11,800
Total Product Losses, $/Yr		105,100		66,300
Total Operating Costs, $/Yr		114,380		72,100
		186,480		
Total Operating Costs, ¢/lb styrene produced		0.037		

*Styrene-ethylbenzene distillation, 500 million pounds per year styrene, 8,200 hours per year

Source: Report EPA 440/1-73/009

The vacuum pump most suitable for this application is a two-stage unit which uses a rotating mass of liquid to draw the vacuum. In this case, the compressant liquid would be essentially ethylbenzene. Most of the organics in the inlet vapor stream from the tower condense in the compressant fluid and can be recycled back to the process. Process flow sheets showing the use of vacuum pumps are presented in Figure 53. The amount of organic substances actually leaving the vacuum system in the exhaust air is extremely small and is itemized in Table 20. The amount shown in this table as recycled is actually discharged from the system via the steam ejector system. The operating costs of using vacuum pumps are summarized in Table 21.

It is evident that a two-stage liquid sealed vacuum pump is more economical than a two-stage steam ejector using surface condensers. The economic advantage is due to the extremely low loss of ethylbenzene and styrene in the exhaust stream from the vacuum pumps. In other words, this modification not only has an economic advantage, but also reduces the RWL of the process. Styrene is used exclusively for homo-, co-, and terpolymers and is produced on the Gulf Coast. Production capacity has grown rapidly to accommodate demand.

FIGURE 53: STYRENE-ETHYLBENZENE DISTILLATION, VACUUM VIA VACUUM PUMPS

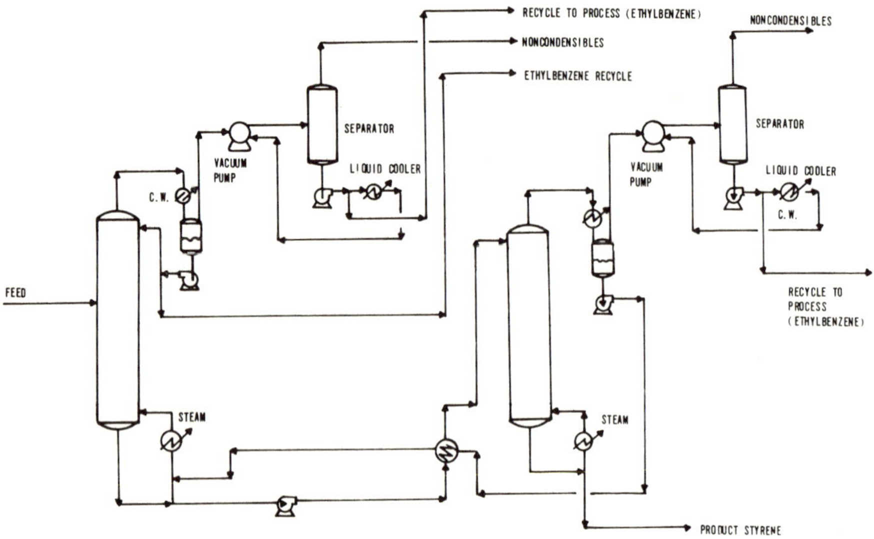

Source: Report EPA 440/1-73/009

TABLE 20: ORGANICS IN EXHAUST AIR FROM VACUUM PUMPS*

	Two-Tower System	
	Tower No. 1	Tower No. 2
Styrene		
In	13**	95**
Out In Exhaust	8	4
Amount Recycled	5	91
Ethylbenzene		
In	340	41
Out In Exhaust	11	5
Amount Recycled	329	36

*500 Million pounds per year styrene, 8,200 hours per year
**Pounds per hour

Source: Report EPA 440/1-73/009

TABLE 21: OPERATING COSTS FOR VACUUM PUMPS*

	Two Tower System			
	Tower No. 1		Tower No. 2	
Investment, $ (vacuum pumps etc.)	30,000		23,000	
Utilities				
Power, 0.800 ¢/kwh	kwh	$/yr	kwh	$/yr
	51	3,350	29	1,900
Cooling Water, 2.5 ¢/Mgal $\Delta T = 20°F$.	GPM	$/yr	GPM	$/yr
	13	160	7	90
Total Utilities, $/yr		3,510		1,990
Investment Related Expenses	$/yr		$/yr	
Maintenance Materials and Labor, 4% of Investment	1,200		920	
Plant Overhead, 65% of Maintenance	780		690	
Insurance, Taxes, 1.5% of Investment	450		920	
Depreciation, 10% of Investment	3,000		2,300	
Total Investment Related Expenses, $/yr	5,430		4,830	
Product Losses	Lbs/yr	$/yr	Lbs/yr	$/yr
Styrene, 7¢/lb.	8	4,600	4	2,300
Ethylbenzene, 3.5¢/lb.	11	3,160	5	1,440
Total Product Losses, $/yr		7,760		3,740
Total Operating Costs, $/yr		16,700		10,560
		27,260		
Total Operating Cost, ¢/lb styrene product		0.005		

*Styrene-ethylbenzene fractionation, 500 million pounds per year styrene, 8,200 hours per year, per letter from Nash Engineering Co. of 5-29-73 to Chem Systems

Source: Report 440/1-73/009

A process developed by *R.E. Hughes; U.S. Patent 3,691,020; September 12, 1972; assigned to The Badger Company, Inc.* involves thermal pollution control and waste heat recovery. The heat recovered from the dehydrogenation reactor effluent of a styrene manufacturing unit is used in the flash evaporator of a water desalination unit to produce pure water from contaminated or high mineral content water.

SYNTHETIC NATURAL GAS (SNG) FROM CRUDE OIL

In a process developed by *N.L. Carr, W.A. Roe, Jr. and H.C. Stauffer; U.S. Patent 3,732,085; May 8, 1973; assigned to Gulf Research & Development Company* a sulfur-free, methane-rich substitute natural gas is produced from a high sulfur crude oil by a nonpolluting process capable of achieving an unexpectedly high thermal efficiency. The process comprises separating whole crude oil into a plurality of fractions comprising a bottoms fraction and a lighter oil fraction, desulfurizing and hydrocracking the oil fraction into a substantially sulfur-free naphtha in the presence of hydrogen, which is provided by converting a residue portion obtained from the bottoms fraction.

The resulting sulfur-free naphtha is converted into substitute natural gas in the presence of steam. Not only is a thermal efficiency of approximately 100% obtained, notwithstanding the use of a portion of the crude oil for refinery fuel, but naphtha is produced at a volumetric efficiency of about 100% based upon the crude oil feedstock. The system is autogenic with respect to hydrogen requirements for desulfurization, nitrogen removal, and hydrocracking.

In addition to the economic advantages related to the thermal and volumetric efficiencies of this process, it is significant that no sulfur (in the form of either SO_2 or hydrogen sulfide) is released by this process to the atmosphere or to a public waterway, and essentially no sulfur is present in the pipeline gas product. Accordingly, neither the process nor the product of this process contributes to air or water pollution.

The only by-products of the process are relatively pure elemental sulfur, carbon dioxide and ammonium hydroxide, each of which has commercial value. For example, the ammonium hydroxide may be converted to ammonium sulfate fertilizer. The only feedstock to the process are crude oil, water and air, and there is but a single hydrocarbon product of the process, namely, a nonpolluting pipeline gas.

2,4,5-T FROM TRICHLOROPHENOL

2,4,5-T is a herbicide (for brush) and a growth regulator of the chlorinated phenoxy type. Its chemical name and formula are as follows.

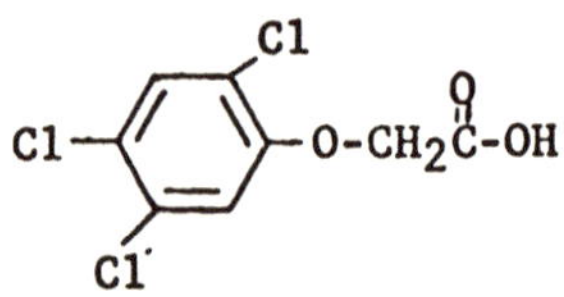

2,4,5-Trichlorophenoxyacetic Acid

It is also used as amine or sodium salt and as esters. It is a solid whose salts are water-soluble and whose acid esters are water-insoluble. It is moderately toxic and nonpersistent.

It is produced by Dow at Midland, Michigan, and by Thompson-Hayward at Kansas City, Kansas, according to the production chemistry set forth on the following page.

FIGURE 54: PRODUCTION AND WASTE SCHEMATIC FOR 2,4,5-T

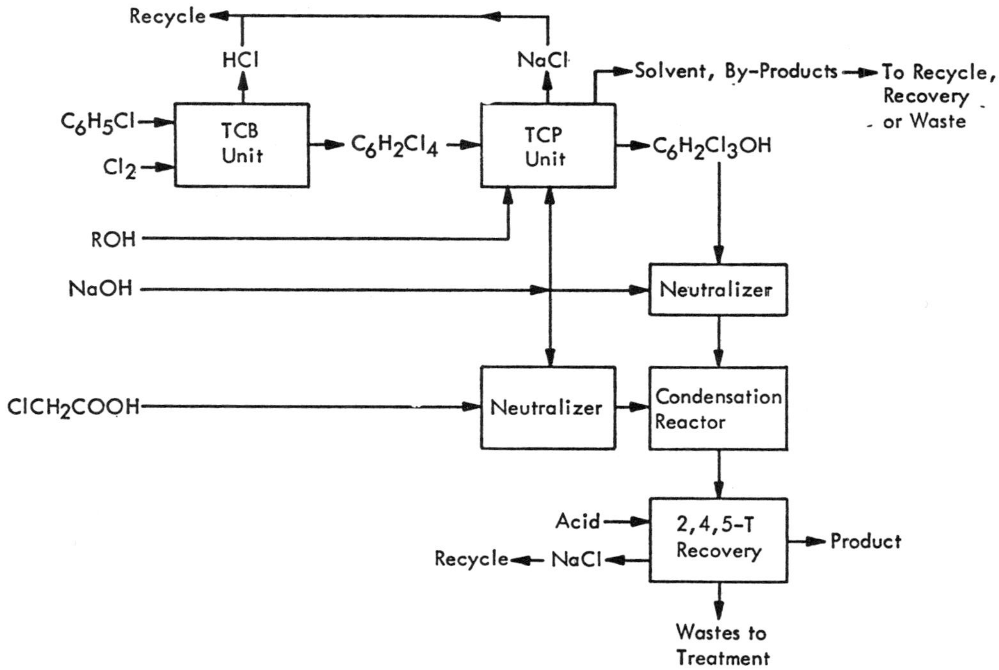

Source: Midwest Research Institute Report PB 213,782; 6/72

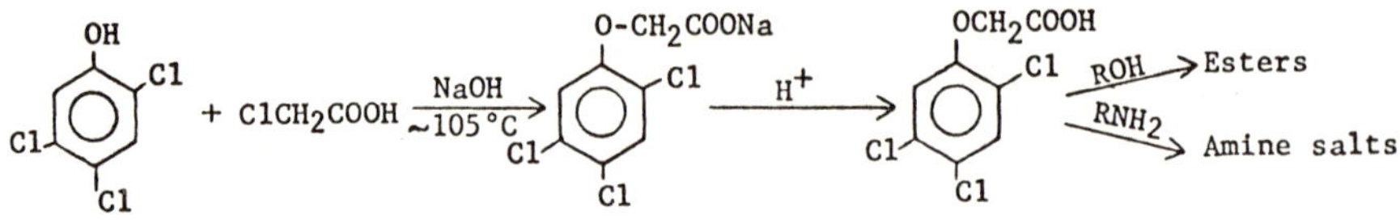

Trichlorophenol — Chloroacetic Acid — 2,4,5-T, Sodium Salt — 2,4,5-T

See the earlier section on 2,4-D for details of Dow's waste treatment facilities. The trichlorophenol plant is new and maintains very careful temperature control during the hydrolysis of tetrachlorobenzene to avoid the formation of dioxins which can cause chloracne in humans. No chloracne cases have occurred with workers in this plant.

During production at the Thompson-Hayward plant, most items are recycled; some of the alcohol goes to the steam condenser and then to the sewer. The biggest problem is disposal of salt. Their liquid waste treatment system is in the process of modification. An oxidation pond presently catches all process, service and surface waters, i.e., it receives some pesticides, detergents, spill residues, etc. Periodically the pond level is partially lowered by pumping water over the levee into the Kaw River. In a new system (under construction at a cost of about $100,000) the clean water streams (boilers, etc.) are being separated from waste streams and will go directly to the river. The waste streams will go to the oxidation pond, and then to a chemical treatment basin (acid or caustic), then via city sewer to the river.

During equipment cleanup, solvent is recycled to the next batch. They usually use the same process and formulation equipment for one product. If the product is changed, they would preclean with solvent which would be recycled; then steam clean with condensate going to the sewer; clean with caustic which detoxifies 2,4-D and rinse. Solid wastes go to a landfill; purge of 2,4,5-T units gives 50 to 75 pounds per month of mixed phenolic wastes which go to a commercial landfill. They have had little problem from air pollution, but some phenolics are troublesome because of odor (detectable at 0.005 ppm). They are collected by wet scrubber and discharged; also use wet scrubbers if they make the amine salts.

As regards pollution problems from accidents, a truck once dropped a trailer and broke containers. Cleanup consisted of absorption with clay and burial in landfill. Mixed lots of shipments on trucks are problems if breakage occurs.

TEREPHTHALIC ACID FROM XYLENE

Terephthalic acid is made by oxidizing p-xylene. This is virtually the sole use for p-xylene. About 30% of the terephthalic acid production is oxidized by nitric acid.

$$C_6H_4(CH_3)_2 + 4HNO_3 \longrightarrow C_6H_4(COOH)_2 + 4H_2O + 4NO$$

The remaining production is made by air oxidation.

$$C_6H_4(CH_3)_2 + 3O_2 \longrightarrow C_6H_4(COOH)_2 + 2H_2O$$

In 1970 about 1½ billion pounds of terephthalic acid were produced. Dacron is polyethylene terephthalate. An annual growth rate of about 30% is expected to continue.

Only one company (DuPont) is using the nitric acid oxidation of p-xylene in the United States. This process is a liquid-phase reaction at approximately 300°F and 125 to 200 psig in dilute HNO_3 (about 30 to 40 weight percent). Oxygen or air is passed into the reaction mixture where oxidation of p-xylene and lower oxides of nitrogen takes place simultaneously. The nitric oxides can be used for nitric acid regeneration.

Terephthalic acid crystallizes out of the solution during the reaction. The slurry formed in the reactor goes to a filter. NO gas, which is generated by the reaction goes to recovery. When the slurry from the reaction step is filtered, the mother liquor (consisting of water and unreacted nitric acid) goes to the NO recovery system. The cake from filtration is dried to become the terephthalic acid product for feed to a Dacron process.

The NO off-gas from the reactor is fed to an oxidizer column, where it is mixed with air. This oxidizes some of the NO to NO_2. The mother liquor from filtration is fed to the top of the oxidizer column and absorbs the oxides of nitrogen to form a dilute nitric acid solution. The spent air from the oxidizer column is vented. The dilute nitric acid produced in the oxidizer column is concentrated to 40 to 50% nitric acid and recycled to the reactor.

There are two sources of air pollution. The vent from the NO oxidizer column will contain air and unabsorbed oxides of nitrogen. The dryer, if it uses air to pick up the moisture, will have a vent which contains air, moisture, and probably product particulates.

The second reaction, represented by at least three commercial processes, utilizes acetic acid as a reaction medium and also involves a heavy metal oxidation catalyst. The most widely used commercial process is the Mid-Century process, in which the oxidation is reported to be based upon a bromine promoted heavy metal catalyst, such as cobalt-manganese. Reaction conditions are 350° to 450°F and 200 to 400 psig. The second process utilizes acetaldehyde as a promoter in place of bromine compounds, and the reaction is carried out at 250° to 350°F and 100 to 200 psig. The third process uses methyl ethyl ketone as the catalyst activator and operates at 200° to 300°F and 50 to 150 psig.

Preheated acetic acid, p-xylene and bromine catalyst, together with high pressure air are charged to a well agitated reactor operating at moderate temperature and pressure. The reactor contents are continuously discharged from the bottom of the reactor as a hot slurry into a crystallizer vessel, where cooling takes place by flashing off part of the acetic acid, unreacted xylene, and some water of reaction. The terephthalic acid slurry is passed to a centrifuge for removal of acetic acid and xylene. The filter cake is washed to remove the remaining reactants and then is dried to give the terephthalic acid product.

FIGURE 55: POLYMER GRADE TEREPHTHALIC ACID FROM p-XYLENE

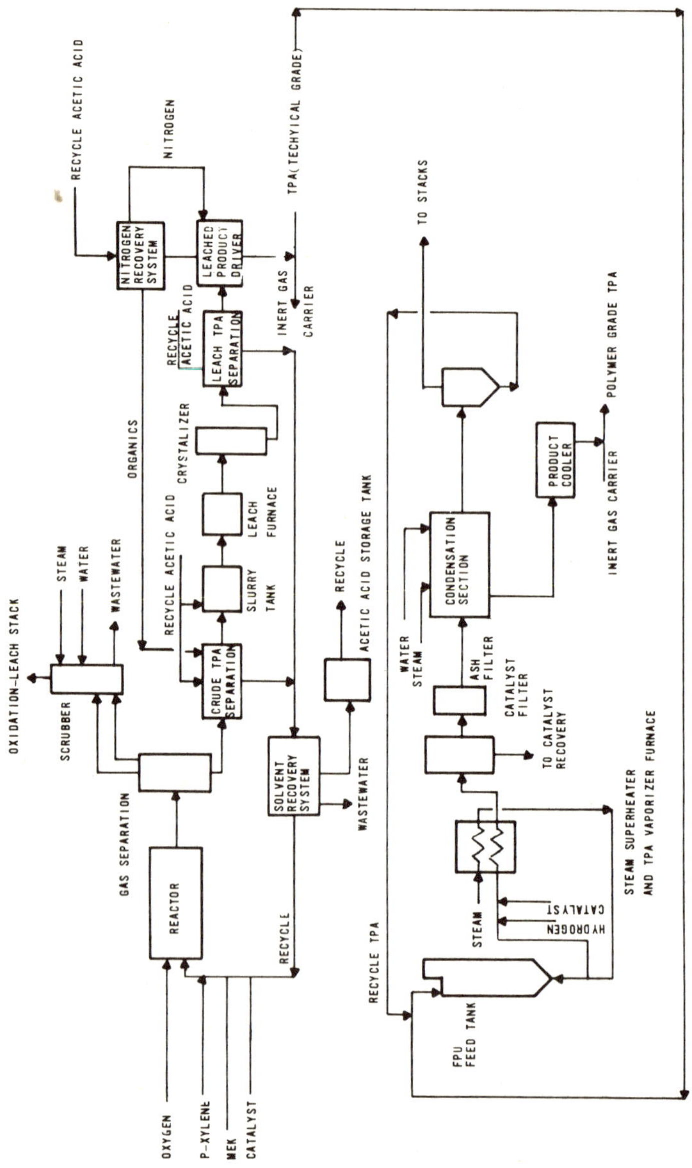

Source: Report EPA 440/1-73/009

The spent reaction liquor and condensate from the crystallizer vessel are distilled to remove water, recover unreacted xylene and acetic acid, and remove any other by-products. The acetic acid is recycled. The off-gas from the reactor is scrubbed with water before being discharged into the atmosphere. TPA obtained from this process is considerably purer than that produced by nitric acid oxidation, usually more than 99 weight percent TPA in contrast to 93 weight percent TPA of the other process.

At some plants, the TPA product is further purified to produce fiber-grade material. The TPA is washed with hot water to remove traces of catalyst and acetic acid. The hot water slurry is then heated and pumped into fixed-bed reactors and hydrogenated. This is followed by crystallization and drying to recover the fiber-grade TPA.

The major wastewater streams in the oxidation process are the bottoms from the solvent recovery unit and the effluent of the off-gas scrubber, and the major waste source in the purification process is the discharged mother liquid from the centrifuge. The characteristics of the wastewater obtained from plant visits are summarized in the following tabulation.

Plant	Process	Flow gal/1,000 lb	COD	BOD5 lb/1,000 lb (mg/l)	TOC
1	Catalytic	43.4	1.95 (5,400)	1.30 (3,600)	1.52 (4,200)
1	Purification	715	8.22 (1,380)	5.15 (865)	3.53 (510)
2	Catalytic				
	10% Occurrence	186	0.915	0.51	0.55
	50% Occurrence	186	1.72	0.82	0.86
	90% Occurrence	186	2.52	1.18	1.16
3	Catalytic	1,090	227 (24,950)	68.3 (7,500)	34 (3,730)
4	Nitric Acid	659	104 (18,900)	58.7 (10,700)	44.9 (8,180)

Plant 2 has five identical modules operating in parallel. Data obtained at this plant over a two-day period were analyzed for probability of occurrence.

Historical RWL data on process wastewater flow and COD were obtained for the catalytic oxidation process at Plant 1. At this plant there are actually two oxidation process modules, which operate in parallel. The data from these two units were subjected to analysis for the probability of occurrence. The tabulation on the following page summarizes the results of this analysis.

	Probability of Occurrence 10%	50%	90%	Ratio 90/50
Flow RWL, gallons/1,000 lb				
Oxidation Unit A	132	174	217	1.25
Oxidation Unit B	95	137	181	1.32
Purification Unit	754	969	1,185	1.22
COD RWL, lb/1,000 lb				
Oxidation Unit A	8.5	12.5	16.5	1.33
Oxidation Unit B	4.9	11.2	25.5	2.28
Purification Unit	12.8	27.4	58.5	2.14

The probability analysis was conducted on monthly average data taken by the manufacturer over a period of twenty-four months. Comparison of the sampling results and historical results for Plant 1 shows that both the measured process wastewater flow and the COD RWL were significantly lower at the time of sampling. This is attributed to the fact that the historical data include surface runoff from the battery limits area. This amounts to approximately 85 gallons per 1,000 pounds of product, with an associated COD loading of 3.5 pounds per 1,000 pounds of product.

The differences in RWL among the plants can be explained. The nitric acid oxidation process produces nitric oxides which are supposed to be used in producing nitric acid. However, it is likely that these nitric oxides are discharged into sewer at the plant which was visited during the survey. This results in a high organic loading in the wastewater. The high RWL of Plant 3 is due to poor process performance. Since both Plant 3 and Plant 4 are scheduled to be phased out in the very near future, further investigation of possible in-process modifications to reduce RWL is not warranted.

Both Plant 1 and Plant 2 utilize steam ejector systems to obtain vacuum for process needs. In contrast to discharging the exhaust stream into the atmosphere, as at Plant 1, Plant 2 employs barometric condensers to condense the exhaust stream. This causes a significant difference in the amounts of wastewater generated.

To define BADCT and BATEA of the oxidation process, vacuum pumps with surface condensers should take the place of steam ejectors and barometric condensers, to reduce the amount of waste flow as well as to preserve the ambient air quality. If a steam stripper like that described in the discussion of aniline should be installed to recover organic contaminants in the wastewater of the purification process, RWL can be reduced approximately by about three-fourths.

Process water usages as well as gross cooling water usages are varied among plants and processes. Information obtained from the plant survey is shown in the tabulation on the following page. Plants are identified with the same identification as that used for RWL.

Plant	Process Water Usage lb/lb product	Cooling Water Usage lb/lb product
1 (Oxidation)	N.A.	N.A.
1 (Purification)	N.A.	N.A.
2	N.A.	188
3	N.A.	N.A.
4	4	20,000

Several other approaches to manufacture of TPA are under investigation but none of them has been commercialized in the United States.

TETRAETHYL LEAD FROM ETHYL CHLORIDE

Over 90% of all tetraethyl lead is produced by some version of a conventional forty-year-old batch process in which an alkyl halide reacts with sodium-lead alloy. The reaction, occurring in a horizontal autoclave provided with a reflux condenser to recover any vaporized alkyl halide, yields a mixture of TEL, salt, and lead. The reaction, carried out at 60 psig and 70°C, is given below.

$$4PbNa + 4C_2H_5Cl \longrightarrow (C_2H_5)_4Pb + 3Pb + 4NaCl$$

The product mixture is fed batchwise to a still, where the tetraethyl lead is separated from the by-product lead and sodium chloride by direct steam stripping. The tetraethyl lead and stripping steam are condensed and sent to a decanter, where tetraethyl lead is drawn off as a bottoms stream. The upper aqueous layer in the decanter, containing unreacted ethyl chloride and dissolved organic by-products, is discharged into a process ditch.

The salty sludge bottoms from the still are sent to a lead recovery unit, and the concentrate is combined with the supernatant from the TEL decanter before being discharged into a settling basin for final recovery of solid lead.

Since recovery of by-product lead is considered an integral part of the TEL manufacturing process, the effluent from the settling basin is considered as the wastewater source of the process. The wastewater characteristics obtained from the plant visit are shown in the following tabulation.

Flow	12,000 gallons per 1,000 pounds
COD	1,100 milligrams per liter
	110 pounds per 1,000 pounds
BOD_5	40 milligrams per liter
	4 pounds per 1,000 pounds
TOC	56 milligrams per liter
	5.6 pounds per 1,000 pounds

The high amount of wastewater is due mainly to the nature of batch processes, which require a large quantity of water in cleaning up the reactor between reaction batches. Another cause of high water use is the vent-gas scrubber at the lead recovery unit. The intermittent dosage of still-aids such as soap or iron

FIGURE:56: TETRAETHYL LEAD FROM ETHYL CHLORIDE

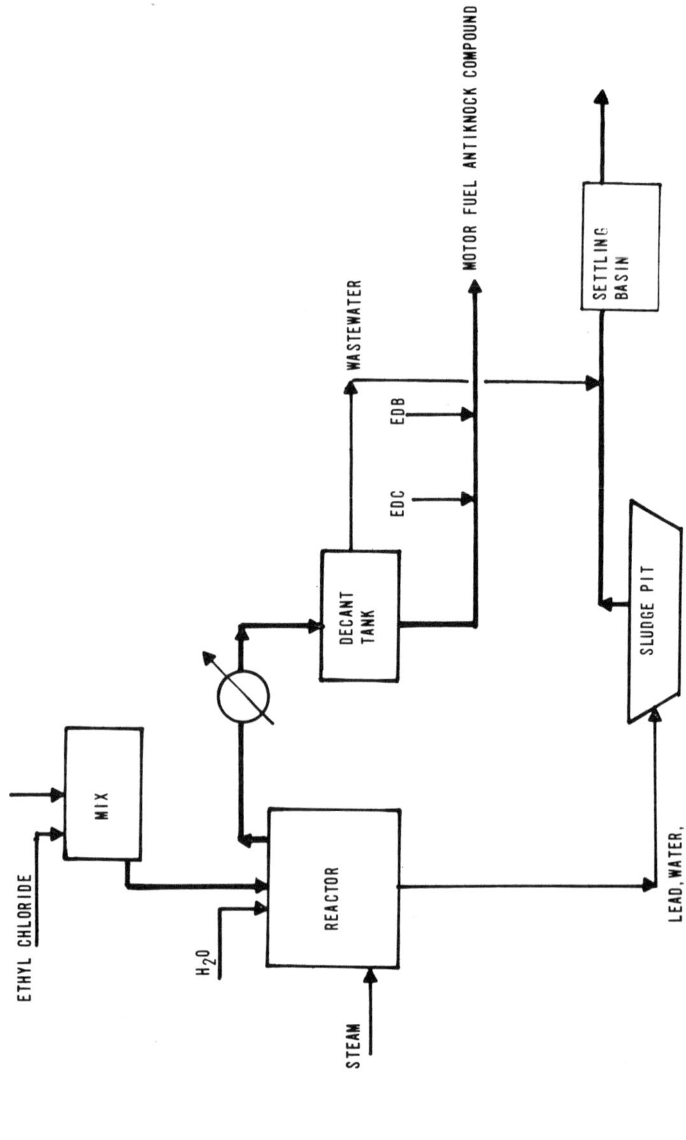

Source: Report EPA 440/1-73/009

to control the plating out of lead on the still walls, as well as unrecovered ethyl chloride, TEL, and metallic lead, all contribute to the high chemical oxygen demand.

In defining levels of control technology, it is suggested that recycling of the aqueous layer in the decanter to reduce fresh water usage, and consequently the amount of wastewater discharged, can be considered for BPCTCA. BADCT and BATEA should have a steam stripper for effective recovery of unreacted ethyl chloride and product TEL from the stream prior to their discharge into the settling basin.

An alternate process, which is based on the electrolysis of an alkyl Grignard reagent, is used by only one company in the world. This involves a totally different approach and offers at least three advantages: (1) it gives higher product yields; (2) it does not make by-product lead, hence eliminating the inefficient recovery and recycle of metallic lead; and (3) it can produce TEL as well as alkyl lead compounds.

The first processing step is the preparation of the Grignard reagent. Agitated propane cooled reactors receive metallic magnesium that reacts exothermically with fresh and recycled alkyl halide in the presence of an electrolytic solvent consisting of a mixture of ethers such as tetrahydrofuran and diethylene glycol dibutyl ether. The yield of alkylmagnesium halide is over 98%. The effluent of the electrolysis cell is sent to a stripper, where a separation of alkyl halide and alkyl lead is performed.

The aqueous effluent from the tetraalkyllead process may be neutral but generally is strongly alkaline, usually having a pH between 11 and 12, due to the reaction of the water with unreacted sodium-lead alloy to produce sodium hydroxide. It contains large amounts of sodium chloride and small amounts of water-soluble organic lead compounds, such as those formed in the purification process described above. In addition to the above, a very small amount of water-soluble inorganic lead compounds are sometimes present in the aqueous effluent.

The disposal of the aqueous effluent from the tetraalkyllead process is a serious problem, since in some locations, the maximum amount of soluble lead compounds allowed in aqueous effluent discharges is 5 parts per million parts of water.

The sodium chloride and sodium hydroxide compounds in the aqueous effluent naturally do not present a serious disposal problem. Moreover, the water-soluble inorganic lead compounds in the aqueous effluent do not pose a serious disposal problem since they can be easily removed from the effluent before discharge by a simple adjustment of the pH to a range of from 8 to 9.5 in the presence of water-soluble carbonates. However, the soluble alkyllead compounds in the aqueous effluent, such as trialkyllead chlorides, trialkyllead hydroxides, dialkyllead dichlorides and dialkyllead dihydroxides, cannot be discharged into lakes and streams without causing a potential contamination problem. Moreover, the disposal of such soluble alkyllead compounds constitutes a significant loss of lead.

While the soluble inorganic lead compounds may be removed from the effluent by adjustment of the pH to about 8 to 9.5 in the presence of water-soluble carbonate as described above, or by precipitation with hydrogen sulfide or other soluble sulfides, the organic lead compounds are not satisfactorily removed by such treatment, and furthermore, any excess soluble sulfide in the effluent would be highly objectionable and would have to be removed from the aqueous effluent before its disposal.

A process developed by *H.E. Collier, Jr.; U.S. Patent 3,308,061; March 7, 1967; assigned to E.I. du Pont de Nemours and Company* involves treating an aqueous effluent from the manufacture of alkyllead compounds and containing about 5 to 5,000 ppm of lead as dissolved organic lead and precipitatable by pH adjustment to 8 to 9.5. The process comprises the steps of: (1) adjusting the pH of the effluent to between 8.0 to 9.5; (2) intimately contacting the aqueous effluent with an ozone-containing gas; (3) precipitating the converted lead compounds and (4) separating the precipitated lead-containing compounds from the aqueous effluent.

This process is based on the fact that ozone is highly effective in removing water-soluble organic lead compounds from wastewater by producing water-insoluble products. The discovery that ozone successfully converted the soluble products to water-insoluble materials was most surprising since air and other oxidizing agents have been found unsatisfactory to effect this removal. The dissolved alkyllead can be removed from the aqueous effluent to such an extent that the aqueous effluent from tetraalkyllead manufacture can be safely discharged into lakes and streams. The process is easy to operate, and the quantity of materials, the time and intimacy of contact are easily coordinated to reduce the dissolved organic lead content to an acceptable level, below 5 ppm or substantially nil if desired.

A process developed by *E.R. Taylor, Jr.; U.S. Patent 3,697,567; October 10, 1972; assigned to E.I. du Pont de Nemours and Company* is a process for removing dissolved organic lead from an aqueous effluent produced in the manufacture of alkyllead compounds by contacting a metal more electropositive than lead but essentially nonreactive with water with the effluent until the dissolved organic lead is converted to an insoluble lead-containing product, leaving a reduced dissolved organic lead content in the effluent. The metal used may be selected from the group consisting of iron, zinc, aluminum and magnesium.

TRICHLOROETHANE FROM VINYL CHLORIDE

The fine solvent properties of 1,1,1-trichloroethane (TCE) have long been known, but in the past TCE has been difficult to make. With cheap vinyl chloride (about 5 cents per pound), the growth of TCE has been about 23% per year. TCE is made in two steps. First, vinyl chloride is hydrochlorinated in the liquid phase.

$$HCl + C_2H_3Cl \longrightarrow C_2H_4Cl_2$$

In the second step, 1,1-dichloroethane is thermally chlorinated.

$$Cl_2 + C_2H_4Cl_2 \longrightarrow C_2H_3Cl_3 + HCl$$

As indicated above, the main use of TCE is as a solvent.

Recycled 1,1-dichloroethane (DCE) vapor and chlorine gas are fed to a tower type reactor, at about 400°C and atmospheric pressure, producing TCE and hydrogen chloride. Yields are over 95%. After cooling the reactor products to about 40°C, they are mixed with fresh vinyl chloride (VC) and fed to a tower type hydrochlorinator in which a liquid mixture of DCE and TCE, containing some ferric chloride catalyst, is recirculated. The hydrogen chloride reacts with the VC producing DCE. The hydrochlorinator is maintained at about 40°C by cooling the recycled DCE/TCE mixture and at atmospheric pressure by venting the inerts. The hydrochlorinator products are decanted from the catalyst and fed to the purification-separation system. Yields are over 95%. The purification-separation system consists of the following steps: adjustment of chlorination; steam distillation; decanting; drying; DCE stripping and TCE fractionation.

The chloride contents of the products from the hydrochlorinator are adjusted by a light chlorination. The products are then stripped from the heavies by steam distillation, decanted from the condensed stream, and dried. From the drier the products are fed to the first distillation column, where the DCE is stripped off and recycled. The first column bottoms are fed to the second distillation column, where the TCE is taken off overhead.

The raw materials are vinyl chloride, chlorine, ferric chloride, and dessicant. The by-products are chloroethanes and heavies. There are two sources of air pollution. First, to keep inerts from building up in the hydrochlorinator, it is vented to the atmosphere. Second, there is a purge on the recycled DCE line, for inerts control.

TRICHLOROETHYLENE FROM VARIOUS RAW MATERIALS

Trichloroethylene is an old compound produced by a well established process based on acetylene. Over 90% of the trichloroethylene is made by this dual step process.

Chlorination of acetylene:

$$2Cl_2 + C_2H_2 \longrightarrow C_2H_2Cl_4$$

followed by dehydrochlorination:

$$C_2H_2Cl_4 \longrightarrow C_2HCl_3 + HCl$$

The remaining 10% of the trichloroethylene is made by the oxychlorination of 1,2-dichloroethane.

$$O_2\text{ (air)} + HCl + C_2H_4Cl_2 \longrightarrow C_2HCl_3 + 2H_2O$$

This process will probably increase in capacity, phasing out the acetylene based plants. In the past, the annual growth rate was about 5%, but this rate is expected to decline to about 3½% per year in the seventies. Over 90% of the trichloroethylene is used in metal degreasing. The rest is used in solvent extractions such as making decaffeinated coffee.

Chlorination, then Dehydrochlorination of Acetylene

High purity chlorine gas, containing no oxygen, is fed with acetylene into a packed tower in which tetrachloroethane is refluxing. The pressure is atmospheric and the temperature is about 50°C. The tower is equipped with a reflux condenser. The chlorine and acetylene react to form tetrachloroethane. The tetrachloroethane is withdrawn from the reactor, vaporized, superheated to 300°C, and fed to the dehydrochlorinator (which is a tower filled with activated carbon), yielding trichloroethylene. The temperature is 300°C, and the pressure is atmospheric.

The hydrogen chloride, trichloroethylene, and unconverted tetrachloroethane from the dehydrochlorinator, are fed to a stripping cloumn where the HCl is stripped off overhead and absorbed in water as hydrochloric acid. The conversion of the tetrachloroethane is about 90%. The bottoms are fed to a second stripper where the product, trichloroethylene, is taken off overhead. These bottoms are fed to a third stripper, where the by-product perchloroethylene is taken off overhead. These last bottoms are fed to a fractionator where the tetrachloroethane is taken off overhead and recycled. The bottoms from the fractionator are discarded.

The raw materials are chlorine (high purity, oxygen-free), acetylene, and water. The by-products are perchloroethylene, hydrogen chloride, and heavies. In the process for producing trichloroethylene, there are two sources of air pollution, the vent on the reflux condenser connected to the reactor, and the vent on the tail gas absorber.

Oxyhydrochlorination of Dichloroethane

Dichloroethane is oxychlorinated by air and hydrogen chloride to produce trichloroethylene. With the hydrogen chloride in excess, air, hydrogen chloride, and dichloroethane are mixed and reacted in the vapor phase in a tubular reactor at 300° to 400°C and about 100 pounds per square inch gauge. The reaction is exothermic, requiring cooling of the reactor. The reaction products are scrubbed with caustic, condensing the product and absorbing the excess HCl. The unreacted air is vented.

The crude product is decanted from the scrubber fluid and fed to a stripper. In the stripper, the product is stripped from the heavies and is fed to a fractionator. In the fractionator, pure product, trichloroethylene, is taken as bottoms. The raw materials are dichloroethane, hydrogen chloride, caustic, and air. The by-products are spent caustic, heavies, and a small amount of lights. The yields are estimated at 90%. There is a single source of air pollutants, the spent air vent on the absorber.

TRIFLURALIN FROM PERFLUOROMETHYL CHLOROBENZENE

Trifluralin is a selective herbicide of the nitroaromatic type. It has the following formula and chemical name.

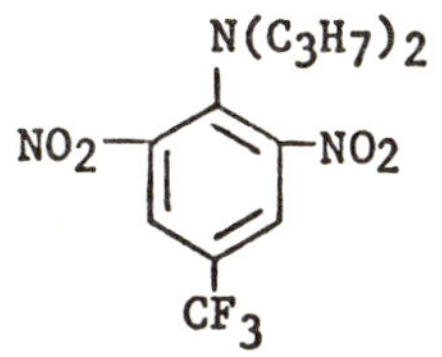

α,α,α-Trifluoro-2,6-dinitro-N,N-dipropyl-p-toluidine

It is a low melting solid of low volatility and water solubility. It has very low toxicity and is nonpersistent. Trifluralin is produced by Eli Lilly & Co. at West Lafayette, Indiana, using the following production chemistry.

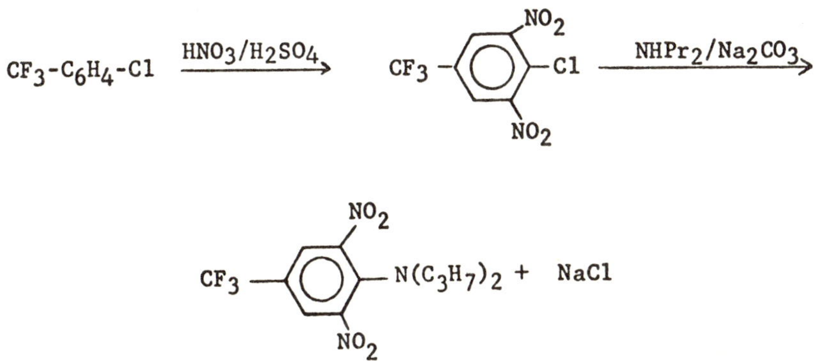

As regards pollution control, NaCl and process water are sent to waste disposal; NO_x to a scrubber; H_2SO_4 upgraded and sold; there are no solid wastes from this process.

The biological waste disposal plant services all of a larger Lafayette facility, but this unit effluent constitutes a significant (perhaps 20 to 40%) part of total. The plant is an \$8 to \$10 million investment with operating costs of approximately \$2 million per year. It uses neutralization, settling, aeration and biological means. The plant is state approved (the Indiana Stream Pollution Board is very strict). The discharge is cleaner than river water and contains no trifluralin product. Lilly has had no fish kills. The fate of the fluoride in product degradation was unknown. (A new Lilly plant at Clinton handles each waste stream at its source in-plant).

Leaky shipping containers have caused only a few problems. They use corrugated layers between cans and strapped pallets for shipment. The product has a bright yellow color which makes leaks obvious.

FIGURE 57: PRODUCTION AND WASTE SCHEMATIC FOR TRIFLURALIN

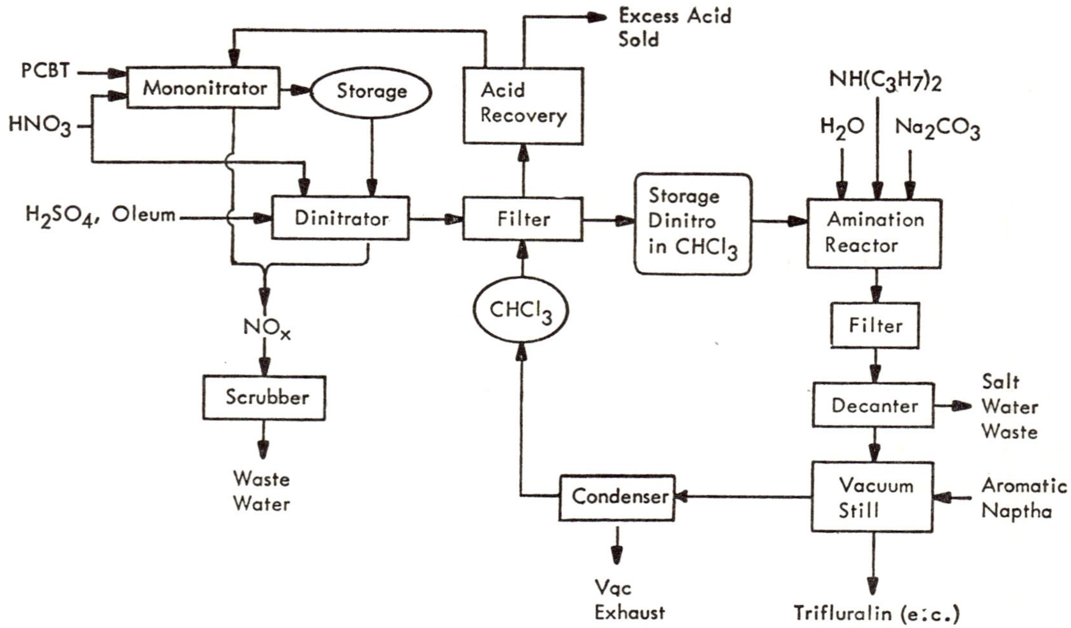

Source: Midwest Research Institute Report PB 213,782; 6/72

Most of the trifluralin produced is formulated directly at Lafayette as liquids. Liquid formulation and filling equipment are used only for trifluralin. The contract formulators of the small amount of granules have no dust problem.

VINYL ACETATE FROM ETHYLENE AND ACETIC ACID

Fresh ethylene, oxygen, and acetic acid are combined with their respective recycle streams, and then are vaporized and fed to a fixed bed reactor. Typical operating conditions are 5 psig and 250°C. Conversion per pass is about 5%, with very high (99%) selectivity. The catalyst is usually a mixture of palladium, copper, and iron chloride on alumina. The acetic acid-to-water mol ratio in the reactor is kept at about 40:1 to suppress acetaldehyde formation. Reactor effluent vapor is partially condensed to recover some of the acetic acid for recycle.

Further cooling and fractionation separate a crude product stream from ethylene, which is recycled to the reactor. The crude product stream is then fed to a series of fractionators for further removal of acetic acid and light ends. Hydroquinone is usually added as a polymerization inhibitor before vinyl acetate is sent to storage.

Since the process is a vapor-phase reaction, wastewater is minimal. The major wastewater stream is generated as bottoms from one of the fractionators. The light ends and heavy ends separated out are either recycled, sold, or disposed of by incineration. Results of survey data are summarized in the following tabulation.

Flow	28 gallons per 1,000 pounds
COD	516 milligrams per liter
	0.13 pound per 1,000 pounds
BOD_5	150 milligrams per liter
	0.04 pound per 1,000 pounds
TOC	220 milligrams per liter
	0.25 pound per 1,000 pounds

This level of RWL can be considered as standards for BADCT and BATEA control technology for this process.

The classical alternate route in the manufacture of vinyl acetate is the simple vapor-phase reaction of acetylene and acetic acid in the presence of a zinc acetate catalyst on a carbon support. Acetylene conversion is about 60% per pass at high (96%) selectivity.

A third route is by liquid-phase synthesis of ethylene and acetic acid. The reaction is carried out in a palladium chloride solution at 450 psig and 250°C. Conversion per pass is about 5% with 97 to 98% selectivity. Acetaldehyde co-product yield is controlled by suitable adjustment of the water content, and this co-product is oxidized in situ to form acetic acid, which is used for the main reaction. The literature indicates that this route produces the best economics.

FIGURE 58: VINYL ACETATE FROM ETHYLENE AND ACETIC ACID

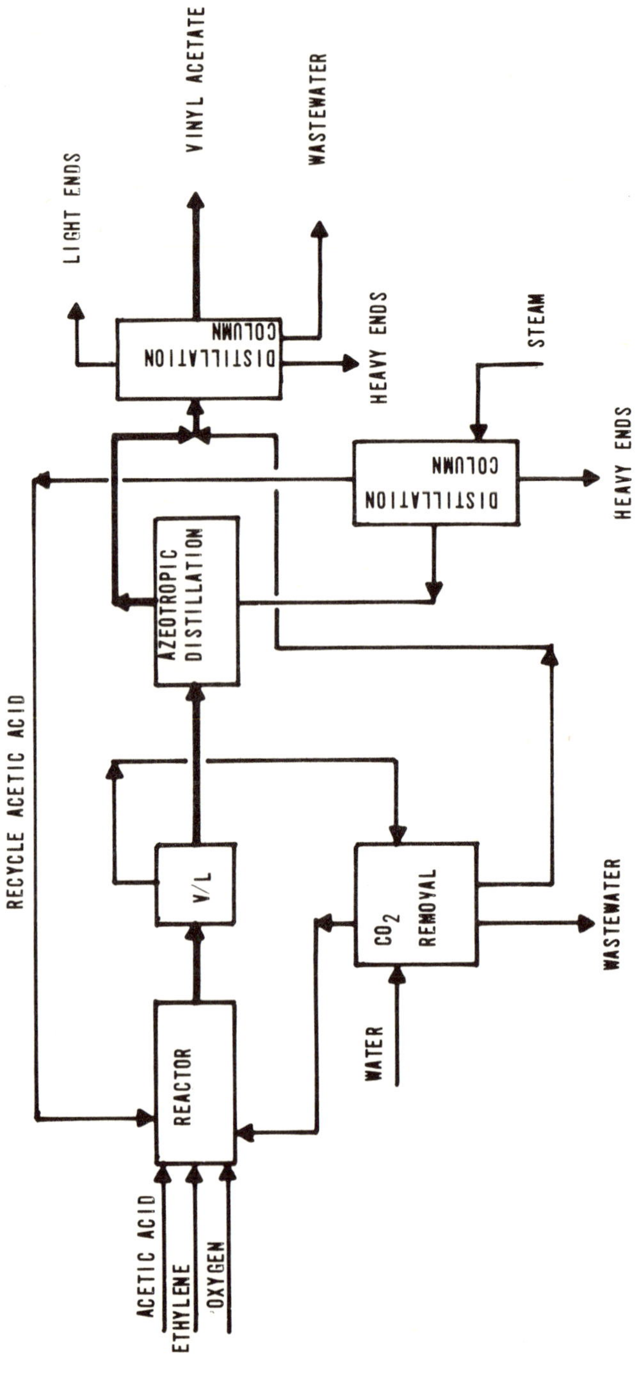

Source: Report EPA 440/1-73/009

VINYL CHLORIDE FROM ACETYLENE

In 1970, less than 20% of the vinyl chloride production was made by this acetylene based process.

$$HCl + C_2H_2 \longrightarrow C_2H_3Cl$$

This older reaction is being replaced by an ethylene dichloride based process, which will also be studied.

$$C_2H_4Cl_2 \longrightarrow C_2H_3Cl + HCl$$

This is a thermal cracking reaction carried out at 900°C and low pressures. Less than 4% of the vinyl chloride is made by the integrated combination of the two processes. The future growth rate is estimated at 13% per year, or a production of 10 billion pounds per year by 1980. Vinyl chloride is used to make polyvinyl chloride and its copolymers.

The vapor-phase reaction between acetylene and hydrogen chloride in the presence of a mercuric chloride catalyst yields vinyl chloride. Anhydrous hydrogen chloride (slight excess) and dry, purified acetylene gas (free from ammonia, hydrogen sulfide, phosphine, and arsine) are mixed and fed to a reactor containing carbon pellets impregnated with mercuric chloride. The reaction is exothermic, so a coolant is circulated around the tubes to hold the reaction temperature at about 200°C and 10 pounds per square inch gauge.

Effluent gases from the reactor are cooled by heat exchange and finally condensed and fractionated in a refrigerated column from which unreacted acetylene and hydrogen chloride go overhead. The acid-free monomer or crude is further fractionated in a second column in which vinyl chloride goes overhead, and by-product ethylidene chloride and aldehydes are removed as bottoms. The condensed vinyl chloride is stabilized with a small amount of phenol before going to storage. Yields are about 80 to 85%.

The raw materials are hydrogen chloride (dry), acetylene, dry (high purity) and mercuric chloride catalyst. The by-products are ethylidene dichloride and aldehydes. There is a purge on the recycled acetylene to prevent the buildup of inerts, which are mainly impurities in the raw materials.

Because no water comes into direct contact with the reactants and products and no reaction water is generated, there is no direct contact process wastewater. However, in the plant visited a "Mercury Treatment System" is associated with this process. This system is used for treating rainfall (which picks up traces of mercuric salts on the surface of the concrete pads and equipment) and periodic distillation column and reactor cleanouts. It is also used to treat the water from surface sprays which cool the outsides of process lines within the process battery limits. Based on flow measurements and sampling of these wastewaters, the following RWL was calculated.

Flow RWL (gallons per 1,000 pounds)	240
COD RWL (pounds per 1,000 pounds)	3.7

The wastewater is collected by a segregated sewer and is pumped into one of two alternate storage tanks. When the storage tank is full, sodium sulfide is added to precipitate mercuric sulfide. Two activated carbon columns, connected in series, are used to polish the filtrate.

The Mercury Treatment System is a batch operation (See Figure 59). The effluent from the activated carbon column is totally recycled to the storage tank until the mercuric concentration has been decreased to approximately 0.5 micrograms per liter. The mercury sludge from the filter press as well as saturated activated carbon is placed in drums and buried or removed by a contractor who recovers mercury. The analytical results for a single batch are presented below.

	Before Treatment	After Treatment
COD (milligrams per liter)	1,836	1,306
TOC (milligrams per liter)	448	33
SS (milligrams per liter)	1,124	24
Hg (micrograms per liter)	2,600	4.1

A more recent process for manufacture of vinyl chloride is by the thermal cracking of ethylene dichloride. This process will be discussed in the following section.

FIGURE 59: MERCURY TREATMENT SYSTEM FOR VINYL CHLORIDE PROCESS WASTES

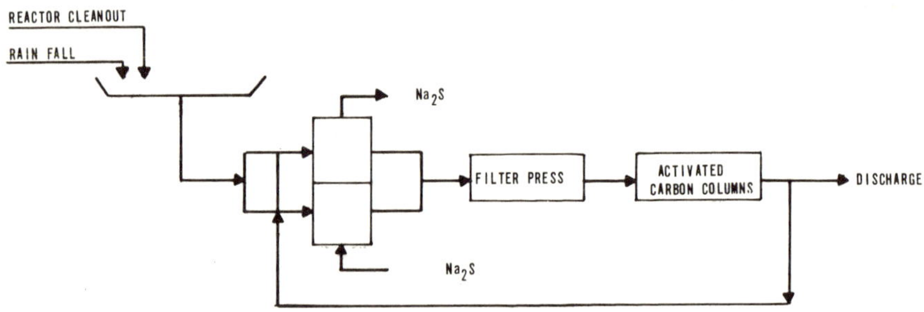

Source: Report EPA 440/1-73/009

VINYL CHLORIDE FROM ETHYLENE DICHLORIDE

Recent developments in ethylene technology, coupled with the low cost and ready availability of ethylene, dictate ethylene as feedstock in all new vinyl chloride plants. Vinyl chloride monomer is produced by cracking purified ethylene dichloride (EDC) in a pyrolysis furnace as follows.

$$C_2H_4Cl_2 \longrightarrow C_2H_3Cl + HCl$$

After quenching by direct contact cooling, the furnace products are separated into HCl and high purity vinyl chloride monomer. The liquid streams from the quencher are fractionated to separate the vinyl chloride product from unreacted EDC, which is then recycled. There is one source of air pollution, the tail gas scrubber on the hydrogen chloride recovery system.

The major wastewater sources are the effluents from scrubbing systems required for hydrogen chloride removal, recycle purification of EDC, and the effluent from associated aqueous by-product production units. The survey data for one plant are presented in the following tabulation.

Flow	336 gallons per 1,000 pounds
COD	2,733 milligrams per liter
	7.661 pounds per 1,000 pounds
BOD_5	Not available
TOC	120 milligrams per liter
	0.33 pound per 1,000 pounds

A large fraction of the RWL shown above is contributed by the aqueous acid production unit. If the by-product were left in an anhydrous form, the anhydrous acid by-product could actually replace the aqueous acid by-product. The RWL of this process will be reduced to 85 gallons of flow per 1,000 pounds of product, 0.203 pound of COD per 1,000 pounds and 0.054 pound of TOC per 1,000 pounds; this level of RWL will be considered as the standard of BADCT and BATEA control technology for vinyl chloride manufactured by EDC cracking.

Total process water usage in existing processes is 2.80 pounds per pound of vinyl chloride, and cooling water usage amounts to 3,464 pounds per pound of product.

An alternate route in manufacture of vinyl chloride is the classical acetylene addition reaction. This has been covered in the preceding section of this book.

AIR POLLUTION CONTROL PROCESSES

HALOGEN ACIDS

In the manufacture of chlorinated hydrocarbons, halogen acids such as hydrogen chloride are produced as by-products. Their recovery is economically desirable and their removal from vent gases is essential from the standpoint of pollution control.

A process developed by *J.E. Seebold; U.S. Patent 2,545,314; March 13, 1951; assigned to Hercules Powder Company* involves the manufacture of hydrochloric acid by absorption from waste gases containing hydrogen chloride.

A process developed by *H.C. Wohlers, F.N. Grover, C.W. Lentz and L.E. Pauling; U.S. Patent 2,730,194; January 10, 1956; assigned to Michigan Chemical Corporation* relates to the separation and recovery of hydrogen chloride, hydrogen bromide and other gaseous substances from waste gases discharged such as from brominators or from chlorinators engaged in the preparation of chloral and related products by alcohol chlorination, paraledhyde or acetaldehyde chlorination.

A process developed by *W. Kunzer and F. Leutert; U.S. Patent 3,036,418; May 29, 1962; assigned to Badische Anilin & Soda-Fabrik AG, Germany* involves the treatment of waste gases containing hydrogen chloride and impurities and originating from organic chlorination reactors. It is known to recover hydrochloric acid from waste gases which contain hydrogen chloride and which have been obtained from organic chlorinations, by allowing the waste gas to flow upwardly through a tower filled with a suitable packing and to which an aqueous absorption liquid is supplied at the top.

When working in this way there is the disadvantage that the waste gas is in general saturated with the vapors of impurities contained therein and originating from the organic chlorination reaction and that there is a risk of these vapors being condensed upon the entry of the waste gas into the absorption zone of the tower. Even when only small amounts of liquid or solid condensate are thereby formed, it is not always certain that at the speed at which the hydrochloric acid flows away from the tower the condensate will completely vaporize again before the hydrochloric acid enters the cooler through which it passes before leaving the tower. The hydrochloric acid obtained may therefore be substantially contaminated especially by substances which evaporate only slowly or of which the boiling point is so high that they are only volatile azeotropically with steam.

It has now been found that the disadvantage can be avoided by heating the contaminated gas containing hydrogen chloride, prior to the absorption, to a temperature which is above the dew point of the impurities. This heating above the dew point of the impurities may advantageously be effected by leading the gas coming from the chlorination apparatus in heat exchange relation to the hot acid flowing from the absorption zone of the tower. For this purpose it is thus only necessary to install in the absorption zone a suitable device for heating up the gases to be worked up, through which the acid produced flows downwardly.

In normal operation there is sufficient heat set free by the absorption for the gas supplied to the absorption zone to be heated above the dew point of the impurities without any appreciable change of temperature taking place in the absorption zone. The separation of the liquid droplets or solid particles entrained with the gases containing hydrogen chloride, from these gases prior to the supply of the gases to the absorption zone is preferably also effected in the device serving for the heating up of the gases.

Whereas according to the methods hitherto known there is obtained from the waste gas containing hydrogen chloride contaminated with benzene from a chlorination plant, a 30% hydrochloric acid which contains 0.2 gram of benzene per liter, the working up of the same gas according to the present process makes possible the recovery of a 30% hydrochloric acid of which the content of benzene amounts to only 0.0015%. In the working up of waste gases which contain carbon tetrachloride in addition to hydrogen chloride, it is possible according to the present process to obtain a 30% hydrochloric acid of which the content of carbon tetrachloride amounts to only 0.006% as compared with a content of 0.01% in 30% acid recovered by the known methods.

HALOGENATED HYDROCARBONS

Off-gases, vent gases and waste gases from organic chemicals manufacture may contain halogenated hydrocarbons. These materials have been recovered in various instances in the past by carbon absorption and by compression-decompression with or without cooling.

These processes are expensive and require large facilities and when cooling is employed and water is present, equipment blockage due to solids formation is common. With the continued construction of plants whose off-gas volumes are astronomical in volumes, the loss of the valuable products in such off-gases can no longer be tolerated. Further, the enormous quantities of photoreactive chemicals which are so often found in off-gases can no longer be permitted to pollute the atmosphere.

A process developed by *J.E. Panzarella; U.S. Patent 3,589,104; June 29, 1971;*

assigned to The Dow Chemical Co. is a process for recovering valuable organic compounds and/or deletorious air polluting chemicals, such as chlorinated hydrocarbons, from hot noncondensable gases containing the condensable organic chemicals and/or pollutants and water vapor. The hot gases are cooled to below the condensation point of the condensable gases by contacting with a cold surface. The steps of the process comprise:

1. establishing a film on a cold surface with a water-miscible organic liquid, such as ethylene glycol;
2. contacting the hot gases with the cold surface upon which there is maintained the film;
3. condensing the condensable gases and water on the surface;
4. venting the noncondensable gases to the atmosphere; and
5. separating the water and miscible organic liquid from the condensed gases.

A process developed by *S. Ezaki; U.S. Patent 3,589,864; June 29, 1971; assigned to Yawata Chemical Engineering Co., Ltd., Japan* is a process for recovering hydrogen chloride from an organo-chlorine compound which comprises burning a spent liquor or a spent gas containing an organo-chlorine compound at an appropriate air ratio, converting the contained chlorine to hydrogen chloride, thereafter recovering the hydrogen chloride in the form of hydrogen chloride gas or hydrochloric acid having an appropriate concentration, and discharging the remaining components as completely harmless and odorless gases.

HALOGENS

A process developed by *M.E.J. Cathala; U.S. Patent 2,868,325; January 13, 1959* involves chlorine recovery by dissolving the chlorine, contained in a mixture of gases, in chloro-sulfonic acid, maintained preferably at a low temperature, degasifying the solution thus obtained in a tower or in a degasifying apparatus of known type, in which a suitable temperature gradient and pressure are maintained, and then expanding the hot chloro-sulfonic acid passing out of the degasifying apparatus in an appropriate apparatus in any known manner.

A process developed by *R.S. Karpiuk; U.S. Patent 2,881,054; April 7, 1959; assigned to The Dow Chemical Company* involves the removal and recovery of chlorine from a gas mixture containing chlorine by the reaction of the chlorine with a dichloroiodide compound.

The process is usually carried out by dissolving a dichloroiodide compound in an acid medium to obtain a dichloroiodide solution containing dichloroiodide anions and contacting the gas mixture containing chlorine with the dichloroiodide solution, thereby reacting the chlorine with the dichloroiodide anions to form tetrachloroiodide anions. The solution containing the tetrachloroiodide anions is then heated to a temperature above the decomposition temperature of the tetrachloroiodide anions to decompose the tetrachloroiodide to the original dichloroiodide anions and chlorine.

Chlorine reacts readily with the dichloroiodide anions to the extent that substantially complete removal of the chlorine is effected from a gas mixture. It is immaterial whether the gas mixture contains a high percentage of chlorine or relatively small amounts. After the mixture is contacted by the usual methods employed in liquid-gas reactions, no chlorine is detected in the treated gas.

HYDROCARBONS

In petrochemical plants hydrocarbons may be emitted to the atmosphere in various ways. Vent gases from various process operations may contain hydrocarbons. Also, large quantities of hydrocarbon-containing gases may have to be vented as a result of upset conditions in portions of a plant or by passage of gases through safety valves. These gases are often passed to a flare for disposal to avoid possible accumulations of explosive hydrocarbons in a plant area.

In a process developed by *W.W. Ford and A.M.L. Kube; U.S. Patent 2,978,063; April 4, 1961; assigned to Phillips Petroleum Company,* provision is made for the absorption of condensable materials, from the gases being passed to the flare line, for further processing and reuse and for passing the noncondensable gases to the fuel line to be used for supplying heat to the plant under consideration.

Further, in accordance with this process, provision is made for a water trap in the line to the flare to prevent venting of gases during normal operation and a by-pass line to carry materials going to the flare through an oil trap maintained at a slightly higher pressure than that of the water trap so that a substantially uninterrupted flow to the flare is assured at all times when a flow of large volume to the flare is required.

A device developed by *R.D. Reed and J.S. Zink; U.S. Patent 3,512,911; May 19, 1970; assigned to John Zink Company* is a burner assembly for the combustion of flared gases at the upper end of a stack wherein steam and air is mixed with the gas in the combustion zone and turbulence is developed promoting mixing of the steam and air with the gases to provide substantially smokeless burning.

A process developed by *E.W. Merrill; U.S. Patent 3,593,496; July 20, 1971; assigned to Charles River Foundation* is a process whereby organic pollutants, such as hydrocarbons, can be removed from air by mixing the air with an aerosol of particles of water containing a surfactant which presents an oleophilic surface to the particles. The air is preferably humidified to 100% RH prior to mixing it with the aerosol, so that evaporation of the aerosol particles will not occur. The aerosol particles are effective in absorbing organic pollutants into their oleophilic surfaces, and removal of the particles by filtration or electrostatic precipitation leaves the air substantially purified of such pollutants.

A process developed by *V. Jasinsky and A.T. Upfold; U.S. Patent 3,667,408; June 6, 1972; assigned to Polymer Corporation Limited* is one whereby flare stack smoke is reduced or eliminated by controlling the temperature of the flare. The temperature is sensed by a sensing element, e.g., one or more thermocouples connected in parallel and located in the vicinity of the flare; the sensing element produces a signal which directly or indirectly activates a steam control valve to adjust the flow of steam to the tip of the flare stack so as to raise or lower the flare temperature to the temperature of optimum combustion of flare stack effluent.

A process developed by *R.D. Reed, J.S. Zink and H.E. Goodnight; U.S. Patent 3,749,546; July 31, 1973; assigned to John Zink Company* involves smokeless combustion of a combustible gas, which is supplied at a flow rate which varies widely from a low to a high value. It involves dividing the total flow rate through a manifold into a plurality of flow lines with each flow line providing gas to a plurality of burners.

The pressure of the gas in the manifold is measured and valves in each of the flow lines are controlled as a function of the manifold pressure, so that they open, one at a time, as the flow increases, thereby supplying gas to each flow line and to its burners at a much narrower range in flow rate. Thus the velocity of the gas is always high enough to get proper air mixing and smokeless combustion.

PARTICULATES

In the art of producing carbon black by pyrolysis a hot gaseous effluent or smoke at a temperature of about 1200° to 3000°F emanates from the pyrolysis reactor. Generally, this carbon black laden smoke is cooled to a temperature in the order of about 400° to 600°F by quenching with water in the reactor and thereafter cooling in an external cooler. Generally the quenching operation consists of contacting the effluent near the exit of the reactor with a spray of water.

The quenched effluent is then conveyed to the cooler and thereafter to a recovery system, wherein the carbon black is separated from the smoke. In order to recover the greatest amount of carbon black from the smoke and, additionally, in order to obviate creating a nuisance by discharging carbon black-containing gas to the atmosphere, it is customary to filter the smoke through cloth or fabric bag-type filters.

One improved type of filter design has been proposed by *D.C. Williams; U.S. Patent 3,491,518; January 27, 1970; assigned to Ashland Oil & Refining Company.* In this process it has been discovered that solids may be effectively separated from solids-laden gas in a novel down-flow filter having tubular open ended bags; and that, contrary to the prior art, the separated solids can be continuously discharged from the filter along with a small portion of the solids-laden gas.

A device developed by *E.K. Caskey; U.S. Patent 3,520,109; July 14, 1970; assigned to Ashland Oil & Refining Company* is one in which hot gaseous effluent containing suspended carbon black is passed through a series of bag filters. The resultant clean gas continues along a generally parallel path to completely surround and heat the collection hopper for the previously entrained carbon black.

The use of the high temperature clean gas to maintain the temperature of the collection hopper above the dew point of the atmosphere within the collection hopper prevents the formation of the condensate therein, and the resultant caking and corrosion associated therewith.

SULFUR COMPOUNDS

A process developed by *T. Nicklin; U.S. Patent 3,649,169; March 14, 1972; assigned to The Gas Council, England* is one in which removal of organic sulfur compounds, carbon disulfide and carbon oxysulfide, from inorganic gases, such as air, or exhaust gases, polluted therewith, is effected by contacting a mixture of steam and the gases to be treated with a catalyst comprising uranium oxide, and, preferably a carrier such as alumina.

A process developed by *S. Prahacs and S.P. Bhatia; U.S. Patent 3,701,824; October 31, 1972; assigned to Pulp and Paper Research Institute of Canada* is one in which objectionably odoriferous sulfurous gases, e.g., H_2S and CH_3SH, are removed from a gaseous stream by intimately contacting the stream with an aqueous alkaline suspension of activated carbon having a pH of at least about 8 in the presence of at least about 12% by volume of CO_2 and not more than about 6% by volume of O_2, the activated carbon thereby promoting the oxidation of sulfide and CH_3SH.

AIR POLLUTION PREVENTION IN COKING OF COAL

With a new look being given to all energy sources in 1974, the coking of coal as a source of industrial organic chemicals may well experience increased emphasis. As noted, earlier, organic chemicals and coal-tar chemicals were once almost synonymous terms.

A process which was developed by *G. Nashan and J. Knappstein; U.S. Patent 3,630,852; December 28, 1971; assigned to Firma Carl Still, Germany* permits the controlled gas-free and dust-free discharge and quenching of coke from horizontally arranged coke furnace batteries. The apparatus includes a closed coke-receiving chamber which is provided with gas and dust exhaust devices.

The chamber extends over the entire length of the coke battery and encloses the doors to the individual furnace chambers. In addition, the receiving chamber is connected with a quenching tower and it includes means for conducting the coke to a conveyor which leads to the quenching tower. The quenching tower itself includes either an arrangement of sprays directed over the conveyor or a quenching pool into which the coke is delivered and which includes a means for conveying the quenched coal into a receiving bin at the exterior of the quenching tower.

A process developed by *E.J. Helm; U.S. Patent 3,647,636; March 7, 1972; assigned to Koppers Company, Inc.* for collecting dust and smoke which are generated as coke is being pushed from a coke oven chamber, comprises a coke guide that is movable along the coke side of the battery and a coke quenching car that is movable along a path substantially parallel to the path traversed by the coke guide.

A hood is movably mounted to the coke quenching car and is positionable in cooperation with the coke guide for receiving coke that passes through the guide. Means is provided to maintain the hood in position relative to the coke guide as the coke quenching car moves relative to the hood in order to distribute coke evenly in the coke quenching car. Equipment is provided on or adjacent the coke guide car to remove particulate matter and gases from the hood.

A process developed by *F. Breitbach and G. Choulat; U.S. Patent 3,661,507; May 9, 1972; assigned to Firma Carl Still, Germany* involves decomposing the ammonia of impure ammonia-containing vapor clouds, particularly ammonia-containing vapor clouds emanating from coke oven plants and gas works and contaminated by hydrogen sulfide, cyano compounds, benzene, naphthalene and the like. The vapor clouds are first heated to the decomposition temperature of the ammonia, whereafter the hot clouds are passed at this temperature through a decomposition zone to cause decomposition of the ammonia into hydrogen and nitrogen.

The combustible matter in the decomposed gases is then burned in an intermediately installed combustion zone by supplying combustion air. The combustible matter including the hydrogen may be burned at a place remote from the decomposition plant or may be added to other gases.

A process developed by *J.J. Kelmar; U.S. Patent 3,661,719; May 9, 1972* utilizes a novel design for a nonpolluting coal carbonization plant for production of by-products from coal. As coal is heated to produce coke and coke oven gas the coke is delivered to a reactor where a bed of it is maintained. Most of the tar and ammonia and oil are removed from the coke oven gas and then the gas is passed through the coke bed.

A tar plant receives the tar removed from the gas and produces coal tar pitch, which is delivered with calcium carbonate to the reactor to desulfurize the coke and gas therein and enriches the gas. The desulfurized gas is continuously removed from the reactor and cooled. Periodically a batch of desulfurized coke is withdrawn from the reactor.

Steam is produced in cooling the coke and is used in heating an ammonia still. Water from the still is processed and used in an ammonia scrubber along with make-up water produced by condensing water vapor from the coke oven combustion gas. Only nonpolluting products leave the system. The rest are consumed or recirculated in it.

A device developed by *W. Kubsch; U.S. Patent 3,675,400; July 11, 1972; assigned to Hartung, Kuhn & Co. Maschinenfabrik GmbH, Germany* is an antipollution apparatus comprising a hood carried on a cantilever arm supported on a hood car mounted on tracks. The hood extends over a coke-quenching car mounted on the same tracks as the hood car and is connected to a suction and dust extraction device such that gases and dust which rise from the coke batch as it is pushed out of a horizontal by-product coke oven and into the quenching car will be conveyed to a gas-stream purifying device rather than being permitted to escape into the air. As the quenching car moves from the coke oven to a quenching station, the hood car and the hood carried thereby follows such that the hood continually covers the quenching car and prevents the escape of pollutants.

A process developed by *R. Kemmetmueller; U.S. Patent 3,697,381; October 10, 1972; assigned to American Waagner-Biro Company, Inc.* involves carrying out pollution-free charging of coke ovens. During introduction of a charge into a coke oven from a larry car gas which escapes through the larry car connections to the coke oven is collected with a suitable gas-collecting pipe. Standpipes which communicate with the interior of the coke ovens communicate with a second gas-collecting pipe so that additional gas escaping from the interior of the coke ovens is collected.

The entire space over the coke ovens is enclosed and the interior of the enclosure communicates with a third gas-collecting pipe in which gas is collected from the space over the coke ovens. All of this collected gas is cleaned before being discharged to the outer atmosphere.

A process developed by *E.E. Pries, K. Errgang, and F.-W. Drebes; U.S. Patent 3,715,282; February 6, 1973; assigned to Dr. C. Otto & Comp. GmbH, Germany* is one in which a tubular coke guide is disposed between the front of a battery of coke ovens and a car that receives coke discharged from the ovens through the guide. Above the car there is a fume-collecting hood, the side of which adjacent the coke guide is formed from a plurality of plates pivotally suspended from their upper ends on a horizontal axis in edge-to-edge relation.

As the car is moved along the battery of ovens, each plate in succession, as it comes opposite the outlet of the coke guide, is swung inwardly to permit coke to be discharged from the guide into the car. The fume that is produced in the hood by the coke is drawn out by an exhaust system and rendered nonpolluting before it is delivered to the atmosphere.

A device developed by *H. Spindeler; U.S. Patent 3,721,609; March 20, 1973; assigned to Dr. C. Otto & Comp. GmbH, Germany* is a pollution-free coke quenching car. A coke quenching car, movable along a battery of horizontal coke ovens, has a pair of spaced vertical walls extending transversely of the car and defining a fall space between them.

The upper part of this space has a front opening for receiving coke from an oven. The space below the opening is closed by a front wall, while the back of the fall space is closed by a back wall. Extending from the bottom of the opening rearwardly in the fall space part way across the car is a ramp as wide as the fall space.

A deck slopes from the back wall downwardly below the inner end of the ramp for receiving coke from it and discharging it beneath the ramp onto a floor that slopes from the front wall downwardly to the back wall. The lower part of the back wall has an outlet opening in it that normally is closed by a door. In the upper part of the fall space and also below the ramp and the deck there are water sprays for quenching coke beneath them.

A process developed by *R. Kemmetmueller; U.S. Patent 3,728,230; April 17, 1973; assigned to American Waagner-Biro Company, Inc.* provides a coke plant and method for operating the same according to which the coke is cooled by dry-quenching and the coal is dried, prior to being charged in a coke oven, with heat extracted during dry-quenching of the coke. The coke plant and method, in addition to carrying out the latter operations, reduce pollution of the atmosphere by releasing only cleaned and desulfurized gas to the outer atmosphere.

A process developed by *J.N. Morrison, Jr.; U.S. Patent 3,746,626; July 17, 1973; assigned to Dravo Corporation* provides pollution control for a coke oven battery during the coke removal operation, and the transporting of the coke to the next processing area. A canopy means is disposed over the area which contains the coke. An exhaust manifold and exhaust means associated therewith are adapted to be connected to the canopy means. A plurality of individually operable dampers are spaced along the travel path of the coke, and a damper control system is provided.

The control system is actuated through predetermined dampers which are disposed to connect the exhaust manifold to the canopy means, with the rest of the dampers remaining closed. Substantially all the evolved pollutants are collected via the dampers and the exhaust manifold for processing.

An apparatus developed by *E. Pries; U.S. Patent 3,748,235; July 24, 1973; assigned to Dr. C. Otto & Comp. GmbH, Germany* for receiving a cake of hot coke from a selected one of a battery of coke ovens and for transporting the coke to a remote coke quenching station in an essentially pollution-free manner includes a vehicle adapted to be driven to a selected coke oven and thence to the coke quenching station.

A housing is carried on the vehicle for receiving a cake of hot coke from an oven, the interior dimensions of the housing conforming substantially to the interior dimensions of the selected oven. The housing has grated side walls to expose a substantial portion of a coke cake disposed therein to the spray generated at the coke quenching station.

A container is mounted on the vehicle for providing a smoke-tight environment around the housing, the container having a door for permitting a cake of hot coke to be pushed into the housing. Guide means are provided for establishing a smoke-tight passage between a selected oven and the container so that a cake of hot coke may be pushed into the housing through the guide means without polluting the surrounding environment.

A device developed by *K. Riechert; U.S. Patent 3,766,018; October 16, 1973; assigned to Dr. C. Otto & Comp. GmbH, Germany* permits quenching coke without polluting the surrounding atmosphere. A coke guide carriage has a conveyor belt-floor for transferring coke from an oven to a quench car through a tubular guide, and the quench car is contained within a hood having top, end, and side walls for confining smoke, fumes, and dust generated by the movement of the coke from the oven to the quench car.

The hood includes a sprinkling system for spraying water onto the coke in the quench car and vent means in the top wall. The vent means includes a stack having a gas inlet end that is enclosed within a gas plenum chamber having a water sprinkling system therein for providing a curtain of water around the inlet end for washing the gas as it enters the stack, whereby the conveyor belt, the sprinkling system and the vent cooperate to minimize any smoke and dust created.

A system developed by *M. Ikeda and M. Sugimoto; U.S. Patent 3,767,536; October 23, 1973; assigned to Sumikin Coke Company Limited, Japan* is a dust collecting system for gases containing dust produced in the case of charging coke ovens with coal. A plurality of divided subducts are arranged on coke ovens and a main duct connected with them is connected with a dust removing apparatus so that a discharged gas from a coal charging operation may be completely introduced into the duct, may have dust well removed with the dust removing apparatus connected with the main duct and may be discharged into the atmosphere as a very clean gas.

A device developed by *W.D. Edgar and A.G. Jonnet; U.S. Patent 3,785,933; January 15, 1974; assigned to Koppers Company, Inc.* is one in which a gas-tight hood is mounted to a coke guide car. The hood is provided with a collar that cooperates with a fixed gas collecting main and with curtains that prevent gases, smoke and dirt from escaping from the quenching car while coke is being pushed.

WATER POLLUTION CONTROL PROCESSES

PRIMARY

There are many impurities in petrochemical or refinery wastewaters which must be removed or altered using pre- or primary treatment processes before subsequent secondary treatment operations can be considered.

Excessive concentrations of suspended solids discharged directly to secondary biological processes normally decrease overall process efficiency, either by reducing the active biological-solids fraction or by adversely affecting the sludge recycle and wastewater contact systems. As a consequence, gravity sedimentation or flotation units are used to reduce suspended materials. The removal of suspended or colloidal materials often can be enhanced by applying chemical coagulation in conjunction with sedimentation or flotation.

The removal of oil or grease by gravity separation is required in many instances as these contaminants have a deleterious effect on most secondary and tertiary treatment processes. Since free oils are easier to remove when the concentration is high, oily waste streams generally should flow through gravity separators prior to dilution with nonoily waste streams. Adding chemicals is often required to enhance separation, particularly when emulsions are present.

Extreme acidity or alkalinity must be neutralized before most secondary treatment processes can be properly applied. Although a biological system provides some buffering capacity (by virtue of such mechanisms as biodegradation of organic acids, respiratory carbon dioxide neutralization of caustic alkalinity, etc.), the presence of free mineral acidity or excessive alkalinity in wastewaters dictates a specific neutralization step.

Extreme variations in organic loadings as well as slug discharges of various organic and inorganic constituents can have an adverse effect on the operation and efficiency of primary, secondary, or tertiary treatment processes. Fluctuations in hydraulic loadings have a similar effect. When such conditions are anticipated, equalization and surge facilities are required.

Specific anions such as sulfides or chlorides affect various unit processes in a variety of ways. For example, high chloride concentrations alter biochemical reactions and adversely affect the gravity separation of biological sludges. Toxic or bio-static effects of inorganic constituents such as ammonia or heavy metals or organic compounds such as phenols are well documented and the presence of such materials often necessitates pre- or primary treatment steps. Approximate threshold conditions which indicate the need for including pre- or primary treatment processes are summarized in Table 22.

TABLE 22: PRE- OR PRIMARY TREATMENT REQUIREMENTS

Constituent	Limiting or Inhibitory Concentration	Treatment
Suspended Solids	>125 mg/l	Lagooning, sedimentation, flotation
Oil or Grease	>50 mg/l	Skimming tank or separator
Heavy Metals	< 1-10 mg/l	Precipitation or ion exchange
Alkalinity	0.5 lbs alkalinity as $CaCO_3$ lb BOD removed	Neutralization for excessive alkalinity
Acidity	Free mineral acidity	Neutralization
Organic Load Variation	> 4:1	Equalization
Sulfides	>100 mg/l	Precipitation or stripping
Chlorides	>8,000 - 15,000 mg/l	Dilution, deionization
Phenols	> 70 - 160 mg/l	Stripping, provide complete mixing
Ammonia	> 1,600 mg/l	Dilution; pH adjustment and stripping
Dissolved Salts	>16,000 mg/l	Dilution, ion exchange

Source: EPA Report 12020 EID 03/71

Oil Separation

Oil/water separation is the first primary treatment process to be considered here. Gravity oil separators are flow-through systems which allow separation of oily substances from the carrier wastewater through a gravity differential. The primary function of the separator is to separate free oil from the wastewater, but it will neither allow separation of soluble substances nor break emulsions. In general, oil separators are rectangular, multichannel structures designed in accordance with the specifications of the American Petroleum Institute (API). Other types of separators which have performed satisfactorily include circular separators, parallel plate separators, and special purpose separators. Design considerations and process variables involved are:

1. flow
2. rise rate of oil globules in wastewaters
3. turbulence correction factors
4. type and concentration of oil
5. temperature of wastewater
6. viscosity and specific gravity of the wastewater
7. geometry of basin

The feasibility of removing or recovering oils using gravity separators can be estimated using a separatory funnel in accordance with the standard API Procedure 734-53 (*Methods for Sampling and Analysis of Refinery Wastes,* API, 1969).

Equalization

Equalization is a method of retaining raw wastewaters in a basin which dampens the fluctuation in its characteristics. The basin should be sized according to the cyclic pattern of the industrial plant operation within practical limits.

If a basin is designed solely to provide for dampening out fluctuations in flow rate, there is no incentive to provide mixing. However, some form of mixing should be provided if wastewater constituents are to be dampened and equalized. This mixing can be provided using proper distribution and baffling techniques, diffused aeration, or mechanical agitation. Design considerations and process variables involved are:

1. flow variations
2. variations in wastewater constituents
3. cyclic patterns of industrial process operations
4. distribution of wastewaters
5. detention time
6. tank geometry
7. degree and type of mixing

The effect of transient organic or hydraulic loads on a given process operation can be simulated on a bench or pilot scale by varying the feed rates or concentration levels of the wastewater to the unit process in question for defined periods of time. An approach to evaluating the effects of transient loads on biological systems and thereby establishing the criteria for equalization requirements has been recently published by C. Adams and W.W. Eckenfelder (Center for Research in Water Resources Report No. 37, Austin, University of Texas, 1969).

Neutralization

It is not uncommon for petrochemical or refinery wastewaters to contain acidic or alkaline materials which require neutralization prior to subsequent treatment. Neutralization can be accomplished by the use of mixing basins, flow-through beds, or gas sparging.

Acid streams can be neutralized by mixing the waste with lime slurries, dolomitic lime slurries, or by adding caustic soda (NaOH) or soda ash (Na_2CO_3). Limestone beds are occasionally used, either as an upflow system or as a downflow system. Alkaline streams can be neutralized with acid (sulfuric, H_2SO_4, or hydrochloric, HCl) or with boiler flue gas (CO_2). Bottled CO_2 is applied to alkaline wastes in a manner similar to that of a diffused air system in an activated sludge plant. It neutralizes alkaline wastes by the same mechanism of flue gas neutralization, i.e., the forming of a weak carbonic acid when dissolved in water. Design considerations and process variables involved are:

1. alkalinity/acidity
2. quantity of sludge produced
3. flow
4. reaction velocity
5. depth and flow rate (limestone beds)
6. degree of mixing (mixing basins)
7. feed rate of neutralizing agent
8. geometry of mixing basin

The neutralization requirements for a mixing basin system can be estimated in the laboratory by taking samples representative of the alkalinity or acidity level in the wastewater and determining the equivalents per day of neutralizing acid or base required to obtain the desired pH. If limestone beds are to be used, bench scale neutralization columns such as those shown in Figure 60 can be used.

FIGURE 60: BENCH SCALE LIMESTONE COLUMN

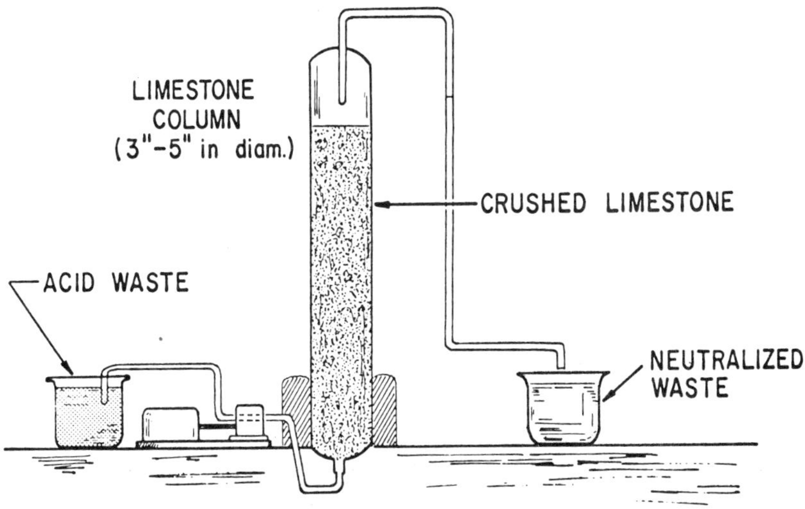

Source: EPA Report 12020 EID 03/71

Sedimentation

Sedimentation is a process employed in wastewater treatment to remove solids from suspension. Most suspended materials in raw petrochemical or refinery wastewaters undergo flocculent settling under quiescent conditions. With this type of settling, particles agglomerate during the settling period with a consequent change in specific gravity and settling velocity.

FIGURE 61: LABORATORY SETTLING COLUMN

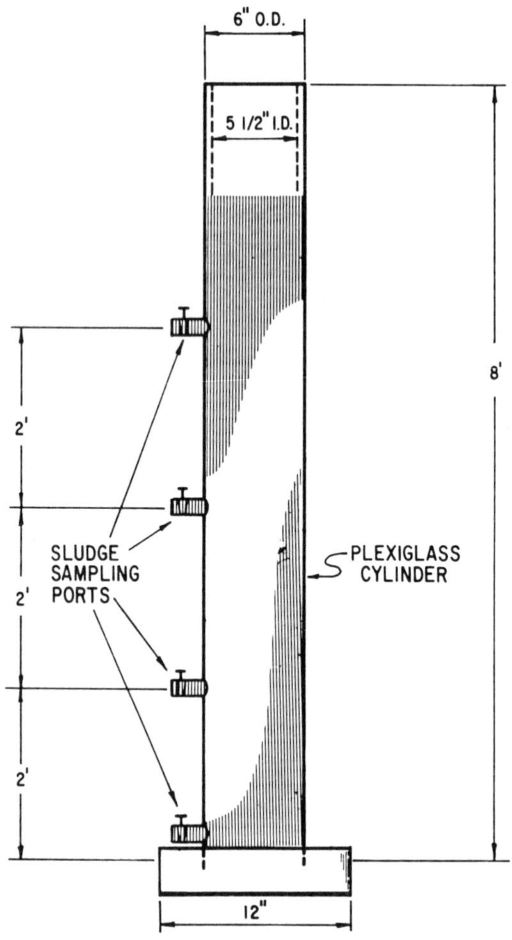

Source: EPA Report 12020 EID 03/71

In the gravity separation of flocculent suspensions, both clarification of the liquid overflow and thickening of the sludge underflow are involved. Sedimentation units are either circular or rectangular and usually have a side water depth of approximately 10 feet. Design considerations and process variables involved are:

1. flow
2. overflow rate
3. suspended solids concentration in raw wastewater
4. desired underflow concentration
5. settling velocity of suspended solids
6. detention time
7. variation in suspended solids concentration
8. structural and weir configurations of clarification basin
9. sludge raking mechanism and raking speed
10. mechanism of wastewater feed and effluent discharge

The basic design criteria for sizing primary clarification units can be developed using bench scale batch settling tests or pilot scale continuous flow clarifiers. The batch settling cylinder as used in the laboratory is shown in Figure 19. The concentrations of suspended materials are observed at various levels for various settling times, and this data can subsequently be formulated into the design.

Flotation

Flotation is a process used to separate the solid phase from the liquid phase, having the same objectives of gravity sedimentation. However, separation of the 2 phases is enhanced by the pressurization of the incoming or recycled wastewater, the release of which provides minute bubbles which attach themselves to or become enmeshed in the suspended particles. The air-solid particles rise to the surface of the flotation unit from which they are removed by a sludge collection mechanism, while the underflow represents the clarifier effluent.

Flotation units are often applicable in removing or reducing the concentration of oily materials present in petrochemical or refinery wastewaters. It is not uncommon to employ gravity separation devices followed in series by flotation units. The efficiency of a flotation unit in removing oil depends on the nature of the oil, its initial concentration, and the level of associated particulate matter. Chemical coagulants are occasionally used to enhance the separation process. Design considerations and process variables involved are:

1. flow
2. recycle ratio
3. air/solids ratio
4. detention time
5. rise rate
6. overflow rate
7. solids loading

The applicability of the flotation separation of solids from liquids can be determined on a bench or pilot scale. Such an evaluation can be made using the pressurized chamber apparatus shown in Figure 62. The wastewater is pressurized and proportionately mixed with additional water in an open container. The rise rate and concentration of the floated material is then noted and used as a basis for the selection and design of flotation systems.

FIGURE 62: BENCH SCALE FLOTATION UNIT

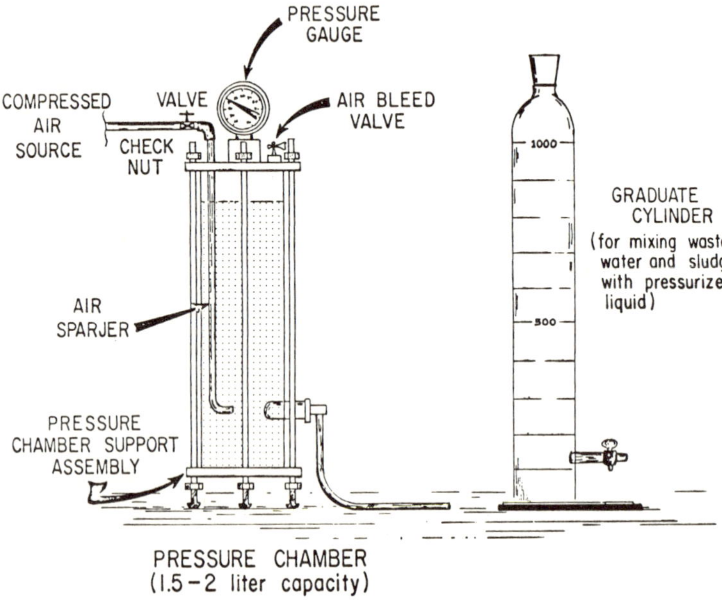

Source: EPA Report 12020 EID 03/71

A process developed by *A.M. Kivari; U.S. Patent 2,876,863; March 10, 1959; assigned to Process Engineers Incorporated* involves separating hydrocarbons from an aqueous waste by pressurizing the waste with a gas so that the gas is dissolved in the waste. The pressure on the waste is thereafter reduced whereby some dissolved gas is freed from solution. The freed gas carries the more volatile hydrocarbons from the waste, and the evolved gas and the volatilized hydrocarbons are separated from the waste. The pressure on the waste is thereafter further reduced and the waste is held in a substantially quiescent state so that additional gas is freed from solution and aids in floating remaining liquid and solid hydrocarbons in the waste to the surface of the waste.

These floating hydrocarbons are then removed from the upper part of the waste. In those aqueous wastes containing settleable solids, the solids are allowed to settle to the bottom of the waste and are continuously removed therefrom.

Flocculation

Reactor-clarifiers are used to combine chemical coagulation-precipitation with gravity sedimentation. These units are particularly applicable when the addition of chemicals significantly enhances the removal of colloidal and suspended materials. Most reactor-clarifiers employ concentric baffles with the center compartment containing chemical feed and mixing appurtenances. A scraping mechanism is used in the bottom portion of the quiescent zone to allow for collecting and concentrating the resulting sludge. Design considerations and process variables involved are:

1. flow
2. overflow rate
3. suspended solids concentration in raw wastewater
4. desired underflow concentration
5. effect of various chemical coagulants on settling velocities and underflow concentration
6. dosage of chemical coagulants
7. pH effect on chemical coagulation and precipitation
8. size of reactor compartment and level of mixing
9. sludge raking mechanism and raking speed
10. mechanism of wastewater and chemical feed and effluent discharge

The efficacy of using chemical coagulation-precipitation can be assayed in the laboratory using the standard jar test procedure, varying the pH, coagulant, and coagulant dosage, to determine the conditions for maximum removals of suspended materials or organic constituents. The effectiveness of adding coagulants to a wastewater is normally measured in terms of color removal, organic reduction (such as TOC, BOD, etc.), and suspended solids removal.

Nutrient Addition

Efficient and successful biological oxidation of organic wastewaters requires a minimal quantity of nitrogen and phosphorus necessary for the synthesis of new cellular tissue. Since many petrochemical and refinery wastewaters are deficient in both constituents, facilities must be provided to supply the wastewater with these critical nutrients.

Nitrogen in the form of anhydrous ammonia, nitrite, nitrate, and some forms of organic nitrogen are available to microorganisms, although it is normally assumed that nitrogen is taken into the cell in the form of ammonia. This means that a reduction of nitrite- and nitrate-nitrogen is required before nitrogen is available as a nutrient source.

Phosphorus is usually added in the form of phosphoric acid or soluble phosphorus salts since they are the most readily assimilable. It is generally assumed that a BOD/N/P ratio of 100/5/1 is required for proper microbial respiration and reproduction, although an alternate way of calculating the requirement is based on the daily synthesis sludge production, viz,

$$\text{Nitrogen requirement} = 0.123\ (\text{sludge production/day})$$
$$\text{Phosphorus requirement} = 0.026\ (\text{sludge production /day})$$

Nitrogen or phosphorus storage tanks are located in the proximity of the raw wastewater influent to secondary biological treatment facilities. Design considerations and process variables involved are:

1. organic load in terms of BOD
2. nutrient feed rates

3. mixing of applied nutrients with incoming wastewater
4. form of nitrogen and phosphorus and level of purity
5. levels of nitrogen and phosphorus in wastewater

The effect of nutrients on biological systems can be evaluated on a bench or pilot scale by measuring the organic removal obtained in biological facilities with varying concentrations of nutrients added.

SECONDARY

Secondary treatment is applied to reduce dissolved organic constituents through chemical or biochemical oxidation to a level acceptable for discharge into a receiving body of water or to the point where tertiary treatment can be effectively employed. Biological systems are the most common processes used for secondary treatment, although chemical oxidation, which has a more restrictive application, can be similarly classified.

Commonly used biological treatment systems include activated sludge, extended aeration, aerated lagoons, trickling filters, and waste stabilization ponds. The biological transformation and removal of nitrogen through nitrification-denitrification processes is classified for the purpose here as a secondary treatment, although such a system sometimes is categorized as a tertiary process.

Screening analyses often are necessary to distinguish biologically treatable streams from those which are biologically resistant or toxicological. This is partially true in a complex petrochemical facility where a treatability evaluation of major waste streams is required in order to plan the total waste treatment system. Biological screening tests can be conducted using either manometric techniques or bench scale batch reactors. These screening procedures will be discussed individually.

Batch Reactor Screening Analyses – A series of small batch-type biological reactors as shown in Figure 63 can be used to accomplish the same objectives as the Warburg approach. An acclimated seed is added to each of a series of batch units, then either various concentrations of a single waste stream or wastewater samples from each stream are added at a volume proportioned by flow to each reactor. Results such as those shown in Figure 64 indicate apparent toxicity or inhibition, using COD or BOD removal as a comparative index.

Manometric Analyses (Warburg Respirometer) — The Warburg Respirometer can be used to obtain a cursory estimate of the effect of various waste streams on biological cultures by measuring the biological response in terms of oxygen uptake for different classifications and concentrations of wastewater. An acclimated biological seed plus the wastewater sample are added to the manometric flask.

As the microorganisms utilize the organic constituents of the wastewater, the corresponding pressure differential in the closed system resulting from oxygen utilization is recorded. Results as shown in Figure 65 indicate the percent by volume of the particular waste stream which is toxic or inhibitory to a biological system (allowance should be made if lag periods of oxygen utilization caused by the incomplete acclimitization of the biological seed occurs).

FIGURE 63: BATCH REACTOR

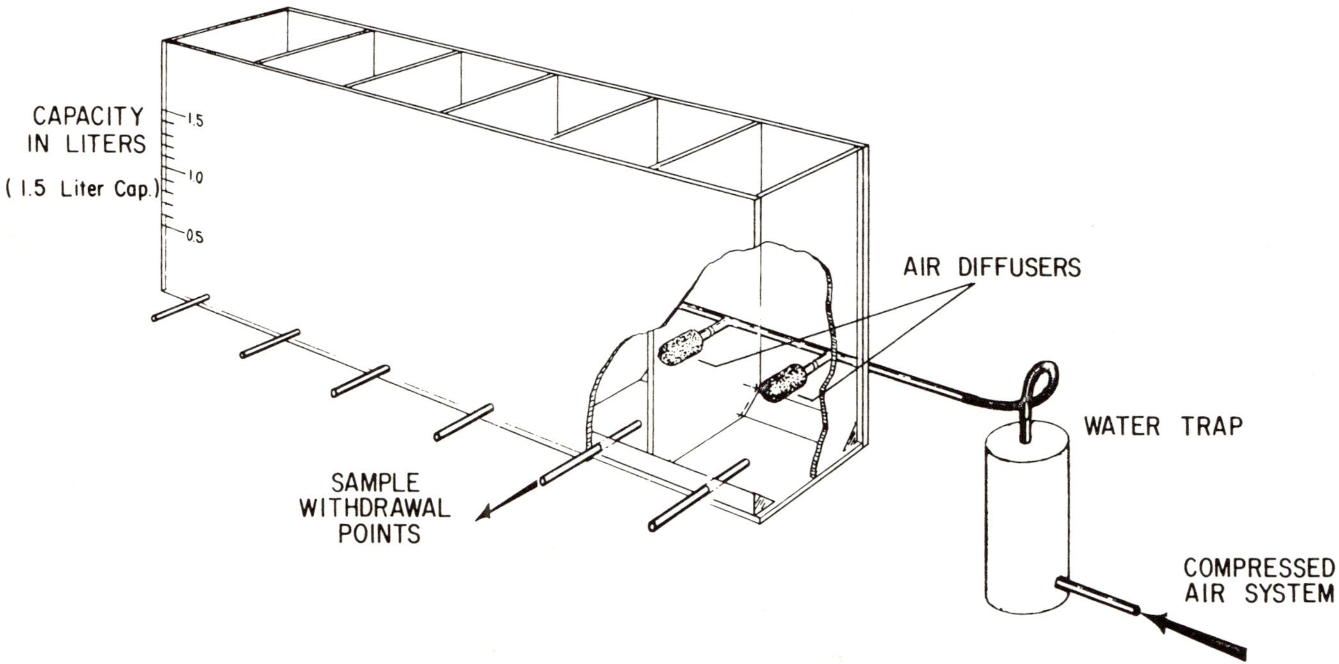

Source: EPA Report 12020 EID 03/71

FIGURE 64: SCREENING PROCEDURES USING BATCH BIOLOGICAL REACTORS

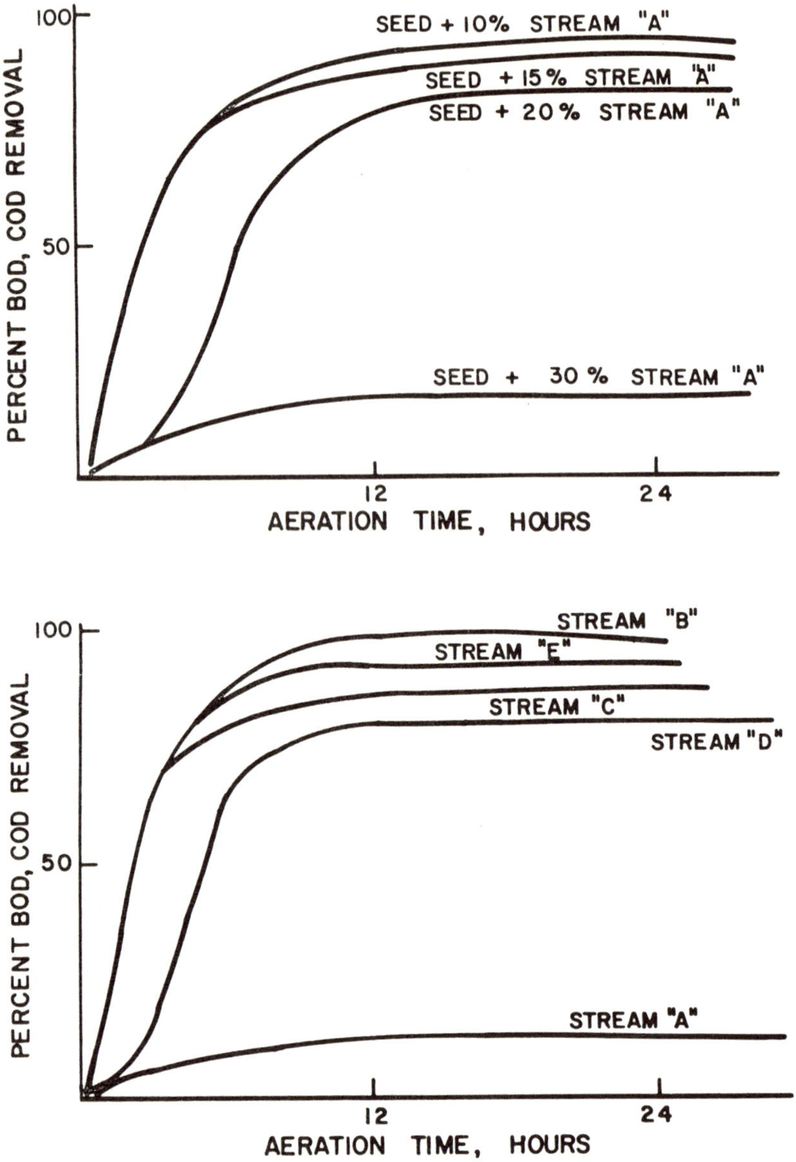

Source: EPA Report 12020 EID 03/71

FIGURE 65: DILUTION EFFECT ON RESPIRATION RATES

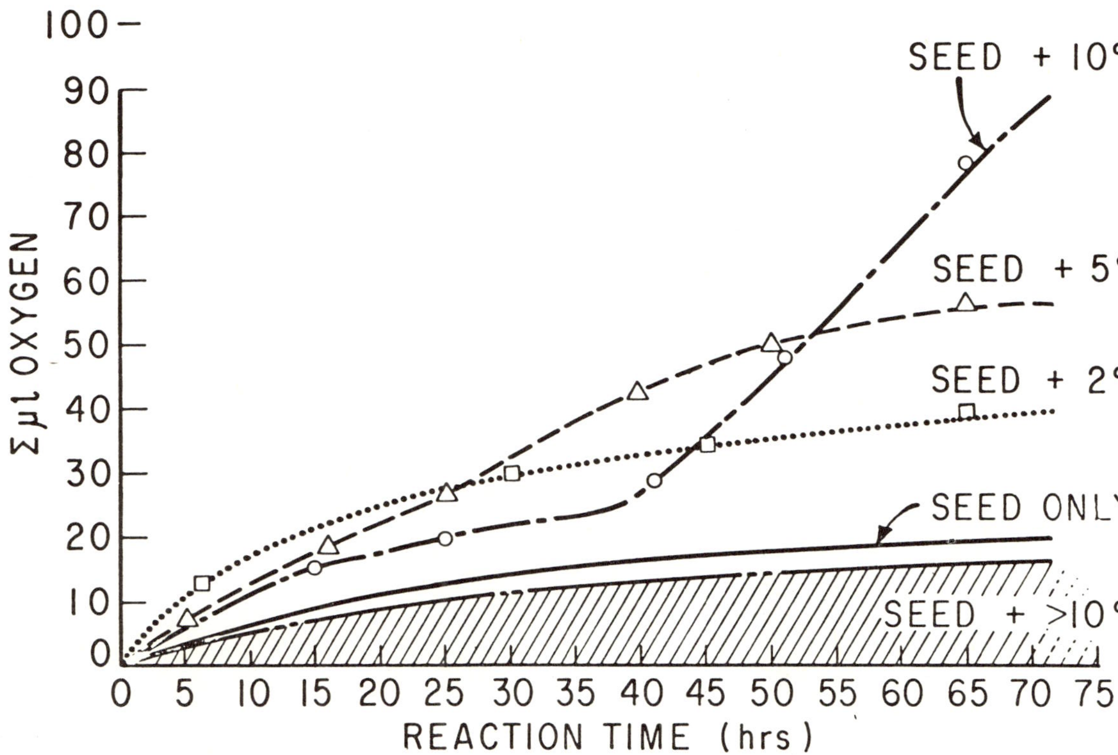

Source: EPA Report 12020 EID 03/71

Chemical oxidation is not economically competitive with biological oxidation for most applications, but it offers several conceptual advantages. It is not overly sensitive to fluctuating environmental conditions, is generally more efficient in the attack on stable or refractory compounds, and can effectively be controlled and regulated.

Activated Sludge

The activated sludge process provides an environment in which flocculated biological growths and wastewater are continuously mixed in the presence of dissolved oxygen. The microorganisms remove organic materials from the waste through physical sorption and aerobic biological oxidation. The biological floc suspensions are then separated by settling, recirculating a portion of the thickened sludge to the aeration basin for contact with additional waste. Excessive sludge accumulation must be wasted and subsequently treated. Oxygen and mixing in the aeration basin are provided by diffused or mechanical aeration. Design considerations and process variables include:

1. organic loading
2. environmental factors (pH, temperature, etc.)
3. nature of wastewater and fluctuation in characteristics
4. level of mixing in aeration basin
5. geometry of mixing basin
6. nature of sludge
7. nutritional requirements

The activated sludge process can be evaluated in the laboratory using the batch reactors shown in Figure 63 or a continuous flow reactor such as the one shown in Figure 66. Although batch-type studies are acceptable for screening analyses and approximating organic removal rates and effluent levels, bench or pilot scale continuous studies are recommended for the development of design criteria. This is predicated on the following factors:

1. Food/microorganism ratios which dictate process kinetics reach equilibrium values in continuous reactor studies, and this information can be more accurately translated to continuous-flow prototype units. Process kinetics are highly dependent on the time variable in batch reactor studies.

2. Oxygen requirements and sludge growth rates observed in continuous reactors have been shown to approximate those values observed in full-scale units when operating conditions are the same. Predictor data derived from batch data has been more inaccurate.

3. The impact of toxic or inhibitory substances, and variation in flow and constituents, on the biological system can be more accurately evaluated for continuous systems when continuous bench or pilot scale reactors are used.

The activated sludge process may be used for the treatment of waste brines from the production of polyhydric organics such as propylene glycol. Details of experience with such a process at the Texas Division of Dow Chemical Co. at Freeport, Texas have been presented by M.A. Zeitoun and W.F. McIlheny in EPA Report 12020 EEQ 10/71, Washington D.C., U.S. Government Printing Office (1971).

FIGURE 66: CONTINUOUS FLOW LABORATORY REACTOR

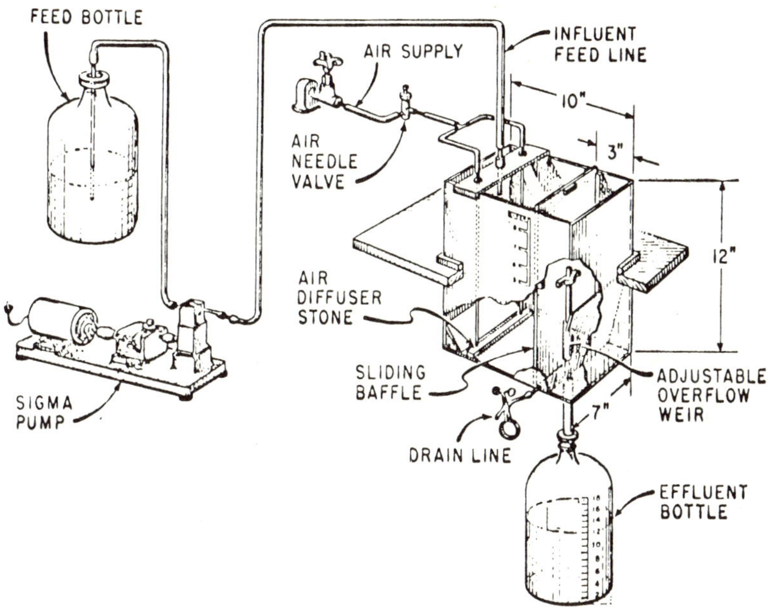

Source: EPA Report 12020 EID 03/71

A process developed by *J.B. Davis and R.L. Raymond; U.S. Patent 3,152,983; October 13, 1964; assigned to Socony Mobil Oil Company, Inc.* involves intimately mixing the oily waste to be treated with a microbial culture or sludge enriched with microbes active on hydrocarbons and maintaining the resultant mixture in continual contact with air or oxygen. The oily substrate, an aqueous medium containing water-insoluble oily or hydrocarbon materials, is agitated to decrease the oily globules to the smallest practical size without addition of emulsifiers. Since the amount of oil is very small, this reduction into globules occurs quite readily for there is a great deal of water around to keep an oil droplet once formed from getting together with another.

Further, the absorption of the oil by the microbial cells assists in this size reduction. Thus, moderate agitation obtained using air spargers, aeration through a perforated pipe under several pounds of pressure or by a mechanical impeller is generally adequate. Prior to or during the agitation an acclimated microbial flora grown on oily wastes or containing hydrocarbon oxidizers is added to the aqueous system. Thus, the mass is rendered as homogeneous as possible; small oil droplets are formed; the surface of oil exposed to microbial action is vastly increased; oxygen is available; and the microbial action at the oil-water interface is vastly increased with resultant oxidation and adsorption of the wanted materials. The oils and hydrocarbons are caused to volatilize in part, to be oxidized in part to carbon dioxide and other harmless materials, to be converted to microbial cell materials such as protein and lipids and to be adsorbed unchanged on the cells. While some extra-cellular material may be produced most of the hydrocarbonaceous material is converted to intra-cellular products. These are contained in the sediment as part of the cells while any solid extra-cellular products are precipitated by their own weight but being a part of the sediment all of which is easily removed.

Periodically, portions of the sludge are removed for charging or recharging vats that are to receive untreated wastes for processing. Since the continuous method affords a more ready maintenance of the concentrate of oxidizers than does the batch process, it is preferred. These acclimated sludges are rich in microbes effective in acting on oily materials. While there may be present microbes which do not oxidize hydrocarbons, there are generally little or no other foods in the hydrocarbon wastes treated by this process. Thus, the hydrocarbon oxidizers used in this process have very little competition for minerals, food or oxygen. Fast disposal results.

A process developed by *G. Jaekel, K. Karrenbauer, E. Goedicke and D. Kirstein; U.S. Patent 3,752,742; August 14, 1973; assigned to Knapsack AG, Germany* is one in which highly polluted effluent from organic chemical production facilities is continuously purified in the presence of microorganisms in a biochemical reaction zone by aeration with an oxygen-containing gas. To this end the oxygen-containing gas is introduced into the biochemical reaction zone having a liquid phase therein, and while flowing therethrough, it is continually transformed with the resultant formation of fresh-surfaced gas bubbles and reduction of the time of contact between the gas in the bubbles and the liquid phase at any surface therebetween.

Extended Aeration

Extended aeration is a modification of the activated sludge process where the organic loading is sufficiently low to allow oxidation of biological solids (endogenous respiration), thus minimizing the excess sludge production attributable to biological synthesis. The sludge is theoretically nonbiodegradable residue but still contains an active biological fraction.

Extended aeration facilities generally are most applicable when the design wastewater flow is less than 2 to 3 MGD because of the tank volume requirements. Design considerations and process variables involved are:

1. organic loading
2. environmental factors (pH, temperature, etc.)
3. nature of wastewater and fluctuations in characteristics
4. level of mixing in aeration basin
5. geometry of mixing basin
6. nature of sludge
7. nutritional requirements

The extended aeration process can be evaluated in a manner similar to that of the activated sludge process. The detention time (or organic loading) required to minimize sludge production can be estimated by evaluating the sludge production per unit time for each of the continuous bench scale reactors subjected to varying organic loadings. The loading at which sludge production is minimized can be considered the point where the system by definition is extended aeration.

Trickling Filters

Trickling filters use a biological slime coated on a fixed media bed to remove dissolved and colloidal organic materials as the wastewater flows through. As the wastewater comes into contact with the slime, the organic material and oxygen diffuse into the growths where biochemical oxidative reactions occur.

FIGURE 67: BENCH SCALE TRICKLING FILTER

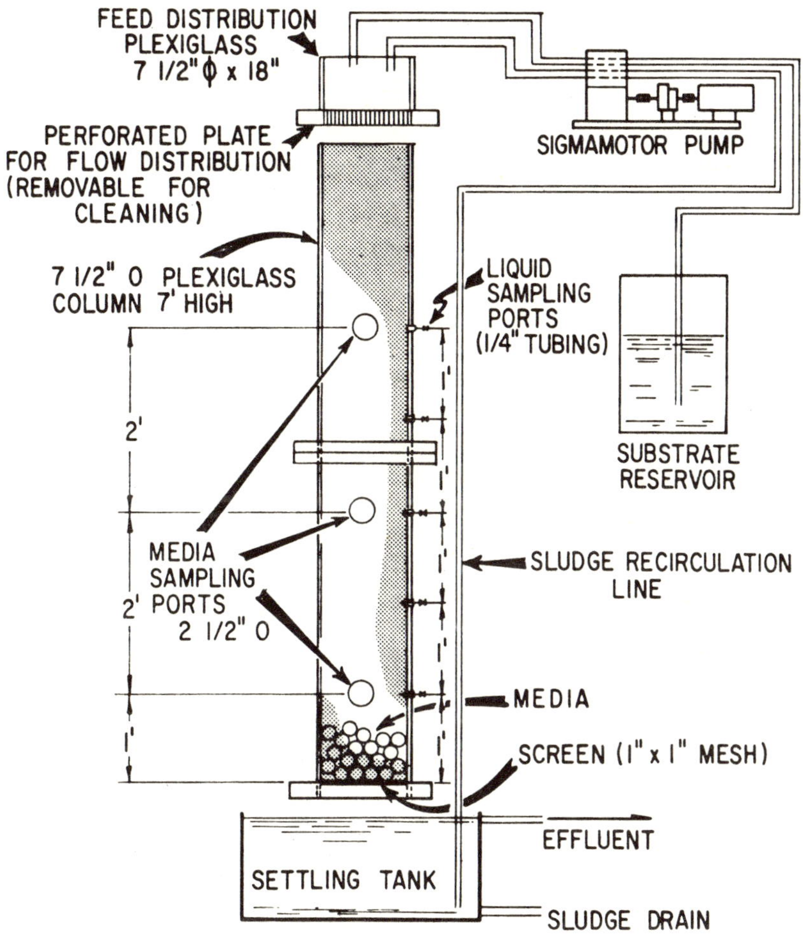

Source: EPA Report 12020 EID 03/71

The characteristics of the media should be such that it provides not only a maximum contact surface area of active biological mass, but also allows maximum wastewater-slime contact time without filter plugging. Good ventilation throughout the filter bed is also required. Design considerations and process variables involved are:

1. filter depth
2. filter media
3. hydraulic and organic loading
4. wastewater characteristics
5. environmental factors (pH, temperature, etc.)
6. recycle pattern
7. nutrient requirements

Trickling filters can be evaluated in the laboratory using a filter model as shown in Figure 67. By varying the hydraulic load, filter depth, and recycle ratio, the parameters necessary for developing the predictor equations can be measured, and the constants which define the substrate removal and filter media can be estimated.

Aerated Lagoons

Aerated lagoons are biological flow-through basins generally applicable when a high quality effluent in terms of suspended solids or organic concentration is not required. Aerated lagoons are more sensitive to temperature changes and biological upsets than activated sludge systems, but serve as good intermediate treatment processes.

Oxygen in aerated lagoons is supplied by diffused air systems or mechanical aeration devices, but the power level of the system is normally insufficient to keep all the biological solids in suspension. Solids deposition with subsequent anaerobic digestion is therefore common in aerated lagoon systems.

These systems may be considered as an interim treatment step in some cases because their organic removal capacity can be enhanced by converting to an activated sludge or extended aeration system by adding clarifiers, sludge return pumps, and additional aeration equipment. Design considerations and process variables involved are:

1. organic loading
2. environmental factors (pH, temperature, etc.)
3. nature of wastewater and fluctuations in characteristics
4. level of mixing in aeration basin
5. nutritional requirements.
6. basin depth and geometry
7. wind, heat transfer coefficient, humidity effects

Aerated lagoons can be evaluated in the laboratory using a continuous reactor similar to that for activated sludge, but providing a strict flow-through system with no facilities for quiescent sludge settling. Aerated lagoons can be simulated on a pilot plant scale by using lined earthen tanks with small mechanical aeration equipment to supply oxygen and partial mixing.

Waste Stabilization Ponds

Such ponds employ biological processes which utilize bacteria and algae to fulfill the process treatment requirements. The available organic materials in the incoming waste are oxidized biochemically while algae, utilizing the simpler bacterial degradation products and sunlight, produce oxygen which is subsequently used by facultative and aerobic microorganisms.

If a high organic loading to a pond exerts an oxygen demand above that provided by photosynthesis and surface reaeration, anaerobic or facultative bacteria will predominate. When this occurs, alcohols, various forms of organic acids, and other miscellaneous degradation products of anaerobic degradation will be produced.

Many complex organic compounds not degraded in the conventional activated sludge process or aerated lagoons can be degraded in a waste stabilization pond because of the prolonged detention times. Therefore, it is feasible in many cases to use an aerated lagoon or activated sludge-waste stabilization pond system, especially where complex petrochemical or petroleum wastewaters are involved. Design considerations and process variables involved are:

1. nature of wastewater organic and inorganic constituents
2. flow, detention time
3. surface loading
4. geometry and depth of pond
5. environmental conditions (pH, temperature, etc.)
6. geographical location
7. color, toxic materials, solids, etc., of wastewater

FIGURE 68: WASTE STABILIZATION POND MODELS

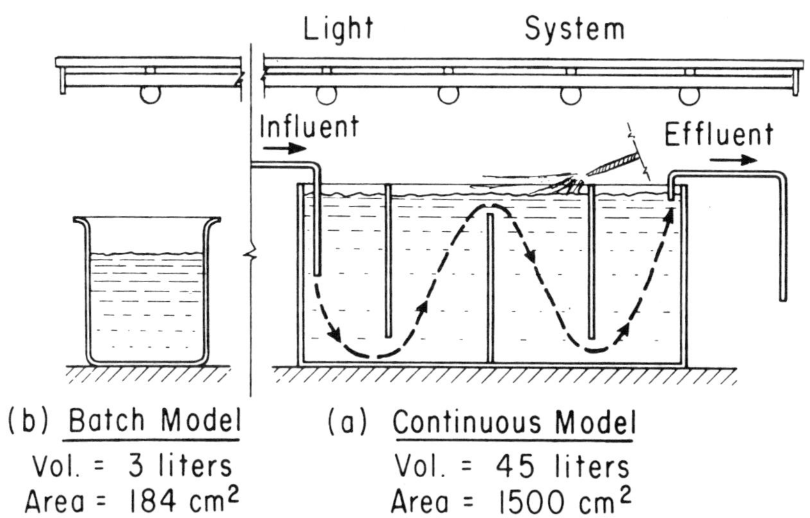

Source: EPA Report 12020 EID 03/71

Waste stabilization ponds can be simulated in the laboratory using models and an artificial lighting system as shown in Figure 68. Both flourescent and incandescent lighting should be provided with an intensity of approximately 600 to 800 foot candles. A compressed air source should be affixed close to the liquid surface to simulate wind action. The organic removal efficiency at defined surface loadings and detention times can be estimated using such a model as well as algal toxicity factors, environmental effects, etc. Larger scale test ponds can be constructed outside for conducting pilot studies if required.

Chemical Oxidation

Wastewaters discharged from many petrochemical or refinery processes are complex and highly resistant to biological degradation. This necessitates some alternate form of secondary treatment. Chemical oxidation using ozone has proved effective for specific wastewaters.

The chemical oxygen demand (COD) of a wastewater containing unsaturated organic compounds can be reduced by ozone oxidation, and the effectiveness of this reduction in terms of ozone utilization can be defined as the number of oxygen atoms reacting with the waste constituents with the consequent reduction in COD per ozone molecule adsorbed. Depending on the extent of oxidizable or unsaturated compounds in the waste, one or all three of the oxygen atoms per ozone molecule can remain chemically attached to the waste constituent during ozonation. Design considerations and process variables involved are:

1. nature of the organic materials (amount of unsaturates)
2. the pH of the wastewater
3. voltage and rate of oxygen flow-through
4. the ozone generator
5. type of ozone distribution system

The amenability of wastewater to ozone oxidation can be evaluated using bench scale apparatus shown in Figure 69. The ozone-oxygen gas is bubbled through the wastewater sample contained in the reaction flask. The ozone utilization and the corresponding COD reduction then can be evaluated in terms of an oxygenation factor:

$$\frac{\text{lbs COD removed}}{\text{lbs ozone used}}$$

A process developed by *T.H. Williams and N.R. Hall; U.S. Patent 2,795,544; June 11, 1957; assigned to The Permutit Company Limited, England, Stewarts & Lloyds Limited, Scotland and Woodall-Duckham Construction Company Limited, England* involves the chemical oxidation treatment of effluent liquors produced in processes in which coal is carbonized. Such a liquor contains in solution various phenols and thiocyanate and thiosulfate which must be removed before the effluent can be discharged.

The liquor, as produced, contains both free and fixed ammonia, and it is commonly passed successively through a free ammonia still and a fixed ammonia still, in the latter of which the liquor is distilled with lime, these stills being either integral with or separate from one another. A coke oven liquor is typical of the liquors in question, and when it emerges from the fixed ammonia still

FIGURE 69: LABORATORY SCALE OZONE TREATMENT

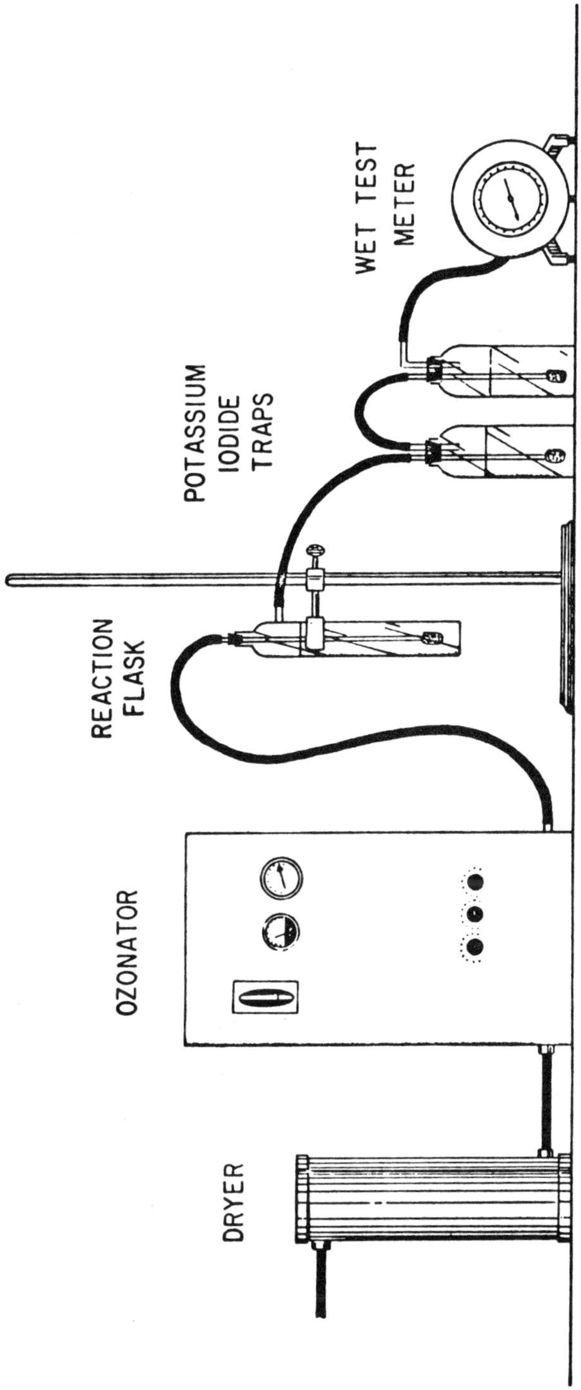

Source: EPA Report 12020 EID 03/71

it contains an excess of lime in solution and may also contain lime in suspension, so that it has a pH of 9 or over.

One process for removing thiocyanate and thiosulfate from a coke oven or similar liquor after it has passed through stills in which both free and fixed ammonia are removed comprises passing the liquor through beds of ion-exchange material, and this treatment is found to effect almost complete removal of these obnoxious compounds. The phenols are subsequently removed by adsorption on activated carbon and form a valuable by-product of the whole process. The treatment with ion-exchange materials is, however, expensive.

It is known that thiocyanate, thiosulfate and phenols are all oxidized by ozone, and on the basis of this knowledge it has been proposed to remove all these compounds from waste liquors by treatment with ozone. In fact, treatment of a typical coke oven liquor as it comes from the two conventional ammonia stills results in immediate oxidation of some of the phenols.

The present process is based on the discovery that the way in which oxidation with ozone proceeds depends on the pH of the liquor, and that below a critical pH there is preferential oxidation of the thiocyanate and thiosulfate, the phenols in the liquor remaining substantially unoxidized. An aqueous solution containing thiocyanate, thiosulfate and phenols and having a pH of 8.4 or less is treated with ozone so as to oxidize the thiocyanate and thiosulfate in it preferentially to the phenols. If the solution initially has a pH above 8.4, that pH is first reduced to 8.4 or less. If the solution is a liquor from a fixed ammonia still, any suspended lime is first removed. The pH is reduced to 8.4 or less.

Either ozonized air or ozonized oxygen is effective. The oxidation by either can be carried on in batch fashion or continuously. It is obviously advantageous that the purification should be continuous. Efficient oxidation can be effected if the liquor is sprayed down a tower in countercurrent to the ozone.

The ozone oxidizes the thiocyanate and thiosulfate to compounds which are either innocuous or can easily be eliminated from the liquor. The oxidation of thiocyanate and thiosulfate with ozone produces acid according to the equations:

$$2NH_4CNS + 2H_2O + 6O_3 \longrightarrow (NH_4)_2SO_4 + H_2SO_4 + 2HCN + 6O_2$$

$$(NH_4)_2S_2O_3 + H_2O + 4O_3 \longrightarrow (NH_4)_2SO_4 + H_2SO_4 + 4O_2$$

This assumes that the thiocyanate and thiosulfate are present in the form of their ammonium salts, as they will be if the liquor has passed only through a free ammonia still. If the liquor has also passed through a fixed ammonia still the salts will be calcium salts, and similar equations will apply. One product of the oxidation is sulfuric acid, which can easily be neutralized.

Another product is hydrocyanic acid, which is partly removed with the spent gas used for the treatment. If the removal is not complete, air can be blown through the liquor after the treatment to remove any remaining hydrocyanic acid. As the oxidation proceeds the pH of the liquor being treated will fall, due to the acid produced, thereby enhancing the preferential oxidation of the thiocyanate and thiosulfate. The lower the pH at the start of the treatment, the less is the risk of the phenols being oxidized, and it is preferred to reduce the initial pH to 7.0 or less. The complete process of purification after the liquor has passed through the ammonia still comprises the removal of any suspended lime which may be present, the adjustment of the pH, the preferential oxidation of the thiocyanate and thiosulfate, the removal of neutralization of products of oxidation, and the removal of the phenols by adsorption on activated carbon or otherwise.

A process developed by *W. Teske and E. Ringel; U.S. Patent 3,337,452; August 22, 1967; assigned to Farbwerke Hoechst AG, Germany* involves the purification of wastewaters from dissolved, oxidizable impurities by oxidative decomposition thereof, which process solves the problem involved in a very advantageous manner and constitutes an advance in the art of purifying wastewaters while avoiding the disadvantages of prior processes.

According to the present process, the wastewaters are treated with oxides or hydroxides of metal occurring in at least two different valence stages, the higher valence state of which has a sufficient oxidation potential with respect to the impurities to be removed, and the lower valence stage of which is capable of being reoxidized to the higher valence stage by oxidants with simultaneous precipitation in the form of a difficultly soluble deposit, at temperatures within the range of about 50°C to the boiling point of the wastewaters, preferably at temperatures exceeding about 80°C.

The oxides or hydroxides that have passed to the lower oxidation stage with reoxidation to the higher valence stage by means of a suitable oxidant are precipitated from the wastewater treated, separated therefrom in known manner and reused for the treatment of further wastewaters. The following is one specific example of the conduct of the present process.

Example: Two liters of wastewater discharged by a large petrochemical installation and containing as main impurities benzene and the homologs thereof was boiled in a flask provided with reflux condenser with 50 grams of hydrated dioxide of manganese for 20 minutes at a pH that had been adjusted to 1 to 2 by means of hydrochloric acid. The manganese that had partially passed into solution as bivalent ion was then precipitated with sodium hydroxide solution as $Mn(OH)_2$ in the alkaline range of 8 to 10 and reoxidized into hydrated dioxide of manganese by passing in air for half an hour.

The sediment of hydrated dioxide of manganese that had deposited was reused for oxidizing the impurities contained in a further part of the wastewater. The hydrated dioxide of manganese applied was used altogether 18 times for the purification of wastewater.

The determinations which were carried out in compliance with German standard methods, gave the average permanganate values, as follows:

(a) Untreated wastewater: 1,573 mg $KMnO_4$/liter of wastewater;
(b) Wastewater treated according to the process: 288 mg $KMnO_4$/liter of wastewater.

The average permanganate value thus decreased by 82%. The wastewater treated in the aforesaid manner did not contain detectable amounts of manganese.

A process developed by *E.L. Cole and H.V. Hess; U.S. Patent 3,642,620; February 15, 1972; assigned to Texaco Inc.* is a process for treating a waste stream containing water-soluble organic wastes as involving heating the stream under turbulent flow conditions to temperatures of 400° to 700°F and pressures of 300 to 3,100 psi in the presence of air or of oxygen thereby splitting off carbon dioxide. The oxidized waste stream, which has a considerably reduced chemical oxygen demand, is continuously fed to a hot contacting zone and is in heat-exchange relationship with the incoming effluent stream. The following example illustrates the effectiveness of the process in the liquid phase oxidation of a maleic anhydride plant effluent.

Example: The composition had approximately 10% dissolved maleic acid and fumaric acid. A simulated effluent composition contained 4.2 grams fumaric acid (saturated at room temperature), 55.8 grams maleic acid, and 600 grams distilled water having 9.1 weight percent dissolved organics and a COD of 78,170.

Two hundred grams of this material were charged to a 1,960 cc bomb. The bomb was pressured to 300 psig with air at room temperature, heated to 650°F for 2 hours (2,900 psig developed), cooled to atmospheric and the liquid sampled. This effluent had a COD of 17,467 for a reduction of 77.6% in this one oxidation stage. Calculations indicate that the oxygen was completely used up.

The remaining liquid in the bomb was again pressured to 300 psig and the bomb was heated to 650°F under rocking conditions for 2 hours (2,400 psig), then cooled and 149 grams liquid (1,807 mg/O_2/liter COD) recovered for an overall COD reduction in the two-stage contacting of 97.7%. An overall recovery of 186 grams of liquid was obtained. The difference between the 186 grams recovered and the 200 grams charged can be accounted for by the oxidation of the maleic and fumaric acids to produce CO_2 and water.

A process developed by *P. Urban and R.H. Rosenwald; U.S. Patent 3,672,835; June 27, 1972; assigned to Universal Oil Products Company* is one in which an aqueous waste stream containing NH_4HS, which is typically produced in a process for converting a hydrocarbon charge stock containing sulfurous and nitrogenous contaminants, is treated to produce elemental sulfur and a treated water stream suitable for recycle to the hydrocarbon conversion process.

This process comprises the steps of (a) catalytically treating the aqueous waste stream with oxygen at oxidizing conditions effective to produce an effluent stream containing NH_4OH, $(NH_4)_2S_2O_3$ and elemental sulfur or ammonium polysulfide; (b) separating sulfur from the effluent stream from step (a) to produce an aqueous effluent stream containing $(NH_4)_2S_2O_3$; and (c) treating the aqueous stream from step (b) with carbon monoxide at reduction conditions effective to form a substantially thiosulfate-free treated water stream.

A key feature of the treatment method is the use of a carbon monoxide reduction step to enable the continuous recycle of the treated water stream back to the hydrocarbon conversion process with consequential abatement of water pollution problems and substantial reduction of requirements for make-up water.

A process developed by *J.-P. Zumbrunn and F. Crommelynck; U.S. Patent 3,711,402; January 16, 1973; assigned to L'Air Liquide, Societe Anonyme pour*

L'Etude et L'Exploitation Des Procedes Georges Claude, France is a process for the complete purification of industrial effluent polluted by phenolated impurities wherein the degradation of the pollutant is obtained by the action of an oxidizing reagent which is a member of the group constituted by monopersulfuric acid and its salts. The oxidizing reagent is an aqueous solution containing HSO_5^- anion and containing furthermore NH_4^+, HSO_4^- ions and, optionally, alkaline cations or hydrogen peroxide.

A process developed by *D.E. McCoy, R.M. McEachern and R.M. Dille; U.S. Patent 3,761,409; September 25, 1973; assigned to Texaco Inc.* is a continuous process for the liquid phase, noncatalytic, air oxidation of sour water in which the sulfur compounds in the water are converted to nonpolluting, nonoxygen demanding compounds which involves heating the water and oxidizing the heated water with countercurrent or cocurrent flow of the water and oxidizing medium. The process rapidly converts objectionable sulfides and intermediate sulfur compounds such as thiosulfates, tetrathionates, polythionates, sulfites and polysulfides to sulfate.

Nitrification-Denitrification

The two-stage conversion of ammonia to nitrate by microorganisms is well known. This nitrification process is influenced by the substrate or carbon loading, the temperature, and the pH. As these and other variables are critical with respect to the oxidation reactions, the design of nitrification tanks must incorporate these factors.

The removal of nitrogen from nitrified effluents can be effected using biological denitrification methods. This is an anaerobic process and is likewise sensitive to many environmental stresses. It is the opinion of many investigators that a series nitrification-denitrification system is a practical method for nitrogen removal. However, when comparing this method of nitrogen removal to alternate systems such as ammonia stripping or ion exchange, it is necessary to consider all the variables which affect the efficiency of the process, the operational problems involved, the quality of the effluent required with respect to total nitrogen, and the process economics. Design considerations and process variables involved are:

1. organic loading
2. environmental factors (pH, temperature, etc.)
3. mixed liquor dissolved oxygen
4. concentration and form of influent nitrogen
5. sludge age
6. aerobic detention time (nitrification) and anaerobic detention time (denitrification)

Bench scale studies involve the use of continuous reactors and trickling filters to obtain nitrification, varying influent carbon, nitrogen, and altering environmental conditions in order to determine the effect of process variables on nitrification rates and capacities. Pilot scale use of filters and aeration tanks have been similarly applied.

The aspects of biological denitrification can be assayed using continuous bench or pilot reactors with mechanical mixing devices. The mixing levels should be sufficient to maintain a relatively homogeneous mixture of biological solids and substrate but not capable of introducing oxygen to the liquid. Again, process variables such as influent nitrate and carbon concentration, mixed liquor solids, etc., are controlled in order to evaluate their effects.

TERTIARY

The treatment of wastewaters by conventional primary and secondary processes does not yield an effluent of sufficient quality to meet the necessary criteria in many instances. Tertiary or advanced wastewater treatment processes consequently are required to remove these residual or refractory contaminants.

Tertiary treatment processes are utilized to remove residual suspended solids, including microorganisms, dissolved solids, complex organics, and other pollutants such as nitrogen and phosphorus compounds, and ABS.

Chemical Precipitation

Chemical coagulation-precipitation can be employed either as a primary treatment or tertiary treatment step, depending on the contaminants present and the desired level of removal. Lime precipitation as a tertiary process for the removal of phosphorus and organic carbon is considered in this section.

The lime precipitation system may be designed either as a two-stage lime process with lime addition to pH 11.5 in the first stage followed by recarbonation to pH 9.5 to 10 with flocculent additives in the second stage, or as a single-stage lime process at pH 11.5 with sodium carbonate added to reduce excess calcium ions. The application of this system in polishing effluents from a secondary plant treating petrochemical or refinery wastewaters would have to be verified either by bench scale or pilot scale studies as the multiplicity of contaminants present in such wastewaters could interfere with the coagulation-precipitation reactions. Design considerations and process variables involved are:

1. flow
2. concentration and nature of the contaminants
3. concentration of coagulant
4. detention time of mixing and clarification zones
5. dissolved oxygen and carbon dioxide concentrations
6. chemical purity of coagulant
7. temperature and pH of the liquid mixture

Bench scale tests using a batch jar test procedure can be used, determining the optimum dosage and pH by varying parameters and monitoring the contaminant removal for each condition. A continuous pilot plant simulation, although more costly, would yield more dependable data. The range of operating temperatures, coagulant dosages, and pH values, with respect to contaminant removal, however, can be determined in bench scale jar tests, using this information to establish pilot plant operating conditions.

A process developed by *A. Renner; U.S. Patent 3,716,483; February 13, 1973; assigned to Ciba-Geigy, Switzerland* is a process for removing dissolved, emulsified or suspended organic substances, for example fat, resin, oil or dyestuff, from water. The contaminated water is brought into contact with a highly disperse, solid, water-insoluble organic polymer, for example melamine-formaldehyde resin or polyacrylonitrile of average molecular weight greater than 1,000 and a specific surface area greater than 5 m^2/g, and the polymer charged with the contamination is separated from the water.

Gas Stripping

Stripping essentially is a form of distillation wherein a small amount of a relatively volatile material is removed from a large volume of less volatile material. In refinery waste disposal, the process is generally used for removing small amounts of volatile impurities, e.g., hydrogen sulfide, ammonia and cyanide from large volumes of wastewaters. The method is not suitable for removing materials of low volatility, such as phenolics, because heating and equipment requirements make it uneconomical. The most frequent use of stripping in refinery waste disposal is for removal of hydrogen sulfide and ammonia from foul waters.

Gas stripping may be considered tertiary treatment or pretreatment. For example, in the refinery industry, it is considered part of pretreatment whereas in domestic wastewater treatment it may be considered tertiary treatment. Stripping is a modification of the aeration process. The process may be utilized to remove both ammonia and hydrogen sulfide under proper conditions of temperature and pH. Air requirements for ammonia removal may be decreased by the inclusion of a preconcentration step utilizing zeolyte absorbents. Lime is normally used as a hydroxide source in the ammonia removal process.

When the removal of hydrogen sulfide is desirable, a stripper can be applied. An acidic condition is necessary for the efficient stripping of sulfides. Ammonia stripping necessitates alkaline conditions as none of the ammonia is in the gaseous state below pH 7.

Gas stripping is generally conducted in a packed tray tower equipped with air blower. The towers usually are designed so that the gas and liquid leaving a desorption plant are in contact with gas containing little or no vapor absorbed from the liquid. The wastewater is pumped to the top of the packed tower where it is distributed uniformly over the packing through which a steady stream of air or gas is drawn. This process may also be categorized as primary treatment in certain instances. Design considerations and process variables involved are:

1. waste flow
2. contaminant concentration
3. hydrogen ion concentration of wastewater
4. wastewater temperature
5. aeration transfer equipment
6. contact time
7. waste loading
8. air flow rate
9. nature and size of packing material
10. diameter of column
11. height of column
12. operating pressure and temperature in column

FIGURE 70: BENCH SCALE GAS STRIPPING TOWER

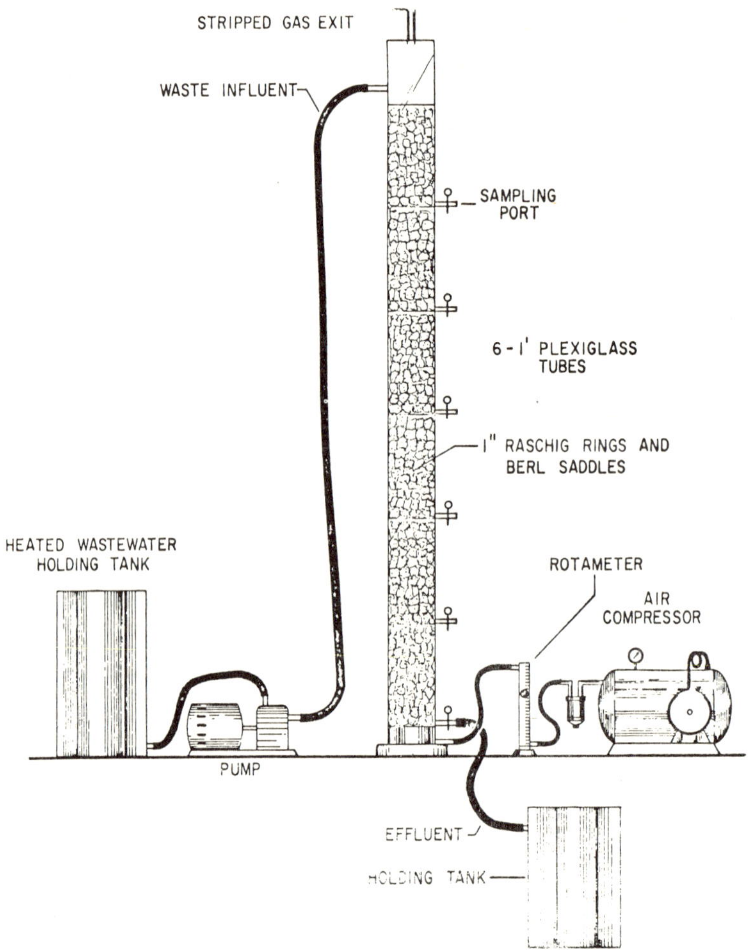

Source: EPA Report 12020 EID 03/71

The first step in the design of a stripping column is to fix its internal diameter. A permissible average gas rate of 400 cubic feet per gallon may be assumed. For packed columns this permissible rate is determined by flooding conditions for a given plate efficiency. The internal diameter of the column can be calculated directly from the permissible gas rate and the volume of waste liquid to be passed through the column. Diameters usually vary between three to six feet with the height being determined by the equilibrium conditions in the column and by the number of contacts necessary for the amount of stripping desired. The relationship between mass transfer coefficients determined in the laboratory or from pilot plant data and corresponding coefficients for full-sized towers have been studied exhaustively, as reported by R.W. Bayley in the *British Effluent and Wastewater Treatment Journal*, February-March, 1967. Provided conditions of flow in both systems are turbulent, the performance of the full-scale unit may be estimated quite rapidly from pilot plant information. A bench scale stripping tower is shown in Figure 70.

A process developed by *P. Kaunert and H. Doering; U.S. Patent 3,096,156; July 2, 1963; assigned to Gewerkschaft Auguste Victoria, Germany* involves the stripping of wastewaters from coke oven plants of hydrocyanic acid and hydrogen sulfide contained therein.

A process developed by *H. Siewers, K. Flasche, A. Stetter and S. Pfeiff; U.S. Patent 3,540,189; November 17, 1970* involves destroying ammonia contained in waters resulting from the operation of coke ovens based on the principle of expelling the gas from the water, which comprises heating the water to be treated with lean gas at an elevated temperature of such a degree that the temperature of the water after treatment thereof is at least 60°C, conveying the lean gas charged with ammonia to regenerators of industrial furnaces, and heating the same therein to such a temperature that the ammonia will become dissociated to nitrogen and hydrogen. By this treatment, ammonia and other harmful gases will be completely eliminated from the waters which are then safe for discharge into the sewers.

A process developed by *E. Feder, K. Deselaers and G. Czehovsky; U.S. Patent 3,669,847; June 13, 1972; assigned to Dynamit Nobel AG, Germany* is one in which steam-volatile organic solvents are removed from process wastewaters by intimately mixing the process wastewaters with steam to form an azeotropic steam mixture, withdrawing the azeotropic steam mixture from the resultant mixture of steam and water, and condensing the azeotropic steam mixture. The major amount of process wastewater, thus freed of solvent content, is discharged as general sewage.

A process developed by *R.D. Kent; U.S. Patent 3,754,376; August 28, 1973; assigned to Texaco, Inc.* is one in which water containing contaminating gases such as hydrogen sulfide, carbon dioxide and ammonia is freed of these contaminants by closed system stripping process which employs an inert gas and steam.

Microstraining

Microscreening, or microstraining, is a form of simple filtration. The objective is the clarification of liquids by the removal of suspended solids, especially those of microscopic size. The process involves the use of a horizontal, rotating drum-type straining element whose periphery is enveloped with perforated or mesh-type screening material. In use, a thin mat forms over the fabric screen and as the wastewater enters the drum and passes through the mesh fabric in radial directions, solids are strained from the wastewater.

The solids are retained on the inside surface of the fabric and continually backwashed to waste. The drum may be installed in a concrete or steel box. The weight or volume of particles intercepted varies with the cube of the screen apertures and therefore significant differences in efficiency are obtained by using varying mesh sizes. This process has been investigated as a potential pretreatment method preceding carbon adsorption, distillation, solvent extraction, reverse osmosis, freezing and electrodialysis. Design considerations and process variables involved are:

1. flow
2. suspended solids
3. nature and sizes of suspended solids
4. variation of suspended solid concentrations
5. filtration area and rotation speed
6. presence of oil, grease, or other floatables
7. filter fabric mesh size

A microstraining plant cannot be designed in the same manner as a rapid sand filtration plant. Rates of flow through the filtering media are much higher than through sand beds; frictional resistance allowable is much less; and the fabric is matted very rapidly by the solids intercepted. The rate of filter blockage is measured in terms of filterability, and this is designated as the filterability index. Given this index from samples of the water in its poorest condition of suspended solids applied and the grade of fabric employed, the general problem of design is to determine the dimensions of the strainers and rotation speeds required to pass the required rate of flow.

The calculation of the filterability index may be performed in many ways. The most accurate method is with a laboratory set-up that measures the head loss across a section of the filter being tested using the waste in question.

Suspended solids removals appear to be relatively unaffected by feed concentration and removal efficiencies on the order of 85 to 90% may be expected. The microscreening process may best be evaluated on a pilot scale.

Carbon Adsorption

Adsorption of molecules from the liquid phase to the solid phase occurs as the result of either electrical attraction of the solute to the adsorption surface, Van der Waals forces, chemical reactions, or possibly a combination of these phenomena. The type of adsorption associated with the removal of organics is generally caused by Van der Waals forces or physical adsorption and usually occurs in three phases: the transport of the contaminant (adsorbate) to the exterior surface of the adsorbing media (adsorbent); the diffusion of the adsorbate into the pores of the adsorbent; and the adsorption of the adsorbate on the surface of the adsorbent.

Activated carbon, carbonaceous material with certain adsorptive and catalytic properties, has an affinity for many contaminants present in petrochemical or refinery wastewaters, and contact between the carbon surfaces and the waste is achieved through fixed-bed or expanded-bed carbon columns. Once breakthrough in the column occurs (the column is in equilibrium with the wastewater), the spent carbon is either wasted or regenerated for subsequent use. Design considerations and process variables involved are:

1. flow rate
2. contact time

3. depth of carbon bed
4. adsorptive capacity of adsorbent
5. influent solute concentration
6. allowable effluent solute concentration
7. particle size
8. environmental factors (temperature, pH, etc.)

Adsorption rates and general relationships can be formulated using batch contact reactors or continuous columns. Carbon is usually added to batch systems in powdered form and removed by flocculation and sedimentation. This can be simulated in the laboratory using test flasks and a shake assembly. The weight of carbon added and the corresponding removal of organic material for varying conditions is then translated into one of the isotherm equations.

Serial continuous carbon columns usually offer the most practical system for the polishing of effluents as the separation of carbon from the wastewater after contact is not required. The concentration of the adsorbed solute is in equilibrium with the influent concentration rather than the effluent solute concentration enhancing removal, and greater flexibility of operation is possible.

A process developed by *P.X. Masciantonio; U.S. Patent 3,373,085; March 12, 1968; assigned to United States Steel Corporation* involves the purification of coke-works wastewaters and the like, and reclamation of their chemical constituents without necessity for special facilities, other than a bed of suitable adsorbent, which is used in combination with conventional by-product coke-works equipment. Although the contaminants are removed by adsorption, the adsorbent used requires no regeneration or discard and the adsorbed chemicals are recovered relatively effortlessly, utilizing usual by-product recovery practices.

Figure 71 is a block flow diagram of the present process. The adsorbent bed **1** is in functional relationship with conventional equipment of a by-product coke works, comprising coal chemical recovery units **2**, coke quencher **3**, and coke ovens **4**.

The present process contemplates use of a carbonaceous adsorbent agent in bed **1** which is a suitable coke-oven charge component, either by virtue of its coke-forming qualities or its compatibility with other coke-making charges, to produce metallurgical coke or other carbonized products of desired characteristics. Specific examples of preferred adsorbents are bituminous coal and bituminous coal treated to increase its adsorptive activity for organics such as phenols.

In the drawing, the adsorbent bed **1** is shown as receiving contaminated waters from the coal chemical recovery units **2** and the coke quencher **3**. In the further operation of the process, upon adsorption of contaminants, including the organic compounds and ammonia, the resulting purified water is separated from the adsorbent and the latter, after having spent its adsorptive capacity, is charged into the coke ovens **4**, carrying with it the adsorbed contaminants.

Upon carbonization of the charge in the coke ovens **4** the contaminants are driven off, along with the usual coal chemicals, whereupon they are available for processing in the coal-chemical recovery units **2** for recovery of useful chemical products.

FIGURE 71: PROCESS FOR PURIFYING COKE-WORKS WASTE LIQUOR

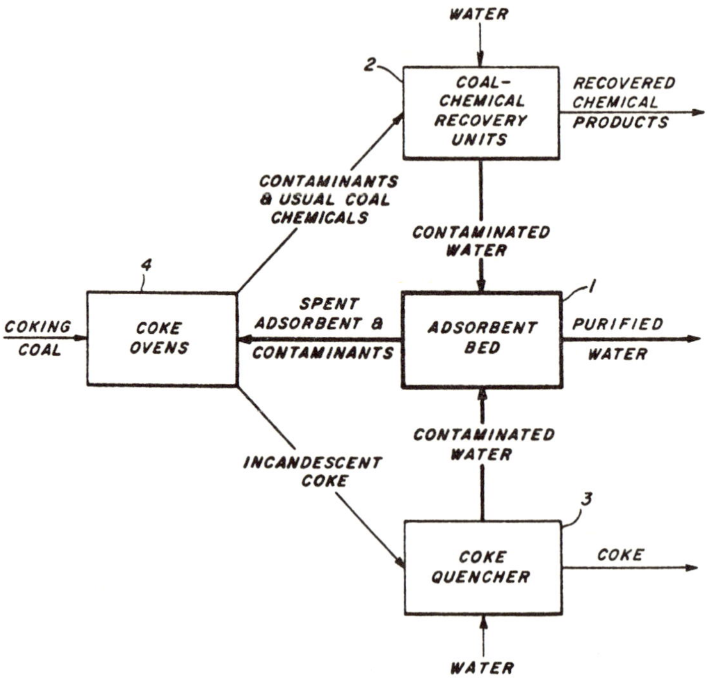

Source: P.X. Masciantonia; U.S. Patent 3,373,085; March 12, 1968

A process developed by *M.M. Mattia; U.S. Patent 3,625,886; December 7, 1971; assigned to Day & Zimmermann, Inc.* permits the adsorption recovery of organic compounds present in aqueous streams at concentrations between 50 and 5,000 parts per million. The process is continuous in that alternate adsorbers are interchangeably used. Economic steam desorption of high-boiling compounds is permitted by use of a thermocompressor to recycle regenerating steam through the adsorbers. Steam condensing in a separator is kept to a minimum in order to obtain the highest possible organic content in the condensate.

Low concentrations of relatively high boiling point compounds in effluent streams constitute one of our major industrial water pollution problems. These compounds frequently prevent reuse of water because they contribute to product contamination and/or lower operating efficiencies. The compounds are not volatile enough to recover by distillation and are too low in concentration to permit economic recovery by evaporation.

Typical examples of such compounds which may be recovered by the present process include ethylene glycol from polyester plant effluent streams, phenol from catalytic cracking operations, phenol in effluent streams from phenolic resin plants, dimethylformamide in wash waters from polyurethane processes, etc.

The chemical nature of the organic compound is not significant, rather only its boiling point, its solubility in water to the extent of 50 parts per million by weight, and that a mixture of the organic compound and water must not form an azeotrope having a boiling point of less than 212°F. By solubility as used herein is meant both true solubility where the organic compound is dissolved in water, and the formation of stable emulsions between the organic compound and water.

Hence, the present process is applicable to a wide variety of liquid organic compounds including mixtures thereof such as aniline, dimethyl sulfoxide, ethanolamine, amyl acetate, dimethyl acetamide, 1,3-dioxane, and caprolactam.

Electrodialysis

This process is based on the migration of positively charged ions and negatively charged ions to respective oppositely charged electrodes and on the ability of semipermeable membranes to selectively permit the passing of ions of one type of charge while repelling ions of the opposite charge. Electrodialysis is specfically for ions and the membranes are essentially impermeable to water. The electrodialysis is a form of tertiary treatment and must be preceded by additional tertiary treatment including filtration. Nonionized molecules and suspended solids are not removed during the process, and, if present, will cause membrane fouling. Sulfuric acid is used to abate membrane fouling. The apparatus is arranged so that concentrated and diluted streams flow continuously.

Only partial demineralization by this method is practical, because electrical power requirements become excessive if the ion concentration is reduced too far. Power consumption is also a function of the amount of ions that must be removed. Single-pass electrodialysis will reduce the total dissolved solids concentration in water by 40 to 50%. Design considerations and process variables involved are:

1. membrane
   a. size
   b. area utilization
   c. current density
   d. thickness
   e. resistance
   f. bursting strength
   g. liquid velocity
   h. space between membranes
   i. headloss through stack
2. stack (cell) design
   a. flow
   b. ionic characteristics of wastewater
   c. influent and effluent ionic concentration
   d. pH and stability of feed water

The removal of inorganics may be defined by Faraday's Law. The determination of electrical resistance and energy consumption requires the knowledge of certain constants related to the particular equipment used, the water, and temperature. These design parameters may be determined through use of pilot plant studies. Such an electrodialysis pilot unit is shown in Figure 72.

FIGURE 72: LABORATORY ELECTRODIALYSIS UNIT

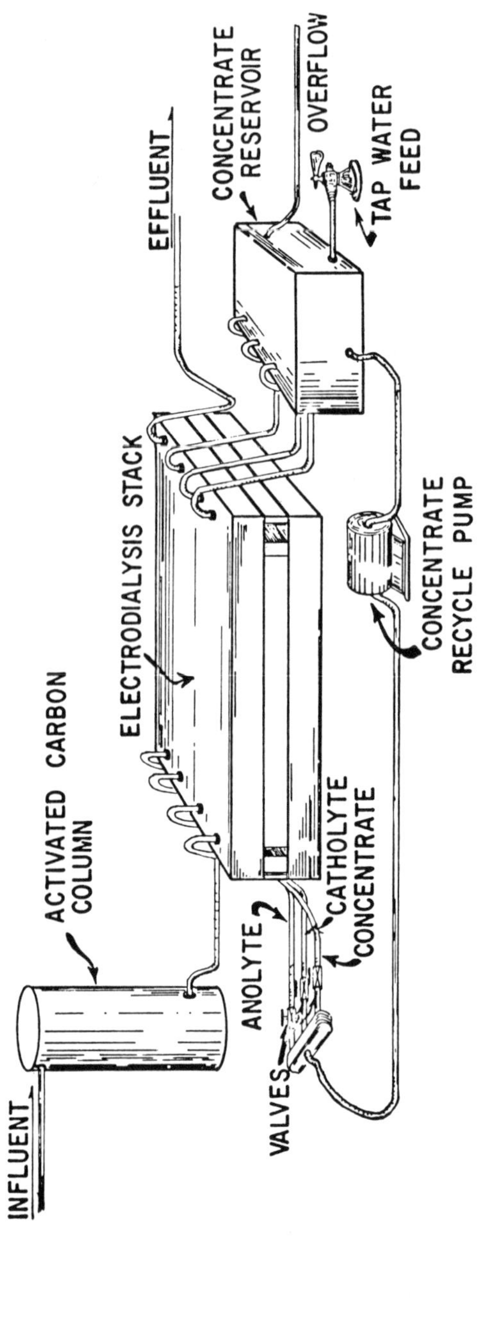

Source: EPA Report 12020 EID 03/71

A process developed by *F.P. Chlanda, H.P. Gregor and K.-J. Liu; U.S. Patent 3,752,749; August 14, 1973; assigned to Allied Chemical Corporation* involves the electrodialytic concentration and removal of acids from aqueous effluents, e.g., malic acid effluent containing less than about 10% malic acid, to produce two streams, a relatively concentrated aqueous concentrate stream of about 30% acid and a relatively dilute stream of less than about 0.3% acid. The process, continuous, semicontinuous or batch offers a practical solution from the standpoint of meeting pollution control standards and for separating or recovering acids which may be returned to the process by direct recycling.

In various manufacturing processes, relatively large amounts of organic acids are discharged in the effluent. While in the past such discharge has been economically tolerable, when the added expense of depolluting such effluents is added to the manufacturing costs, concentration and recycling of such acids become an important economic factor.

For example, in the manufacture of malic acid, large quantities of residual malic acid are discharged in the effluent to the extent that a moderately sized plant has an effluent having an average volume of about 10,000 gallons a day and containing on the order of an average percentage of about 1.3 to 6.0% malic acid and about 0.1 to 0.3% maleic acid and 0.1 to 0.5% fumaric acid, together with lesser amounts, e.g., 10 to 20 ppm chlorine as sodium chloride and iron in amounts of 8 to 19 ppm.

Of course, acid streams with greater acid concentrations may also be treated, e.g., streams having an average concentration of malic acid from about 1.0% to about 10.0% and maleic acid in similar amounts, i.e., of 1.0% to about 10.0%. Fumaric acid may be present in amounts of maximum solubility at room temperature, i.e., about 0.5%.

An effluent of this kind is a substantial pollutant to the extent that its biological oxygen demand by federal regulation must be lowered by 85% in order to comply with acceptable standards. Apart from such regulations, however, by concentrating the acid content of the stream in a plant of the above size to about 30%, a recovery of nearly one million pounds of malic acid per year would result in the overall process.

While such effluents may be processed in a conventional manner such as by use of lime and ponding in an effort to avoid stream pollution, the cost of such treatment is high, i.e., it is estimated to be more than $100,000 a year for a 10,000 gallon/day effluent size plant, and no acid recovery would result from such disposition.

Other treatments such as ion-exchange methods, reverse osmosis, and evaporative concentration have been suggested for treatment of the effluents; however, none of these methods have appeared suitable because of the excessive costs involved and/or because the conditions surrounding the disposition of the effluent were unsatisfactory.

Ion Exchange

Ion exchange is an exchange adsorption process in which ions associated with the solid adsorbent are exchanged for ions in solution and is used to remove undesirable anions and cations from wastewater. Most ion exchange resins presently used are synthetic resin materials consisting of a network of hydrocarbon radicals to which are attached soluble ionic functional groups. The total number of functional groups per unit weight of resin determine the exchange capacity and the group type determines the exchange equilibrium and the ion selectivity.

Ion exchange resins may be strong or weak acidic cation exchangers or strong or weak basic anion exchangers. Generally, hydroxyl ions are exchanged for the anions removed from solution while hydrogen or sodium ions are replaced by the cations which are taken up by the exchange media. Design considerations and process variables involved are:

1. applied feed rate per unit volume of resin
2. exchanger depth
3. backwash expansion
4. flow rate per unit area of resin
5. regenerate flow rate
6. rinse water flow rate
7. regenerate concentration
8. nature of resin

FIGURE 73: LABORATORY ION EXCHANGE COLUMNS

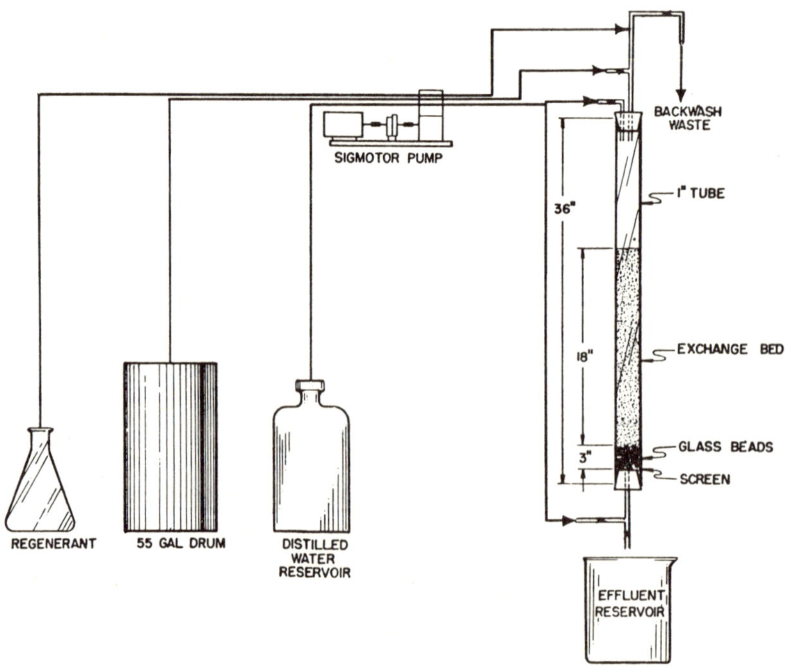

Source: EPA Report 12020 EID 03/71

An ion exchange column as shown in Figure 73 can be used to develop data necessary for design of prototype ion exchange units. Based on the observed number of ion equivalents removed per unit time, the resin volume requirement can be estimated. Additionally, the rinse water and regenerate requirements can be predicted.

Evaporation

Evaporation is a process by which liquid is changed to vapor. Although molecules of water are continuously leaving the water surface, others are returning, and the rate of evaporation is determined by the net accumulation of those leaving the water surface. If the temperature of the surface is to be maintained, large quantities of heat must be supplied by radiation and conduction from the overlying air or at the expense of any heat stored below the surface.

Evaporation may be accomplished in special equipment designed for this specific purpose. However, the process is relatively expensive and is limited primarily to the recovery of volatile by-products from waste liquids and to the treatment of wastewaters where no alternate methods are available. This discussion will be concerned with evaporation by natural environmental effects.

The evaporation process in its simplest form consists of providing shallow reservoirs and allowing the wastewater to be detained until the transformation of the liquid to vapor has been completed. Design considerations and process variables involve the following:

1. effective surface area
2. temperature of the water and of the air
3. the movement of air above the water surface
4. the relative humidity
5. nature of the wastewater
6. depth of the reservoir
7. salinity of the wastewater
8. concentration of surfactants or other evaporation suppressants in the wastewater
9. precipitation
10. geographic location

Basic design criteria can be established by use of evaporation pans. The standard Weather Bureau Class A Pan is the most widely used in the United States. The pan is filled to a depth of 8 inches and the water surface level is measured daily for evaporation determinations. Other measuring equipment should be available at the pan site for meteorological determinations and evaluation of other factors contributing to evaporation.

Heretofore, severe atmospheric corrosion of structures, buildings, and process equipment has occurred at coke processing plants for many years. The problem has recently become more critical as a result of the installation of additional chemical processing facilities thus placing a high concentration of complex process equipment in an area exposed to this atmospheric corrosion. A major cause of this problem is the use of aqueous waste streams to quench the coke. These waste streams contain dissolved inorganic salts that become air-borne with the steam produced in the quenching operation, and subsequently upon cooling, rain out of the atmosphere on adjacent buildings and equipment. Unfortunately, these salts are very corrosive to structures, buildings, and process equipment that are in the path of this salt fallout. This situation has been tolerated because no economically attractive alternative method of disposing of these waste streams was available.

Disposal of the aqueous waste streams in a deep well has been considered and is being practiced in some coke plants. However, deep well disposal at some coke processing plants requires a disposal well of about 19,000 feet; and the probability of such a well being able to handle the large volume of water has been reported to be fifty percent. Further, a danger exists that underground streams will be contaminated by this method waste disposal.

A forced evaporation process has therefore been developed by *W.J. Didycz and D. Glassman; U.S. Patent 3,790,448; February 5, 1974; assigned to United States Steel Corporation* for purifying wastewater containing corrosive salts produced in a coke producing apparatus and normally utilized to quench coke in a coke quenching apparatus, with waste steam from the coke quenching apparatus to produce substantially salt-free condensate.

The process consists of:

1. heating the waste fluid in heat exchange relationship with the steam to a first predetermined temperature at a first predetermined pressure;
2. condensing the high temperature steam to form substantially salt-free condensate;
3. vaporizing the heated waste fluid at the first predetermined temperature at the first predetermined pressure to vaporize a portion of the waste fluid and to concentrate the corrosive salts to form a concentrate, the first predetermined pressure being lower than the pressure of the waste steam from the coke quenching apparatus;
4. condensing the heated vaporized portion of the waste fluid to form substantially salt-free condensate; and
5. utilizing the substantially salt-free condensate in an industrial application, thereby eliminating the disposal of the waste fluid and the contamination of the atmosphere with the waste fluid if the waste fluid were employed to quench the coke in the coke quenching apparatus, and utilizing the waste steam for a useful purpose.

Reverse Osmosis

A natural phenomenon known as osmosis occurs when solutions of two different concentrations are separated by a semipermeable membrane. With such an arrangement water tends to pass through the membrane from the more dilute side to the more concentrated side. The driving force that imparts this flow through the membrane is related to the osmotic pressure of the system, and this pressure is proportional to the difference in concentration between the two solutions.

When pressure is applied to the more concentrated solution, the flow of water can be retarded or even reversed, i.e., water moves from the more concentrated compartment to the less concentrated compartment. The properties of a membrane that permit water molecules to pass through but impede the flow of contaminant molecules are not clearly understood, but membranes composed of modified cellulose acetate prepared with various inorganic additives exhibit this separability. The reverse osmosis process provides an affirmative barrier to viruses, inorganics and refractory organics, and removes unionized as well as ionized materials.

Reverse osmosis, which is a form of tertiary treatment, should be preceded by additional treatment including filtration. Secondary effluent is pumped under a pressure exceeding the osmotic pressure into a high pressure vessel separated from a collection vessel by a semipermeable membrane. The product water is collected on the low pressure side of the membrane, usually at atmospheric pressure, and the concentrated brine from the high pressure vessel is discharged to waste. Synthetic osmotic membranes are available which can reduce the total dissolved solids concentrations of saline feed waters from 50,000 mg/l to 500 mg/l. Design considerations and process variables involved are:

1. water flux (gallons/square feet/day through the membrane)
2. membrane area, square feet/cubic feet
3. salt rejection, percent
4. operating pressure, psi
5. flow
6. total dissolved solids concentration of feed
7. acidity and fouling characteristics of waste
8. salt removal efficiency

Laboratory systems data are only partially representative of the performance of larger reverse osmotic systems. Membrane surface area is smaller in laboratory systems and less likely to contain imperfections. Flow regimes and patterns in the laboratory desalination cell may differ from those present in larger systems.

Rejection increases with pressure and therefore comparisons between laboratory and pilot plant performance may be misleading if the operating pressures are not considered. Process requirements may readily be determined from pilot plant operation and with care, laboratory studies may be used to evaluate the feasibility for such operations. Four types of reverse osmotic units have been tested: plate and frame, spiral, tubular, and hollow fine fiber. A tubular laboratory unit is shown in Figure 74.

A reverse osmosis process for the recovery of glycerin from a petrochemical waste stream containing inorganics and polyglycerins has been studied in detail with the results applied to the design of an effective process scale unit as reported by E. Klein, S.V. Desai, J.K. Smith and R.E.C. Weaver in EPA Report 12020 DQC 03/71 entitled "Polymeric Materials for Treatment and Recovery of Petrochemical Wastes", Washington, D.C., U.S. Government Printing Office (March, 1971).

Membranes employed were asymmetric cellulose acetate butyrate and cellulose acetate. The pilot scale experimental studies were performed with tubular membrane modules which readily accommodated the sample plant stream being studied. Good separation was achieved operating between 600 and 800 psig for best selectivity. The product throughput rate appeared the limiting consideration and proved sensitive to increased turbulence and reduced feed viscosities,

FIGURE 74: LABORATORY TUBULAR REVERSE OSMOSIS UNIT

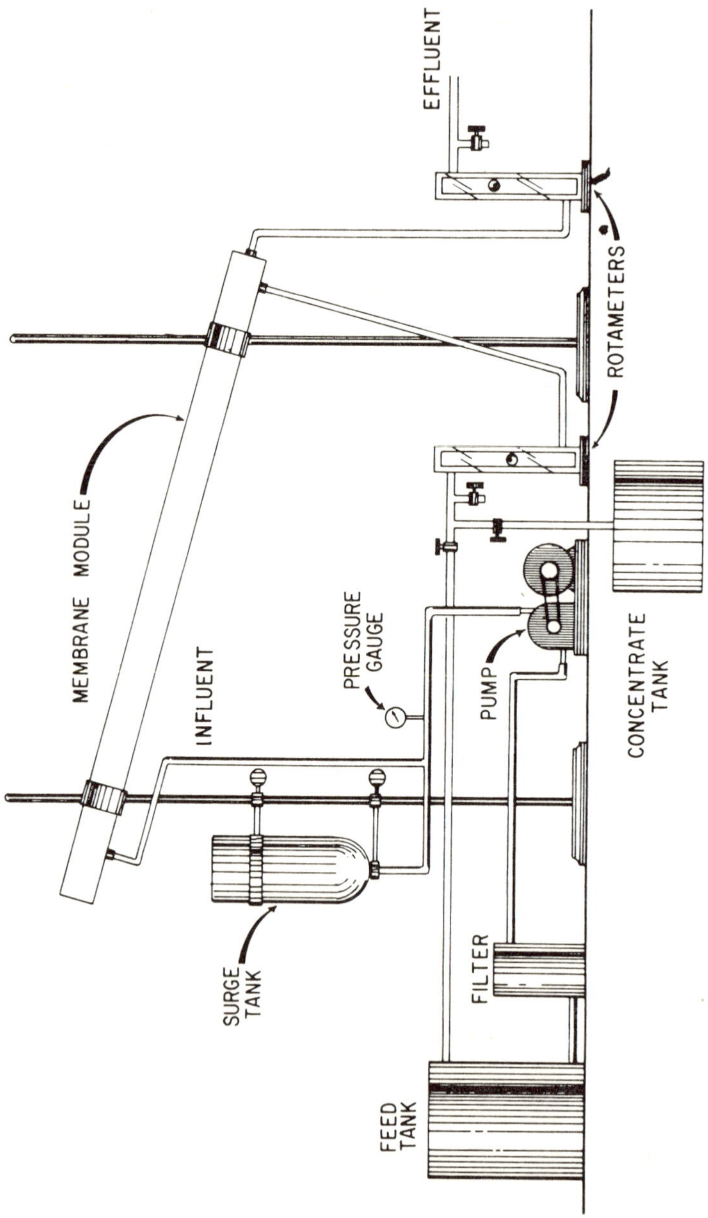

Source: EPA Report 12020 EID 03/71

the latter achieved by dilution. The pilot unit data were used to design a countercurrent multistage battery to achieve even closer separations. It is shown that sufficient glycerin could be recovered to provide an attractive return on the required investment.

Chlorination

Chlorination has long been considered to have the greatest practical potential of all disinfection systems for freeing wastewaters of pathogens. Other agents have been proposed for disinfection such as bromine and iodine. Ozone has been effective for disinfection of water supplies and disinfection may also be accomplished by the use of electromagnetic radiation.

Although chlorine is primarily used as a disinfectant, its use may also be considered a form of tertiary treatment in that residual organics are oxidized by free chlorine. As a result of the strong oxidizing power of chlorine, this element will destroy or inhibit the growth of bacteria and algae. Metal ions which are in their reduced state react with chlorine and are also oxidized.

The type and amount of chlorine compound in solution is pH and temperature dependent. Initially, chlorine added to wastewater is reduced by compounds which react rapidly with the chlorine. Finally, when all chlorine-reducing substances are oxidized in the wastewater, the additional quantity of chlorine added to the water results in an equivalent residual chlorine. The residual chlorine is the form which is necessary for effective disinfection.

Chlorination may be accomplished by the addition of gaseous chlorine or by the addition of calcium or sodium hypochlorite. Disinfection is time dependent and it is normally necessary to provide a detention period of 20 to 30 minutes to provide effective disinfection and also to account for short-circuiting in the detention basin. Disinfection of a wastewater effluent is indicated by a chlorine residual which may be detected in the effluent after sufficient contact time. Design considerations and process variables involved are:

1. flow
2. hydrogen-ion concentration of the wastewater (pH)
3. extent of chlorine reducing compounds in the wastewater
4. form and concentrations of nitrogen compounds in the wastewater
5. chlorine demand
6. chlorine feed concentrations
7. contact time
8. temperature
9. nature and number of organisms in the water

Basic design criteria for disinfection by chlorination can be established through laboratory techniques. The objective of the laboratory procedure is to relate contact time, chlorine dosage, and chlorine residual to effective disinfection. The coliform family of bacteria is normally used as an indicator of contamination and their degree of removal may be related to disinfection efficiency.

The contact time determined from the laboratory procedure is normally increased to account for short-circuiting and other difficulties encountered in actual plant operation. Wastewater treatment facilities normally require between 3 and 15 mg/l of chlorine for disinfection, and it is customary to carry a 1 mg/l residual measure in the effluent after a 20 minute chlorine contact period. The effects of chlorination on fourteen selected organic chemicals remaining in effluents from biological treatment have been described by E.L. Barnhart and G.R. Campbell of Hydroscience, Inc. of Westwood, New Jersey. The results are presented in EPA Report 12020 EXG 03/72, Washington, D.C., U.S. Government Printing Office (March, 1972).

Rapid Sand Filtration

Rapid sand filtration can be used effectively to polish secondary wastewater effluents. Rapid sand filtration may be accomplished by gravity and under pressure, but gravity sand filters handle large flows more economically.

The conventional rapid sand filter consists of 18 inches of gravel overlain by 30 inches or less of sand. Natural silica sand is a common filtering medium, but other fine grained filter materials such as crushed anthracite have been used. The rapid sand filter is normally rectangular in shape and reinforced concrete construction is commonly used for housing the filter beds. The filter beds are underlain with a piped lateral collection system and the installation is valved to permit backwashing and to control the application of wastewater to the beds.

Conventional filters are operated at hydraulic loadings in the range of two to four gpm/ft^2. Rapid sand filtration has been enhanced by the use of multimedia filters incorporating, for example, sand-coal-garnet beds and hydraulic loadings of four to eight gpm/ft^2 are commonly used in these multimedia beds. Design considerations and process variables involved are:

1. flow rate
2. media grain size
3. filter depth
4. properties of suspended solids
5. temperature
6. use of chemical treatment and filter aids
7. filter hydraulic loading, gpm/ft^2
8. allowable headloss to the bed

Rapid sand filtration is affected by a large number of variables and at the present time, there is no accepted mathematical model that correlates all the variables affecting the filtration phenomenon.

In general, it seems that because so many variables affect the removal of solids and sand filters, all of them cannot be included in one filtration equation. Floc size, size distribution, shape, composition and concentration, together with sand size and shape, size distribution, rate of filtration, temperature, etc., all play a part in determining the manner and extent to which suspended solid removal is accomplished in a given filter layer. Therefore, filter design still remains the art of relaying past experience to the prediction of filter performance.

The rapid sand filtration process lends itself to pilot plant study and many of the variables can be evaluated in the laboratory. Figure 75 illustrates a laboratory sized rapid sand filtration unit.

FIGURE 75: LABORATORY SAND FILTER

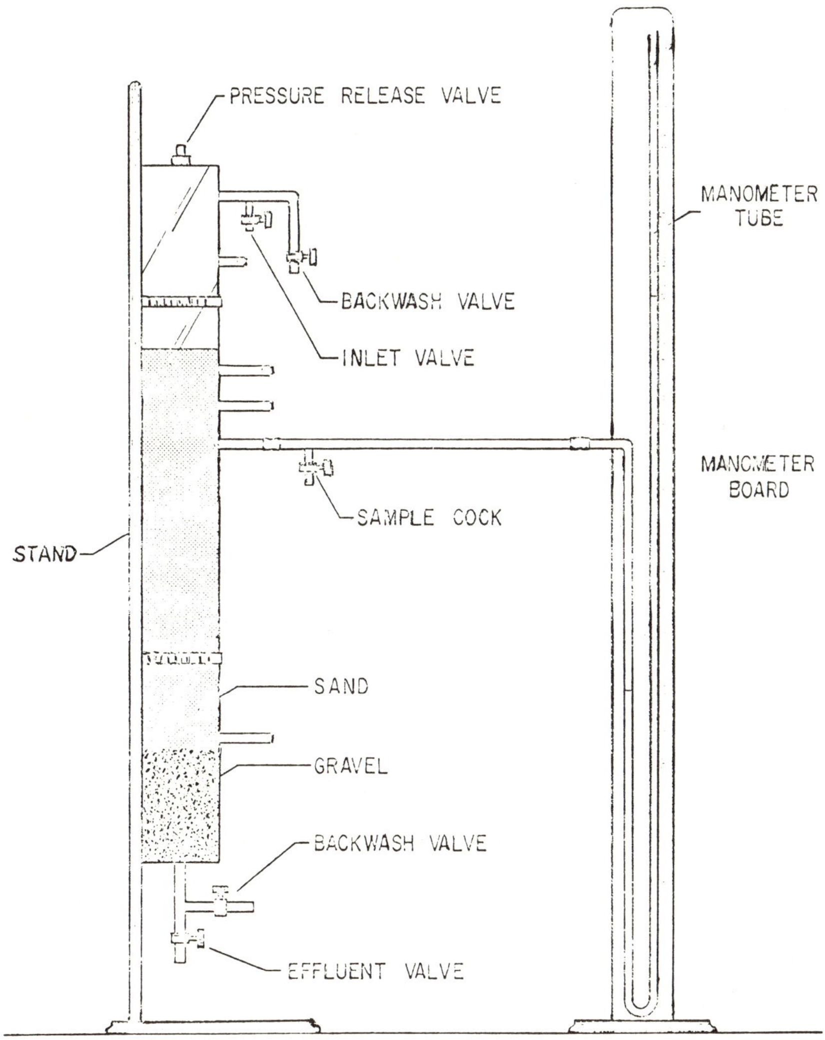

Source: EPA Report 12020 03/71

SLUDGE HANDLING

Sludge dewatering and disposal is one of the more important components of the total waste treatment system. Not only are large quantities of waste solids accumulated daily, but the day-to-day cost of sludge disposal frequently exceeds the cost of any other single process in a wastewater treatment plant. Primary sludges from refinery wastewaters are among the most difficult to handle due to the presence of oils and emulsions.

FIGURE 76: ANALYSIS FOR SLUDGE CHARACTERIZATION

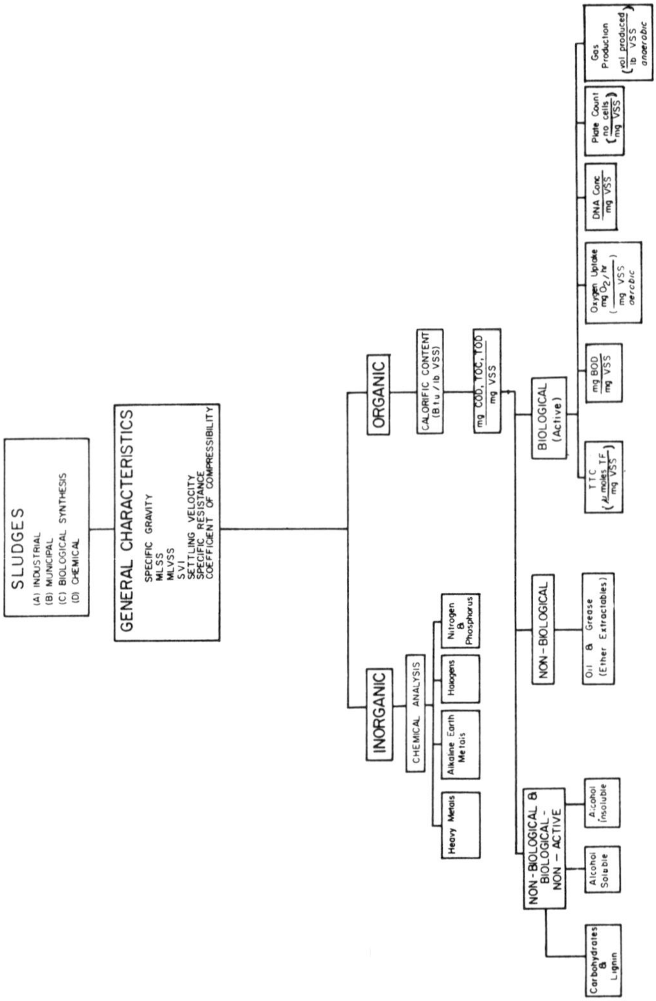

Source: EPA Report 12020 EID 03/71

Sludge handling and disposal includes concentrating the sludge for process use, dewatering and processing the sludge for conversion into a form suitable for disposal, transportation of the sludge, and final disposal.

Characterizing sludge handling methods is extremely difficult as two or more processes may be used conjunctively for one operation. In the past, anaerobic digestion was widely practiced in the treatment of primary sludges and thickened secondary sludges. However, few anaerobic digesters are presently used in refinery or petrochemical wastewater treatment facilities.

Mechanical dewatering of sludges is being adopted by increasing numbers of cities and industries due to increasing land and labor costs. Barging of thickened sludge to the ocean is being practiced by municipalities and industries near coastal areas, but this method may be restricted in the future. Sludge incineration, therefore, is considered the process having the most potential.

Often sludge handling and disposal represents 25 to 50% of the total capital and operating cost of a wastewater treatment plant and the volume of waste sludge will continue to increase. Sludge handling and disposal is an integral part of the total waste treatment process and required the same degree of study given to other plant processes. Predictor equations may be more difficult to derive for sludge handling operations; however, through the use of laboratory and/or pilot plant equipment, basic design criteria can be established.

Analyses helpful in characterizing wastewater sludges with regard to handling and disposal are schematically presented in Figure 76. The sludge fractions are categorized as inorganic and organic, with a suggested subcategorization of the organic component into its biological and nonbiological fractions. As the effects of benthic sludges, effluent suspended sludges, and excess sludge for disposal are environmentally significant, it becomes necessary to quantitatively and qualitatively evaluate its biological composition.

Aerobic Digestion

Aerobic digestion is a process in which excess biological sludge is aerated for long periods of time, resulting in the cellular oxidation or destruction of volatile solids. The concepts of this destruction are similar to those applied in the extended aeration process, although the sludge age values are such that a less viable and more inert sludge is prevalent.

If the bacterial cells are represented by the chemical formula $C_5H_7NO_2$, then the cellular destruction through aerobic digestion is represented by:

$$C_5H_7NO_2 + 5O_2 \longrightarrow 5CO_2 + 2H_2O + NH_3$$

This occurs when substrate in an aerobic system is insufficient for synthesis of energy, and the rate of destruction exceeds that of growth. Primary sludges often are combined with excess activated sludge, although the VSS destruction rate usually decreases as higher food to microorganism ratios (F/M) prevail. Design considerations and process variables involved are:

1. nature and concentration of solids
2. solids loading

3. sludge age
4. oxygen and mixing requirements
5. environmental conditions (temperature, pH, etc.)

The applicability of digesting sludges aerobically will depend on the nature of the sludge, the applied loading, and the sludge age. If aerobic digestion is considered as a candidate process in a waste treatment system, bench or pilot scale tests should be undertaken to determine the VSS destruction rates and oxygen requirements necessary for design purposes.

The destruction rates of digestible sludges can be evaluated in the laboratory using batch reactors such as those shown in Figure 63. The sludge mixture is aerated and the VSS measured at various periods of time. Once the VSS concentration becomes stabilized, the degradable fraction can be estimated. The data are then plotted and the empirical relationship determined.

The oxygen uptake rate of the sludge mixture is also monitored throughout the aeration period. This rate at the aeration time required for the desired VSS destruction level can then be used in estimating the system oxygen requirements.

Anaerobic Digestion

The anaerobic treatment of wastewaters or digestion of sludges involves a multitude of sequential biochemical reactions whereby facultative and anaerobic microorganisms ultimately convert organic material to carbon dioxide and methane. These reactions basically involve the conversion of substances such as organics, nitrates, and sulfates to reduced organics, carbon dioxide, methane, ammonia, and hydrogen sulfide.

As the anaerobic conversion of various substrate materials yields little energy to the microorganisms, only a small portion of the waste is converted to new cells. This is an inherent advantage of the anaerobic process in treating wastewaters, as large volumes of sludge are not synthesized as in the aerobic process. Other advantages include low nutrient requirements, no oxygen requirements, and production of a useful product, methane. Design considerations and process variables involved are:

1. organic loading
2. solids loading
3. temperature and pH
4. acidity/alkalinity
5. hydraulic detention time and microbial growth rate
6. gas production
7. mixing
8. nature and concentration of solids

Anaerobic digestion and anaerobic treatment of liquid wastewaters can be evaluated using bench scale units such as those shown in Figures 77 and 78 or on a larger pilot scale. The process conditions can be changed as required in order to evaluate the corresponding process efficiency. It should be recognized that operation and control becomes time consuming and critical as the scale is increased.

FIGURE 77: LABORATORY BATCH-FED DIGESTOR

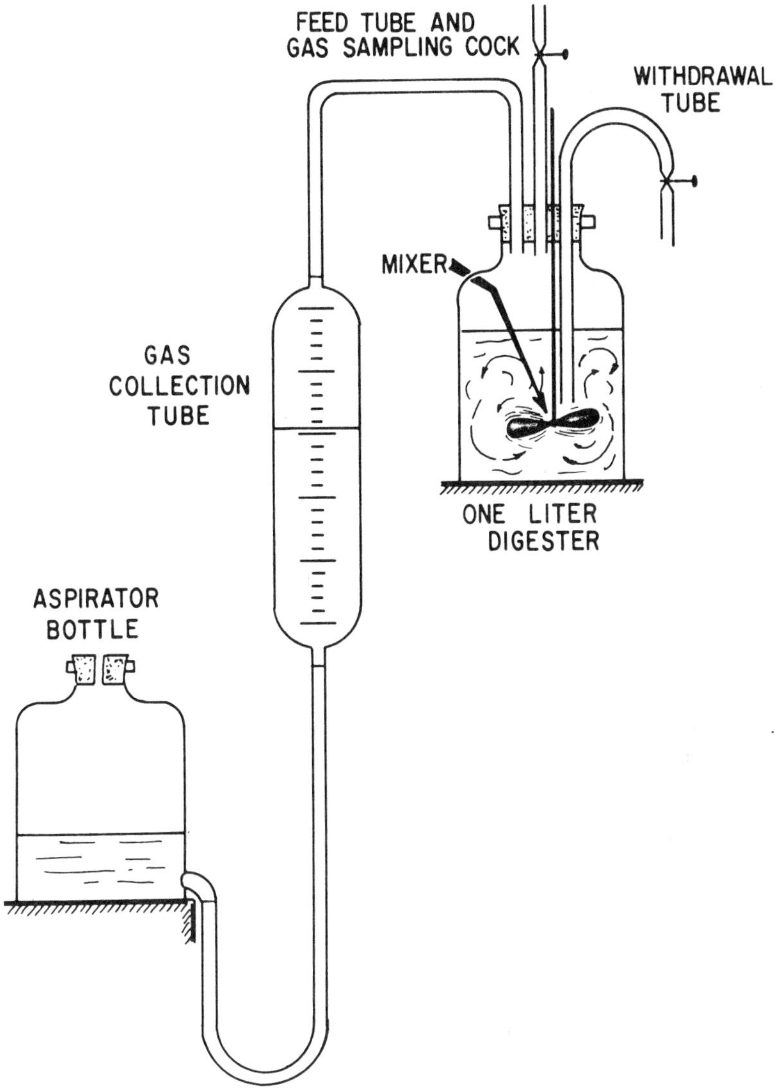

Source: EPA Report 12020 EID 03/71

FIGURE 78: CONTINUOUS-FEED DIGESTOR

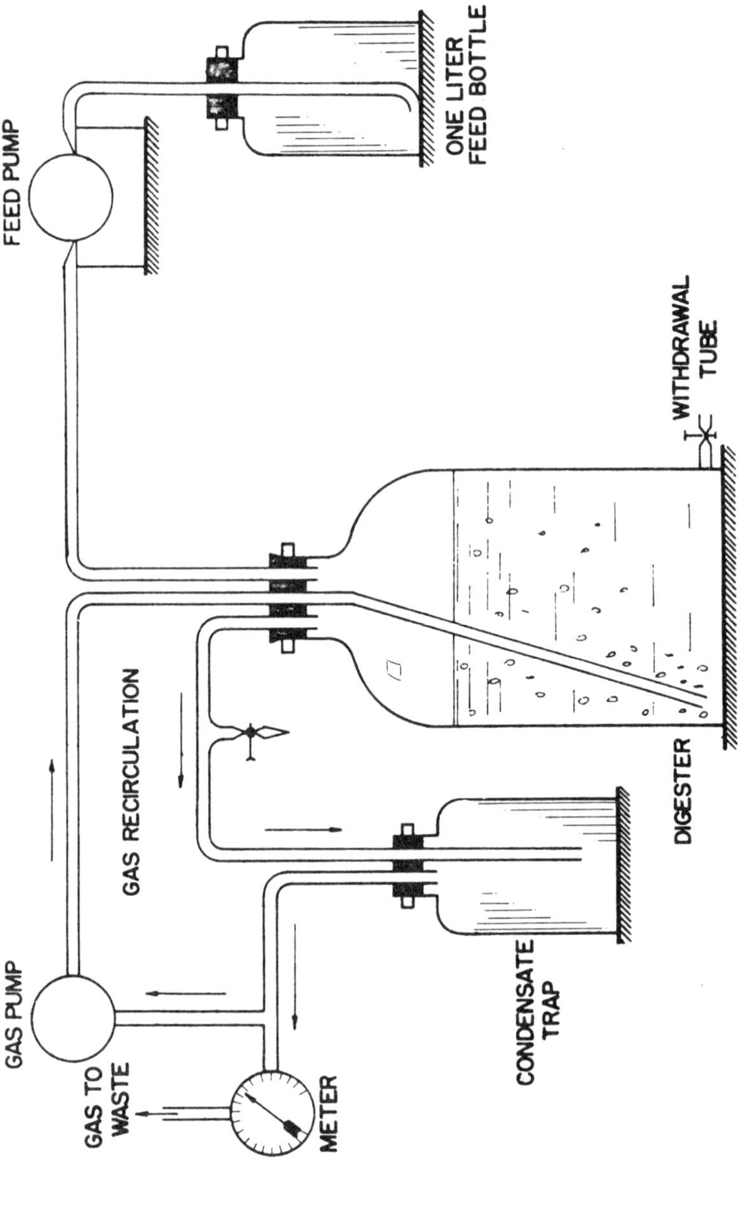

Source: EPA Report 12020 EID 03/71

The anaerobic treatment of synthetic organic waste has been described by J.C. Hovious, J.A. Fisher and R.A. Conway of the Research and Development Department, Chemicals and Plastics Division, Union Carbide Corporation, South Charleston, West Virginia. The results have been presented in EPA Report 12020 DIS 01/72, Washington, D.C., U.S. Government Printing Office (January, 1972).

Wet Oxidation

Heat treatment of wet sludges may be classified as a combustion process. However, much lower temperatures are used in heat treatment than used in the incineration process. The process may be expressed in terms of a pressure cooker system and is known as wet air oxidation. The process developed in Europe is known as the Porteous Process and the American version developed by Zimmerman is appropriately referred to as the Zimmerman Process.

In both methods the feed sludge temperature is elevated by passage through a heat exchanger and then the sludge is pumped to the reactor where steam is added. The reactor temperature is maintained at approximately 380°F. The wet air oxidation process differs from the Porteous because of the addition of air and scrubbing of exhaust gases in the former. Both systems require that the sludge be ground to about 0.25 inch size particles and be pumpable.

Exposing the sludge to heat and pressure coagulates the solids, breaks down the gel structure, and reduces the hydration and hydrophilic nature of the solids. The liquid portion of the sludge can then be separated for additional dewatering and final disposal. Treating sludge, on leaving the reactor, passes through the heat exchanger, giving up most of its heat to the incoming raw sludge.

After leaving the exchanger, the supernatant is decanted from the solids. The thickened sludge is dewatered further by one or more of several methods (Porteous, 1968). Heat-treated sludge may be dewatered by filter presses or vacuum filters or used for landfill. The Zimmerman Process has effectively treated sludges containing 3 to 6% solids without preliminary dewatering; however, economics seem to justify prethickening. Design considerations and process variables involved are:

1. sludge flow
2. solids concentration
3. volatile solids concentration
4. chemical oxygen demand
5. chemical efficiency, COD removal
6. power recovery desired
7. sludge variability
8. ultimate disposal method
9. nature and composition of sludge

The basic design criteria for heat treatment may be determined by performing physical and chemical analyses of the sludge. In a study by the U.S. Public Health Service, it was concluded that data obtained from pilot plant studies were not amenable to simple kinetics and relations should be developed empirically for specific situations.

Thickening

Sludges from primary or secondary processes usually require thickening prior to dewatering by air drying, vacuum filtration, or centrifugation. If a sludge can be thickened, the process offers the advantages of improving digester operation, reducing sludge volumes for direct land or sea disposal, and enhancing the efficiency of process sludge dewatering systems.

Biological sludges usually can be thickened to a solids concentration of 4 to 6%, depending on the type of thickener and the nature of the sludge. As the ability of many sludges to thicken can be defined only in general terms, bench or pilot thickening studies should be initiated prior to finalizing the thickener design. Air flotation, mentioned previously, is also used for sludge thickening in certain applications. Design considerations and process variables involved are:

1. sludge loading
2. diameter and configuration of thickening unit
3. depth of thickener
4. temperature
5. raking mechanism
6. detention time
7. sludge depth in thickener
8. general sludge characteristics

The most common approach in evaluating thickening requirements for sludges is to use batch settling cylinders such as the one shown in Figure 19. The sludge in question is put into the batch settling cylinder and the settling and thickening characteristics are evaluated under quiescent conditions, serving as the basis for scale-up and design.

Lagooning

Lagoons may be divided into 3 classes: (a) thickening, storage, and digestion lagoons; (b) drying lagoons; and (c) permanent lagoons. Lagooning is the most popular sludge disposal technique at industrial wastewater treatment plants. Lagooning may be considered as a stage process in the handling of sludge or as a final sludge disposal process.

The first type of lagoon is used when conventional digestion units are overloaded or sometimes as substitutes for conventional processes. Drying lagoons are used as substitutes for sand drying beds and a permanent lagoon where sludge is never removed is one of the cheapest methods of sludge disposal.

Permanent lagoons may be considered as a land filling method of sludge disposal. A supernatant decanting system enhances the operation of permanent lagoons by providing additional storage. Sludge stored in lagoons may be dewatered from about 95% moisture to 55 or 60% moisture in 2 to 3 year periods. The standard operating procedure for permanent lagoons is to discharge digested sludge to the lagoons at regular intervals, allowing a drying and cleaning period.

Waste stabilization ponds can be used effectively to treat excess activated sludge. Lagoons are often used for sludge dewatering prior to ultimate disposal and in effect the lagoon acts as a gravity thickener. When a lagoon is used for dewatering and not permanent storage, recommended filling depths are 2.5 to 4 feet. However, ponds 5 to 6 feet in depth have been used to stabilize waste activated sludge. Design considerations and process variables involved are:

1. available land area
2. climatic atmospheric conditions
3. subsoil permeability
4. lagoon depth
5. lagoon surface loading
6. sludge characteristics
7. moisture content of sludge

Sludge lagoons have been designed on the basis of surface or volumetric loadings or based upon ultimate disposal requirements. Model ponds of the type illustrated in Figure 68 have proven useful in evaluating several variables of pond design. These model ponds were developed for studies of waste stabilization pond treatment of wastewaters, but can also be used to study sludge stabilization. Development of design criteria is dependent on lagoon type, e.g., whether the process is to be designed for thickening, drying, digestion or storage.

Sand Drying Beds

Dewatering of digested sludge is commonly accomplished by the use of open sand drying beds. Sewage sludge should be properly digested before application to atmospheric drying as digestion enhances sludge drainability and raw sludge may cause undesirable odors. Most beds are open and completely exposed, while others are glass-covered to reduce the effects of inclement weather on the drying process.

Drying beds usually consist of 4 to 9 inches of sand over 8 to 18 inches of graded gravel or stone. 6 to 12 inches of digested sludge are applied to the bed and allowed to drain through the sand beds where the filtrate is collected by underground laterals, conveyed to sumps, and subsequently recycled through the plant. Experiments have shown that tile-drained sludge beds dry 25% faster than beds with impervious bottoms. However, other experiments and studies indicate that nonunderdrained beds can be built and operated at one-fifth the cost of a tile-drained bed.

The sludge drying process incorporates drainage of the sludge moisture through the bed and the simultaneous removal of water by evaporation. The amount of moisture lost by evaporation during the initial stages is insignificant but becomes very important 1 to 2 days after the sludge is applied to the bed.

The sludge is allowed to dry until it reaches a liftable condition at which time the dried sludge is removed from the beds for final disposal. This condition normally occurs when the moisture content is 70 to 80%. The moisture content of the dried sludge is on the order of 60%. Design considerations and process variables involved are:

1. climatic and atmospheric conditions
2. depth of sludge application
3. presence or absence of chemical conditioning
4. sludge moisture content

5. source and type of sludge
6. extent of sludge digestion
7. sludge age
8. sludge composition
9. solids concentration when applied
10. sludge bed construction

Only recently have attempts been made to formulate rational design standards for sludge drying beds. In the past, bed requirements have been based only on empirical relationships or experience factors. Investigators have attempted to derive design criteria from laboratory experiments and pilot plant operations. However, correlation between experimental data and actual plant data has been relatively poor.

Selected variables of the sludge drying processes have been successfully studied in the laboratory as well as on a pilot plant scale as described by B.A. Carnes in a master's thesis, University of Texas (1966).

Vacuum Filtration

Vacuum filtration is used to dewater wastewater sludges in which water is removed under an applied vacuum through a porous media which retains the solids. In the operation of vacuum filters, a rotary drum passes through a slurry tank in which solids are retained on the drum under an applied vacuum.

When the drum emerges from the slurry tank, the deposited cake is further dried by the transfer of liquid to the air which is drawn through the cake by the applied vacuum. The filter cake is then removed to a collection hopper for hauling or incineration. Design considerations and process variables involved are:

1. sludge feed concentration
2. sludge conditioning (chemical additives)
3. degree of thickening preceding filtration
4. sludge viscosity
5. filtrate viscosity
6. operating vacuum
7. chemical and physical composition including partial size, shape, etc.
8. type and porosity of filter media

The most rapid means of estimating the vacuum filter yield for a given sludge is by use of the leaf test apparatus. The test variables such as an applied vacuum, sludge concentration, sludge conditioning, etc., are established and the sludge yield for each condition is estimated. Continuous pilot scale tests are recommended if design information for large and costly vacuum filtration installation is required.

Centrifugation

Basically, centrifuges separate solids from the liquid through sedimentation and centrifugal force. The most effective centrifuges for dewatering sewage sludges are horizontal, cylindrical-conical, solid bowl machines. In a typical unit, sludge is fed through a stationary feed tube along the center line of the bowl to the hub of the screw conveyor. The screw conveyor is mounted inside the rotating conical bowl and rotates at a slightly lower speed than the bowl. Sludge leaves the end of the feed tube, is accelerated, passed through the ports of the conveyor shaft, and is distributed to the periphery of the bowl. The solids are settled through the liquid and are moved along the bowl wall by the blades of the screw conveyor.

The solids move out of the liquid bowl and onto a drainage deck on which they are continuously conveyed by a screw to the end of the machine, at which point they are discharged. The liquid effluent is discharged through effluent ports after traveling the length of the pool under centrifugal force. The depth of the liquid or pool volume can be varied by adjustment of weir plates located at opposite ends of the bowl.

Centrifugation has been used for sludge thickening and sludge dewatering, and has been used with and without chemical conditioning or the addition of polyelectrolytes. The centrifuge is most effective in recovering primary solids, and least effective when recovering waste activated sludge. Design considerations and process variables involved are:

1. feed rate
2. sludge solids characteristics
3. feed consistency
4. temperature
5. chemical additives
6. bowl design
7. bowl speed
8. pool volume
9. conveyor speed

Unlike filtration, it is not easy to predict centrifugal performance by the use of laboratory techniques. Pilot studies must be made and appropriate scaling factors determined for process design. When scaling pilot plant data to larger units, it is necessary to consider the residence time of the liquid in the bowl, the solids thickness and the residence time of the solids on the deck, and particularly the effective centrifugal force on the liquids and solids (Albertson and Guidi, 1969).

It may be possible to determine the feasibility of centrifugation from simple laboratory tests. The sludge to be dewatered may be spun in a laboratory hand centrifuge for 30 seconds at 2,000 rpm, and after centrifugation, the tube filtered to determine the stability of the deposited cake.

Filtration

Filter pressing presently is not widely practiced for sludge handling in the United States, probably because of the high labor and maintenance costs involved in the batch operation. However, the applicability of filter pressing sludges is becoming more promising. The press does not dewater solids by squeezing, but rather acts as a pressure filter similar to a rotary vacuum filter with the exception that higher pressures are used. Like the vacuum filter, the filter press usually requires chemical conditioning, bench scale filtration studies, and careful selection of the filtering media. Economics of the process dictate sludge concentrating by some other means prior to pressing. The Sludge All System developed by Beloit-Passavant Corporation is a completely automated process and the customary shutdown for cake removal is not necessary. The system includes thickening, pressing, and incineration; and the accumulated filter cake from the press contains about 50% solids.

Thickened sludge is ground prior to filtration and often a precoat is added and mixed with the feed sludge before pressing. The sludge enters the press through a center feed and is forced into the filter modules. The filtrate is continuously pressed through the fabric and discharged for recycle. Design considerations and process variables involved are:

1. sludge volumes
2. raw sludge solids concentration
3. nature of the sludge
4. chemical conditioning
5. ultimate disposal of filter cakes
6. filter pressing pressure
7. filter mesh aperture size
8. filter pressing time

The basic design criteria for sizing a filter press unit can be developed by use of a pilot plant unit. Pilot plant data should also provide thickening and chemical conditioning requirements. Means for ultimate disposal of the cake solids often will dictate the required solids concentration in the final sludge cake. A final cake solids of 40% may be expected in the sludge cake provided feed solids are maintained near 10%.

Land Disposal

The disposal of liquid digested sludge on open land surfaces is quite common among smaller wastewater treatment plants. In England, the disposal of liquid sludge to farmland is very popular and in the arid and semiarid parts of the United States, the reclamation of water from municipal sewage is becoming increasingly recognized as an important water conservation measure. Liquid digested sludge and supernatant have been applied to lands for final disposal to fertilize grass or agricultural crops for soil conditioning.

Digestion, aerobic or anaerobic, is almost always required before spreading liquid sludge on land. Sludge is distributed on the land and processed in a variety of ways. The sludge may be injected into the subsoil under pressure or simply pumped or gravity fed through a pipeline to agricultural fields or land to be reclaimed. A common technique is disposal directly to the land by spraying from tank wagons. Design considerations and process variables involved are:

1. proximity of surface waters and distance to groundwater table
2. toxic constituents of the sludge
3. nutritional value of the sludge
4. availability of disposal sites
5. transportation costs
6. suitability of soil for sludge disposal
7. nature of any aesthetic nuisance
8. effects on vegetation

9. application rates
10. atmospheric and climatic conditions
11. method of sludge application

Design criteria can only be developed through the use of fairly large demonstration sites and extensive time involving studies is required. The effects of land disposal on crops and ground and surface waters is of paramount importance.

Incineration

Various incineration processes have been applied to the disposal of residues resulting from wastewater treatment. These processes may be classified as multiple hearth fluidized bed flash combustion and atomized suspension. The wet oxidation process (classified previously under heat treatment) and incineration are considered combustion processes. All these methods depend to some extent on the organic material in the waste as a source of fuel.

However, regardless of the heat value of the sludge, some auxiliary fuel system is necessary during start-up of the incineration process. The incinerator may be classified as an open flame water evaporator since the actual incineration does not take place until the sludge has lost most of its moisture through evaporation. The temperature in the burning zone must be at least 1350° to 1400°F for complete combustion of the sludge solids and for complete oxidation of the existing gases to innocuous compounds.

In the multiple hearth furnace, preconditioned dewatered sludge is conveyed to the upper hearth of the furnace, and mechanical rakes move the sludge from one hearth to the next lower hearth, where the sludge is sequentially dried, burned, and the ashes cooled.

In the fluidized bed incinerators, air is utilized to maintain hot sand in an expanded or fluidized state. The violent boiling of the expanded sand allows intimate contact of the sludge and other fuel with oxygen without the aid of mechanical mixing. Dewatered sludge is fed directly into the fluidized bed to which preheated air is introduced.

The atomized suspension process is essentially a high temperature, low pressure thermal process for the oxidation of fine particles of sludge to innocuous ash. Sludge is concentrated, ground, and sprayed as an atomized suspension into a stainless steel cylinder. The walls of the cylinder are maintained at a temperature between 1000° to 2000°F and heat is transferred to the falling droplets of sludge by radiation.

In the cyclone flash dryer and incinerator process, dewatered sludge is mixed with previously heat dried sludge to produce a fluffy material. In this form, it is carried through the drying stage by the high velocity of hot gases and the dried sludge is burned in the furnace. Design considerations and process variables involved are:

1. sludge flow
2. moisture content of sludge
3. volatile solids content of the sludge
4. heat value of the sludge

5. type of operation, continuous or intermittent
6. source and type of auxiliary fuel
7. method of final disposal of solid residue
8. sludge variability
9. sludge loading
10. possibility of alternate methods

Design criteria are established through incinerating the sludge in a bench or pilot scale burner. The auxiliary fuel requirements, nature of the ash, and composition of the off-gases are the important parameters to be ascertained.

Sea Disposal

Sludge may be transported by rail, truck, barge, pipe, or conveyor. For hauls up to 50 miles, it is likely that truck transportation would be more economical than rail. Freight rate structures in the United States are extremely diverse, causing comparisons of transportation methods to be difficult at best.

In addition to the cost of hauling, there will be an incurred cost for loading sludge into hopper cars in the case of rail haul or pumping sludges to barges in the case of barge haul. Barge rates are lower than rail rates since the tractive force involved is less and barging normally does not bear the cost of the waterway.

For coastal communities or communities adjacent to the well-developed inland waterway systems on the Mississippi, Ohio, Missouri, Illinois, and Tennessee Rivers, sea disposal is frequently the most economical and simplest method available for ultimate sludge disposal, although this entire concept is presently under review by the regulatory agencies. It should be recognized that present and future laws governing sea disposal practice will in large measure dictate its economics. Design consideration and process variables involved are:

1. proximity of final disposal site
2. pretreatment required
3. transportation unit costs
4. type of transportation available
5. distance to loading point
6. requirements for final disposal
7. sludge characteristics

The decision to use transportation in conjunction with sludge disposal is based upon an analysis of alternate dewatering and disposal methods. Laboratory analysis of the sludge are required for such an investigation. There are many factors that may rule out the possibility of sludge transportation, such as the unavailability of economical transportation or the distance to a final disposal site. Barging and sea disposal should be considered as a means for ultimate sludge disposal for plants located near rivers close to the sea.

ULTIMATE DISPOSAL

The final disposition of residual or refractory wastewaters and sludges resulting from various processes is generally termed ultimate disposal. Solutions to ultimate disposal problems include subsurface storage, conversion of wastes to innocuous end products, storage in ponds or land spreading, and ocean disposal. Land spreading, lagooning, and ocean disposal processes have been presented elsewhere and only deep well injection and thermal oxidation are discussed in the paragraphs which follow.

Thermal Oxidation

The thermal oxidation of liquid wastewaters involves the transfer of heat from an auxiliary fuel to the wastewater. In some cases, the wastewater has calorific value which decreases the total auxiliary fuel requirements.

The wastewater is first atomized as finely as possible in order to present the greatest surface area for mixing with combustion air. The temperature of the system is maintained at a level necessary for the reaction to proceed and sufficient residence time is provided to allow complete oxidation. The system is also designed so that there is sufficient turbulence for air, liquid, and fire contact.

The equipment normally consists of a horizontal or vertical refractory lined furnace with an auxiliary fuel burner firing at one end or tangential to a cylindrical shell. The size of the incinerator depends on the heat release of the system, the amount of combustion air to be used, and the quantity of wastewater injected. Design considerations and process variables involved are:

1. flow
2. calorific value
3. residence time in burner
4. air flow
5. ignition temperature of organic material in the wastewater
6. degree of wastewater atomization
7. degree of system turbulence
8. auxiliary fuel calorific value

Before any liquid incineration unit is installed, pilot scale studies should be undertaken. This requires that representative samples either be obtained from the waste stream or synthesized. The wastewater is then fed to the pilot liquid incineration unit, atomized, and burned. The stack gases are monitored for products of incomplete combustion while process variables such as auxiliary fuel feed, wastewater pumping rates, temperatures, etc., are varied as required throughout the test series. The necessary design criteria then can be established based on observed performance.

Deep Well Disposal

Subsurface disposal of liquid wastes is not a new concept as the oil and gas producers have been using this method for disposal of oil field brines for half a century. Only recently have the process industries realized the applicability of deep well injection for disposing of concentrated and relatively untreatable waste streams. The depth at which these wastes are discharged vary from a few hundred feet to over 2,000 feet and well head injection pressures for some wells have approached 4,000 psi. A suitable disposal formation is the very heart of a good disposal system and if water can be disposed of with a vacuum on the injection well head, operating expenses would be much less than when injection pressures are required. Deep well disposal requires the injection of liquid waste into a porous subsurface stratum which contains noncommercial brines. The wastes are merely stored below ground in strata which are sealed by impervious strata, thus isolated from usable underground water supplies or mineral resources. The disposal system consists of a well and surface equipment, such as pumps and pretreatment units. Design considerations and process variables involved are:

1. subsurface geology
    a. permeability
    b. porosity
    c. fracture gradient
2. subsurface hydrology
3. fluid compatability
4. pretreatment considerations
5. flow
6. characteristics of waste, including temperature

Initially, a feasibility study is required which includes acquisition of basic information concerning subsurface geology and hydrology. Based on these data, the operating well head pressure may be estimated and for steady-state single-phase flow, the injection rate may be predicted. Pilot wells are often constructed prior to the design and construction of the waste injection well system.

Preliminary estimates of injection rates and hole pressure may be verified by these pilot studies. These studies are particularly important for the design of wastewater pumps and appurtenances. The test hold will also serve to identify subsurface geology and allow for sampling of the formation fluids. The duplication of formation hydrologic and geologic conditions in the laboratory is highly impractical and in most cases impossible.

Fluid compatability can be evaluated in the laboratory by mixing the waste and formation waters and observing any changes in the physical appearance. Preliminary investigations should also include bacteriological analyses of the formation water and wastewater as microorganisms may also cause plugging of the well formation.

OVERALL WATER POLLUTION CONTROL MODELS

BEST PRACTICABLE CONTROL TECHNOLOGY CURRENTLY AVAILABLE (BPCTCA)

Best practicable control technology currently available (BPCTCA) for the organic chemical industry is based on the utilizations of both in-process controls and end-of-process treatment technologies.

Alternative in-process controls commensurate with BPCTCA include the implementation of process observation and sampling to determine the quantity, compositions, concentration, and flow of the process waste streams. Such waste characterization studies logically lead to the selection of various process waste sources for segregation.

Exemplary plants within the industry segregate contaminated contact process water streams from noncontaminated streams such as cooling water. This practice appreciably reduces the waste volume to be treated in a centralized waste treatment plant. In addition process water streams are segregated on the basis of the ease with which certain constituents can be recovered as well as the ease with which the wastes can ultimately be treated.

Process modification consistent with BPCTCA include the substitution of nonaqueous media in which to carry out the reaction or to purify the products. In some cases aqueous waste by-products are eliminated by changes in the reactants, reactant purity, or catalyst system. Where waste is used in the process, its use should be restricted and the possibility of using recycled or reused water should be investigated. Examples of this practice include recycle between an absorber and a steam stripper, countercurrent washing techniques, and the collection of vacuum-jet condensate, rain water and floor water for reuse.

Equipment associated with the separation of an organic phase from an aqueous phase, such as decanters, are provided with backup coalescers or polishing filters for the aqueous phase. Direct vacuum-jet condensers are replaced with indirect condensers or vacuum pumps.

In addition to waste reductions obtained through segregations and process change, exemplary plants using BPCTCA combine recovery of products and by-products with wastewater purifications. The recovery of chemicals from the wastewaters includes both the physical separation of chemicals from the wastewater as well as subjecting the wastewater to additional chemical reactions that will render them more amenable to recovery and purification.

Physical separation processes utilized by exemplary plants include adsorption, solvent extraction, and distillation. Adsorbents in use include activated carbon, zeolites, and synthetic resins. The adsorbed chemicals are recovered by desorption which also serves to regenerate the saturated adsorbent.

One system for the nondestructive, in-place, regeneration of activated carbon is the use of pH change to cause the adsorbed chemicals to desorb. Such a system has been used successfully to recover phenol and acetic acid by the addition of caustic.

Solvent extraction is used for the recovery of phenol from the wastewater of the cumene process for phenol manufacture. Solvent extraction is practiced when the chemical can be extracted into a solvent already in use in the process. Excess solvent is steam stripped from the effluent. Effluent phenol concentration is expected to average 0.1 to 0.5 mg/l from the treatment system.

Distillation is used to recover by-products from reduced volume wastewater streams by steam stripping. This concentration step produces an overhead condensate containing the strippable organic substances and water. This condensate is then reused in the process. Exemplary plants utilizing either solvent extraction or steam distillation of wastewaters usually apply additional polishing treatment to the effluent to remove the small remaining quantities of organic substances.

Chemical reactions such as chlorination, hydrolysis, cracking, dechlorination and dealkylation have been used to convert impurities into forms suitable for subsequent physical separations. A typical example is the hydrolysis of aromatic tars with caustic with subsequent acidification and physical separation of the organic and aqueous phases. It is not possible to delineate a specific sequence or combination of in-process controls which could be considered as an across the board definition of BPCTCA. However, methods taken from those previously described should enable all processes within each category to attain the following standard raw waste loads. These values are listed in Table 23.

TABLE 23: SUMMARY OF MEDIAN RAW WASTE LOAD DATA AS THE BASIS FOR CALCULATING EFFLUENT LIMITATIONS

Subcategory	Process Wastewater Flow (liter/1,000 kg)	(gal/1,000 lb)	BOD_5 (kg/1,000 kg)	COD (lb/1,000 lb)
A	500	60	0.12	0.31
B-1	1,320	158	0.35	1.1
B-2	3,580	429	1.77	6.2
C-1	2,340	280	1.90	6.5
C-2	10,800	1,300	53.0	118.0
D	175,800	21,050	79.0	1,075.0

Source: Report EPA 440/1-73/009

End-of-pipe treatment technologies commensurate with BPCTCA are based on the utilization of biological systems including the activated sludge process, extended aeration, aerated lagoons, trickling filters, and anaerobic and facultative lagoons. These systems include additional treatment operations such as equalization, neutralization, primary clarification with oil removal, and nutrient addition. Because the removal of certain organic materials may require the utilization of high concentrations of biological solids, effluent polishing steps such as coagulation, sedimentation, and filtration are considered as commensurate with BPCTCA. Effluent suspended solids are expected to be maintained below 30 mg/l average concentration. The following waste reductions are considered consistant with BPCTCA:

Subcategory	Percent Reduction of BPCTCA Raw Waste Load Median Values	
	BOD_5	COD*
A	90%	75%
B-1	90%	75%
B-2	98%	75%
C-1	95%	75%
C-2	99%	75%
D	95%	75%

*COD effluent limitations are not specified for BPCTCA

These reductions have been applied to the standard raw waste loads for each subcategory. The effluent limitations for BPCTCA are listed in Table 24.

It should be noted that because biological systems have been proposed as the mode of treatment consistant with BPCTCA, the BOD_5 parameter is controlling and is the only one for which the effluent limitations are to be applied. It may be desirable in certain cases to establish limitations for COD or TOC instead of the BOD_5 parameter. The feasibility of such a substitution can only be determined on an individual basis after adequate correlation has been established. Effluent limitations are specified on the bases of the maximum for any one day and the maximum average of daily values for any period of 30 consecutive days.

TABLE 24: BPCTCA EFFLUENT LIMITATIONS

Effluent Characteristics	Maximum Average of Daily Values for Any Period of Thirty Consecutive Days kg/1,000 kg*	Maximum for Any One Day kg/1,000 kg*
Subcategory A		
BOD_5	0.025	0.045
TSS	0.023	0.038
Phenols	0.00025	0.0005
Subcategory B		
B-1 Product-Processes		
BOD_5	0.06	0.10
TSS	0.06	0.10
Phenols	0.00066	0.0013
B-2 Product-Processes		
BOD_5	0.17	0.30
TSS	0.16	0.27
Phenol	0.0017	0.0034

(continued)

TABLE 24: (continued)

Effluent Characteristics	Maximum Average of Daily Values for Any Period of Thirty Consecutive Days kg/1,000 kg*	Maximum for Any One Day kg/1,000 kg*
Subcategory C		
C-1 Product-Processes		
BOD_5	0.17	0.30
TSS	0.16	0.27
Phenols	0.0017	0.0034
C-2 Product-Processes		
BOD_5	0.9	1.5
TSS	0.49	0.8
Phenols	0.005	0.011
Subcategory D		
BOD_5	9.0	15.0
TSS	7.88	13.0
Phenols	0.088	0.17

*kg/1,000 kg production is equivalent to lb/1,000 lb production.

Source: Report EPA 440/1-73/009

The single stage activated sludge system was chosen as a model system for BPCTCA and a general flow diagram for the wastewater treatment facilities is shown in Figure 79. A summary of the general design basis is presented in Table 25.

TABLE 25: BPCTCA MODEL TREATMENT SYSTEM DESIGN SUMMARY

Treatment System Hydraulic Loading
(capacities covered, in gpd)

7,200	360,000
43,200	720,000
72,000	1,440,000
216,000	2,160,000

Pump Station
Capacity to handle 200% of the average hydraulic flow.

Equalization
One day detention time is provided for Subcategories A, B, and C, and three days for Subcategory D. Floating mixers are provided to keep the contents completely mixed.

Neutralization
The two-stage neutralization basin is sized on the basis of an average detention time of twenty minutes. The lime-handling facilities are sized to add 2,000 lb of hydrated lime per mgd of wastewater, to adjust the pH. Bulk-storage facilities (based on 15 days usage) or bag storage is provided, depending on plant size. Lime addition is controlled by two pH probes, one in each basin. The lime slurry is added to the neutralization basin from a lime slurry recirculation loop. The lime handling facilities are enclosed in a building.

Nutrient Addition
Facilities are provided for the addition of phosphoric acid and aqua ammonia to the biological system in order to maintain the ratio of BOD:N:P at 100:5:1.

(continued)

TABLE 25: (continued)

Aeration Basin

Platform-mounted mechanical aerators are provided in the aeration basin. In addition, concrete walkways are provided to all aerators for access and maintenance. The following data were used in sizing the aerators:

Oxygen utilization	1.5 lb O_2/lb BOD removed
α factor	0.9
β factor	0.9
Wastewater temperature	20°C
Oxygen transfer	3.5 lb O_2/hr/shaft hp at 20°C and zero DO in tap water
Motor efficiency	85%
Minimum basin DO	1 mg/l

Oxygen is monitored in the basins using DO probes.

Secondary Clarifiers

All secondary clarifiers are rectangular units with a length-to-width ratio of 3:4. The side water depth is 10 ft and overflow rate varies between 100 and 500 gpd/ft^2 depending on plant size. Sludge recycle pumps are sized to deliver 100% of the average flow.

Air Flotation

The air flotation units recommended for Subcategory C plants are sized on a solids loading of 20 lb/ft^2/day. In addition, liquid polymer facilities are provided to add up to 50 mg/l of polymer to enhance solids separation.

Sludge Holding Tank-Thickener

For the smaller plants, a sludge-holding tank is provided, with sufficient capacity to hold 5 days flow from the aerobic digester. The thickener provided for the large plants was designed on the basis of 6 lb/ft^2/day and a side water depth of 10 ft.

Aerobic Digester

The aerobic digester is sized on the basis of a hydraulic detention time of 20 days. The sizing of the aerator-mixers was based on 1.25 hp/1,000 ft^3 of digester volume.

Vacuum Filtration

The vacuum filters were sized on a cake yield of 2 lb/ft^2/hr, and a maximum running time of 18 hr/day. The polymer system was sized to deliver up to 10 lb of polymer/ton dry solids.

Final Sludge Disposal

Sludge is disposed of at a sanitary landfill assumed to be 5 miles from the wastewater treatment facility.

Design Philosophy

The plant's forward flow units are designed for parallel flow, i.e., either half of the plant can be operated independently, thus providing reliability as well as flexibility in operation. The sludge facilities are designed on the basis of series flow. All outside tankage is reinforced concrete. The tops of all outside tanks are assumed to be 12" above grade.

Source: Report EPA 440/1-73/009

FIGURE 79: BPCTCA WASTE TREATMENT MODEL

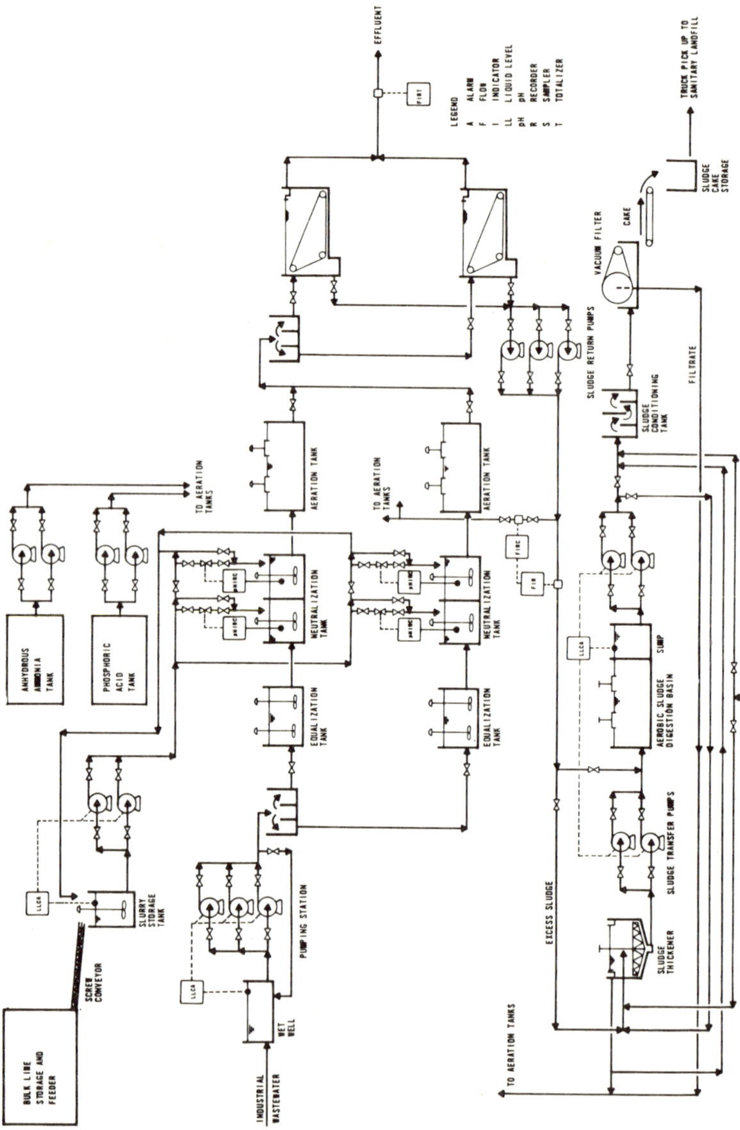

Source: Report EPA 440/1-73/009

The following is a brief discussion of the treatment technology available and the rationale for the selection of the previous unit processes to be included in the BPCTCA treatment system. An optional API oil-water separator was sized for an average flow of 720,000 gal/day. This was done in order to indicate the percentage increase in the total BPCTCA cost which would be involved as a result of excess floating oil.

The pump stations are generally located after the API separator so as to avoid emulsifying the floating oil. Topography of a particular plant site will dictate whether pumping equipment is required. Equalization facilities are provided in order to minimize short-interval (e.g., hourly) fluctuations in the organic loading to the treatment plant, as well as to absorb slug loads from reactor cleanouts, accidental spills, etc. and to minimize the usage of neutralization chemicals, etc.

A three-day detention time based on average flow is provided for Subcategory D flows, in contrast to one day's detention time for Subcategories A, B and C. This increased detention time is provided to allow for the hydraulic as well as the organic variability inherent in batch chemical production. Also, this detention time will provide for continuous seven days per week operation of the wastewater treatment facilities when the manufacturing facilities operate only five days a week.

After equalization and depending on the individual plant's product mix, it may be necessary to neutralize the wastewater, to make it more amenable to biological treatment. Alkaline neutralization is provided in the form of hydrated lime storage and feed facilities. Primary clarification facilities were not included in the facilities for the cost estimate, because the TSS RWL data indicated that it would not be necessary to remove TSS before biological treatment. It should be noted that a plant's particular product mix should be evaluated before a decision is made to omit this process.

An activated sludge process was selected for the biological portion of the system. However, single stage activated sludge is not the only system which may be applicable nor should it be construed as being totally applicable to all process wastewaters which may be generated in the industry. In addition, aerated lagoons are also applicable to meet the proposed effluent requirements. The following is the rationale for selection of activated sludge and its inclusion in the cost estimates.

Both aerated lagoons and activated sludge processes involve aeration basins, the major difference being that aerated lagoons normally discharge directly without a clarification step. This results in a much lower concentration of microorganisms in the aerated lagoon than in an activated sludge basin because the microorganism mass is not recirculated back to the aeration basin.

Therefore, for comparable organic loadings, a much larger aerated lagoon would be required to provide treatment equivalent to that of an activated sludge plant. In actual practice, the aerated lagoon process has found wide application where it is not necessary to provide a method of sealing the lagoon or where the soil characteristics are such that they form a natural seal. The activated sludge process was selected for cost estimating purposes in order to provide an estimate of the economic impact of the proposed effluents limits on any treatment facility including those for which it is necessary to make the basin linings impervious.

Beside activated sludge and aerated lagoons, various other combinations of biological treatment processes may be utilized. The combination of available unit processes are numerous and the treatment scheme should be selected only after a thorough engineering as well as economic analysis.

In the biological process, for every pound of BOD removed from a wastewater, approximately 0.6 lb of biological solids is produced, and this must be removed from the system. In many areas where aerated lagoons are applicable, settling lagoons are used to separate these biological solids.

These settling lagoons are periodically dewatered and dredged, and the dredgings are pumped to sludge-holding lagoons. (In the activated sludge process sludge wasting is done daily to dewatering facilities.) These sludge lagoons may act as solar evaporation ponds or as drying lagoons if the water and precipitation are decanted off.

The sludge will generally dewater naturally if applied at depths less than 15" The dewatered sludge may then be landfilled, or the existing lagoon can be covered with earth and a new lagoon constructed. Which alternative is used depends upon individual state sanitary landfill requirements and the possibility of ground water pollution through contamination with leachate.

Many plants in the United States are so geographically situated that aerated lagoons provide a viable alternative. However, in order to make the subsequent cost estimates more meaningful and universally applicable, activated sludge was selected. In addition, because the Subcategory C wastes are so concentrated and the TSS mixed liquor levels are so high (as previously discussed), an air flotation unit is included to facilitate solids separation after the secondary clarifier.

Activated sludge facilities pose a distinct sludge disposal problem. Because biological sludge must be wasted on a daily basis from most larger plants, it creates a handling problem which often cannot be solved as expeditiously as with an aerated lagoon.

The relatively small amounts of sludge generated by the BPCTCA plants dictated the selection of the most viable sludge disposal alternative. The quantities of sludge are such that sludge incineration cannot be justified because of the very small equipment sizes that would be involved. Therefore, it was decided, for smaller plants, to aerobically digest the sludge, decant the supernatant, and then vacuum filter the sludge. The sludge cake (which would be at least 20% solids) would then be acceptable at a sanitary landfill.

For the larger plants, sludge thickening is provided to concentrate the waste activated sludge from 0.8 to 2.0% before digestion. Thickening is not applicable for the smaller plants because the sludge quantities are such that odors could develop because the small equipment would result in long detention times. For these cases, a sludge holding tank is provided in order to add flexibility to the operation. During vacuum filter down time, digested sludge may also be held without upsetting the solid handling facilities.

BEST AVAILABLE TECHNOLOGY ECONOMICALLY AVAILABLE (BATEA)

The best available technology economically achievable is based upon the most exemplary combination of in-process and end-of-process treatment and control technologies.

The full range of treatment and control technologies which are applicable to the major organic chemicals segment of the organic chemicals manufacturing industry has been previously described. This level of technology is primarily based upon significant reductions in the chemical oxygen demand (COD), as well as the biochemical oxygen demand pollutant parameters.

End-of-process treatment has been determined to be biological plus additional activated carbon treatment. It must be noted that this does not preclude the use of activated carbon as an in-process treatment in lieu of its use at the end-of-process. This may be desirable when product can be recovered or when harmful pollutants must be removed prior to treatment. Two model systems are presented for cost estimation purposes:

(1) Activated sludge treatment followed by filtration and activated carbon adsorption in fixed-bed columns (applied to Subcategories A and B).
(2) Activated sludge treatment followed by carbon adsorption in pulsed bed columns (applied to Subcategories C and D).

These systems or equivalent combinations can provide the reduction in BOD_5 and COD pollutant parameters as listed below:

	BOD_5	COD	
		% Removal Basis	
Subcategory	% BPCTCA RWL	% BPCTCA RWL	% BADCT Effluent
A	99	90	70
B-1	99	94	70
B-2	99.5	94	70
C-1	99.5	94	70
C-2	99.7	94	70
D	99	90	70

The applicable reductions were used as a basis for determining effluent limitations. Low concentrations for TSS and phenols are also attainable via BATEA treatment and control technologies. The maximum average for any 30 consecutive day period, and daily maximum effluent concentrations for TSS are 15 mg/l and 25 mg/l respectively. For phenols these values are 0.1 mg/l and 0.2 mg/l for the 30-day and daily values respectively.

BATEA treatment and control technologies are expected to provide maximum control of effluent variability by process controls and end-of-process investment. Effluent limitations for BATEA are presented in Table 26.

TABLE 26: BATEA EFFLUENT LIMITATIONS

Effluent Characteristics	Maximum Average of Daily Values for Any Period of Thirty Consecutive Days kg/1,000 kg*	Maximum for Any One Day kg/1,000 kg*
Subcategory A		
COD	0.02	0.04
BOD_5	0.002	0.004
TSS	0.004	0.006
Phenols	0.000025	0.00005
Subcategory B		
B-1 Product-Processes		
COD	0.065	0.13
BOD_5	0.004	0.008
TSS	0.01	0.017
Phenols	0.000065	0.00013
B-2 Product-Processes		
COD	0.37	0.74
BOD_5	0.01	0.02
TSS	0.0025	0.0042
Phenols	0.00017	0.00034
Subcategory C		
C-1 Product-Processes		
COD	0.39	0.78
BOD_5	0.01	0.02
TSS	0.005	0.0083
Phenols	0.00034	0.00068
C-2 Product-Processes		
COD	7.2	14.4
BOD_5	0.2	0.4
TSS	0.16	0.27
Phenols	0.0011	0.0022
Subcategory D		
COD	65.0	130.0
BOD_5	0.4	0.8
TSS	1.30	2.19
Phenols	0.0085	0.017

*kg/1,000 kg production is equivalent to lb/1,000 lb production.

Source: Report EPA 440/1-73/009

As previously discussed in in-process recovery systems, activated carbon has possible applications in the organic chemicals industry for in-plant recovery of specific chemicals. In addition, activated carbon has also been demonstrated in many cases that it can be used as an end-of-pipe wastewater treatment technology.

During the plant survey program, seven samples of the individual industrial treatment plant effluents from industrial Subcategories B, C and D were obtained and carbon isotherms were performed using powdered activated carbon. A summary of the analytical results are presented in Table 27. The complexity of the organic chemical industry is evidenced by the results of the carbon isotherms. The initial COD varied tenfold, while the variation in carbon exhaustion rate was over one hundredfold.

The carbon adsorption isotherm is widely used to screen various activated carbons and to quantify overall removal efficiencies. However, the exhaustion capacity, generated by a carbon isotherm is not sufficient to be used for design purposes. In pilot scale column operations, the following factors should be recognized:

the limitations inherent in extrapolating laboratory data to multicolumn systems, and

the problems of channeling and wall-effects that limit the utility of data taken in small diameter laboratory columns.

Ideally, pilot-plant continuous column studies should be run to generate reliable design data.

The carbon isotherm data shown in Table 27 indicated that, except in the particular case of Plant 6, COD removal efficiencies by carbon adsorption of biological treatment plant effluents are well above 70%. In addition, it is known that additional organic substances can be degraded through biological activity which occurs in the carbon bed.

In order to develop BATEA effluent criteria, the activated carbon system in addition to BPCTCA treatment systems was chosen as a model system for BATEA for Subcategories A, B, C and D.

TABLE 27: CARBON ISOTHERM DATA PERFORMED ON INDIVIDUAL BIOLOGICAL TREATMENT PLANT EFFLUENTS

					- - Exhaustion Capacity - -	
Plant ID	Industrial Category	Initial Soluble COD conc mg/l	Final Soluble COD conc mg/l	% Removed	lb COD Removed per lb Carbon	lb Carbon per 1,000 Gallons
1	B	146	19	87	4.5	0.27
2	D	304	21	93	1.35	1.1.87
3	C	525	150	71	0.35	12.2
4	C	573	146	75	0.36	13.3
5	B	774	97	87	0.50	18.7
6	B	972	758	22	0.035	232
7	B	1,297	397	70	0.34	32.2

Source: Report EPA 440/1-73/009

The treatment plant effluents were filtered to insure the removal of all insoluble COD.

BATEA effluent criteria may be attained in actual practice via a number of possible routes. In order to quantify the impact of BATEA criteria on the individual categories, two treatment systems were designed and subsequent cost data reported previously. A general flow diagram for BPCTCA treatment system for Subcategories A and B is shown in Figure 80 and for Subcategories C and D in Figure 81. A summary of the general design basis is presented in Table 28.

Dual media gravity filtration is provided since activated carbon typically requires that the concentration of TSS be 50 mg/l or lower in order to maximize carbon adsorption and minimize the filtration function. High TSS would involve shortened filter runs and increased amounts of backwash water usage.

FIGURE 80: BATEA WASTE TREATMENT MODEL CATEGORIES A AND B

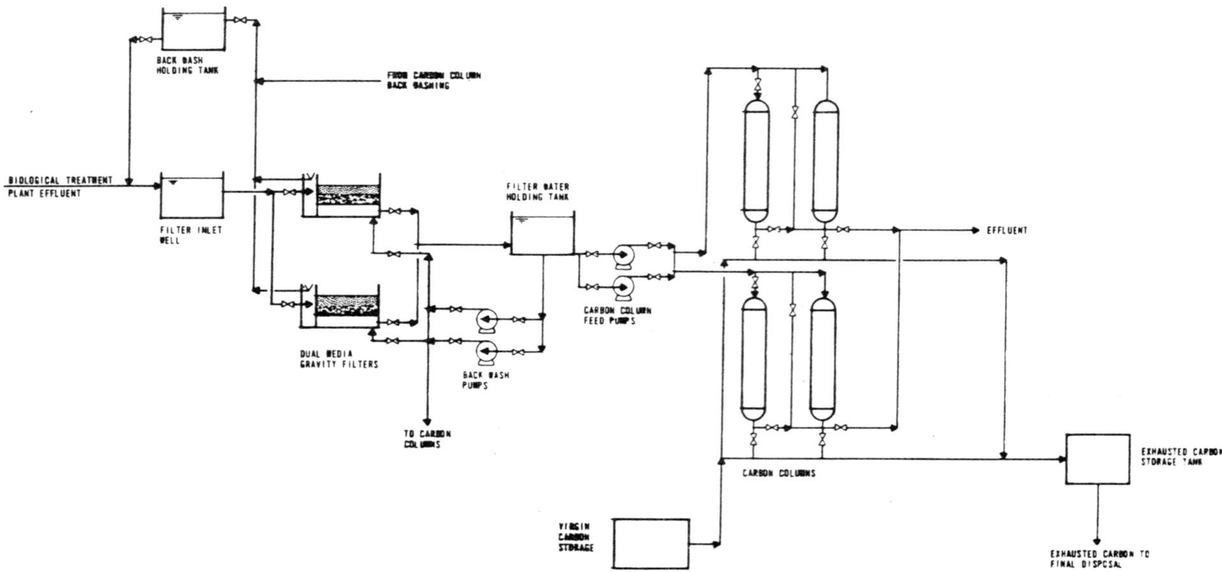

Source: Report EPA 440/1-73/009

FIGURE 81: BATEA WASTE TREATMENT MODEL CATEGORIES C AND D

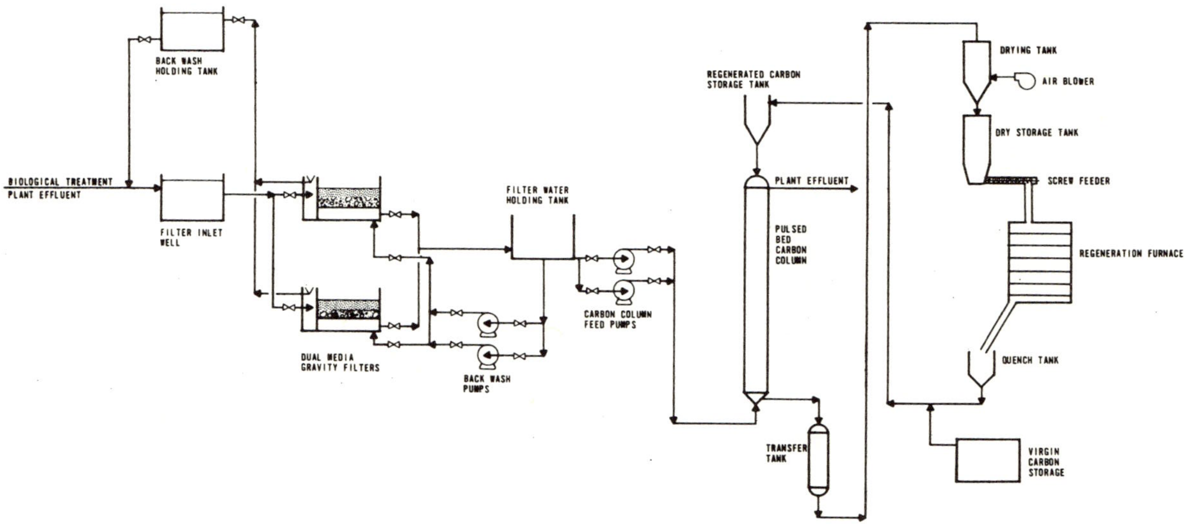

Source: Report EPA 440/1-73/009

TABLE 28: BATEA END OF PIPE TREATMENT SYSTEM

Design Summary

Gravity Dual Media Filtration

The filters are sized on the basis of an average hydraulic loading of 3 gpm/ft^2. Backwash facilities are sized to provide rates up to 20 gpm/ft^2 and for a total backwash cycle of up to 20 minutes in duration. The filter media are 24" of (No. 1½) and 12" of sand (0.4 to 0.5 mm sand).

Granular Carbon Columns

The carbon columns are sized on a hydraulic loading of 4 gpm/ft^2 and a column detention time of 40 minutes. A backwash rate of 20 gpm/ft^2 was assumed for 40% bed expansion at 70°F.

Design Comments:

Subcategory A and B are fixed-bed downflow units, while the Subcategories C and D systems are pulsed-bed upflow unit, with the carbon being wasted over a prescribed time sequence, e.g., wasted for 15 minutes every 2 hours.

Filter-Column Decant Sump

Tanks are provided to hold the backwash water and decant it back to the treatment plant over a 24 hour period. This will eliminate hydraulic surging of the treatment units.

Regeneration Furnace

The following exhaustion rates were used for the sizing of the regeneration facilities:

Subcategory	Influent COD mg/l	Exhaustion Capacity lb COD/lb Carbon
A	100	4.5
B	120	4.5
C and D	1,200	0.35

These exhaustion capacities were selected, based on the carbon isotherm data previously presented in Table 27. A multiple-hearth furnace is employed for regeneration of the carbon only for Subcategory D. The quantities of carbon exhausted based on the previous exhaustion capacities for Subcategories A and B are not sufficiently large to warrant the investment in a regeneration furnace.

Regenerated Exhausted Carbon Storage

Tanks are provided to handle the regenerated and exhausted carbon both before and after regeneration.

Source: Report EPA 440/1-73/009

BEST AVAILABLE DEMONSTRATED CONTROL TECHNOLOGY (BADCT)

Determination of the best available demonstrated control technology (BADCT) for new major organic sources involves the evaluation of the most exemplary in-process control measures with exemplary end of process treatment. Some major in-process controls which are applicable to new sources are as follows:

(1) The substitution of noncontact heat exchangers using air, water or refrigerants for direct contact water cooling equipment (barometric condensers).

(2) The use of nonaqueous quench media, e.g., hydrocarbons such as furnace oil, as a substitute for water, where direct contact quench is required.

(3) The recycle of process water, such as between absorber and stripper.

(4) The reuse of process water (after treatment) as makeup to evaporative cooling towers through which noncontact cooling water is circulated.

(5) The reuse of process water to produce low pressure steam by noncontact heat exchangers in reflex condensers or distillation columns.

(6) The recovery of spent acid or caustic solutions for reuse.

(7) The recovery and reuse of spent catalyst solutions.

(8) The use of nonaqueous solvents for extraction of products.

Although these control measures are generally applicable, no attempt was made to identify all of these or any single one as universally applicable. The end of process treatment model has been determined to be biological treatment with the additional suspended solids removal by clarification, sedimentation, sand and/or dual media filtration. The following system is proposed for cost estimating purposes and does not limit the use of equivalent systems: two stage activated sludge plus dual media filtration. Although biological treatment has been described as the basis for the BADCT, it is recognized that chemical-physical systems such as activated carbon may also be employed as an end-of-process technology or as an in-process or by-product recovery system. It may also be necessary to remove certain wastes which are toxic to or interfere with biological waste treatment systems by in-process chemical-physical control processes. The reduction in major pollutant parameters as defined by BADCT is listed by category in the following tabulation:

	Percent Reduction of BPCTCA Raw Waste Load Median Values	
Subcategory	**BOD_5**	**COD**
A	95	80
B-1	95	80
B-2	97	80
C-1	97	80
C-2	99.5	80
D	97	80

Total suspended solids and phenol effluent concentration with BADCT technology are equivalent to those for the BATEA. Daily and any 30 consecutive day period maximum concentrations for suspended solids are 25 and 15 mg/l respectively. Phenol concentration on a daily and any 30 consecutive day period maximum are at 0.2 and 0.1 mg/l respectively.

Effluent limitations are presented in Table 29, which is shown on the following page, for new sources for major organic sources.

TABLE 29: NEW SOURCES PERFORMANCE STANDARDS (BADCT)

Effluent Characteristics	Maximum Average of Daily Values for Any Period of Thirty Consecutive Days kg/1,000 kg*	Maximum for Any One Day kg/1,000 kg*
Subcategory A		
BOD_5	0.012	0.020
COD	0.10	0.15
TSS	0.0075	0.012
Phenols	0.00005	0.00010
Subcategory B		
B-1 Product-Processes		
BOD_5	0.035	0.06
COD	0.40	0.55
TSS	0.02	0.033
Phenols	0.00013	0.00026
B-1 Product-Processes		
BOD_5	0.085	0.15
COD	2.2	3.0
TSS	0.05	0.083
Phenols	0.00034	0.00068
Subcategory C		
C-1 Product-Processes		
BOD_5	0.085	0.15
COD	2.3	3.3
TSS	0.05	0.083
Phenols	0.00034	0.00068
C-2 Product-Processes		
BOD_5	0.40	0.75
COD	40.0	60.0
TSS	0.16	0.27
Phenols	0.0011	0.0022
Subcategory D		
BOD_5	0.85	1.5
COD	390.0	540.0
TSS	2.60	4.38
Phenols	0.017	0.034

*kg/1,000 kg production is equivalent to lb/1,000 lb production

Source: Report EPA 440/1-73/009

POLLUTION CONTROL ECONOMICS

The necessity of holding this volume to some reasonable size precludes the possibility of discussing pollution control economics here. However, there are reports which do discuss this topic in detail and they are cited here for the reader's convenience:

(1) Datagraphics, Inc., *Projected Wastewater Treatment Costs in the Organic Chemical Industry,* Report 12020 GND 07/71, Wash., D.C., Supt. of Documents (July 1971).

(2) Little, Arthur, D., Inc., *Economic Analyses of Proposed Effluent Guidelines – Organic Chemical Industry,* Report EPA-230/1-73-019, Wash., D.C., Environmental Protection Agency (August 1973).

FUTURE TRENDS

A series of recommendations has been made for improving the technology necessary for successful pollution abatement programs in the petrochemical and refining industries by W. Jewell and D.L. Ford of Engineering-Science Inc. of Austin, Texas in *EPA Report 12020 EID- 03/71.*

(1) The quality of wastewater discharged from standard process units within a refinery such as crude units, catalytic cracking units, coking units, utility areas, etc., can be categorized with respect to major contaminants. Therefore, attempts should be made to survey and statistically report data available from operating refineries. Such information would greatly enhance the programming of wastewater surveys in refineries where this information is not now available as well as provide some design bases for treatment facilities serving refineries not yet constructed. It is recognized that such a tabulation should be limited to "standard" production units. Applicable characterization parameters for each standard unit also should be listed.

(2) Particular emphasis should be given to the development of a monitoring and alarm system within a refinery or petrochemical complex. This includes the detection of specific substances resulting from batch dumps or spills which adversely affect treatment process efficiencies and/or the biology of the receiving environment. Such substances include specific organics, refractory hydrocarbons, heavy metals, etc.

(3) As more treatability studies are performed and documented and treatment facilities constructed therefrom, more sophisticated scale-up factors should be defined. This is particularly true when considering biological and chemical-physical treatment systems. This can be achieved by obtaining operating data from the bench, pilot, and full-scale units and analyzing and relating the data statistically.

(4) A more comprehensive evaluation of the impact of storm flow runoff on treatment facilities and surrounding areas should be included in any wastewater

survey. This information then should be included in the treatability evaluations if such flow is judged to significantly influence test system responses.

(5) A better-defined bench or pilot scale approach for simulating and determining the effects of occasional pollutional discharges such as those from ballast handling areas to a treatment system should be developed.

(6) The design of continuous flow wastewater treating processes should always be based on data developed from continuous reactor treatability studies. Semicontinuous or batch reactor studies should only be conducted when a similar operation is expected of the full-scale system. When carefully interpreted, studies made using batch and semicontinuous reactors can serve as screening tests for detecting toxicity, or for roughly estimating continuous flow process performance.

(7) It is recognized that sludge handling and disposal facilities may constitute a significant portion of both construction and operating costs for a petrochemical or refinery waste treatment plant. These sludges include oily sludges, chemical sludges, or excess biological sludges. Consequently, an integral and important part of the treatability study should be the development of data to allow intelligent selections and accurate sizing of processes for handling these sludges.

(8) There is a need to develop a rational and more uniform approach toward bench or pilot scale analysis of tertiary treatment processes, particularly as more tertiary plants are put into operation providing operational data available for treatability evaluation.

(9) A more uniform approach in conducting surveys and treatability studies is suggested through the preparation of "Standard Methods"- type treatability manuals. Such documents should incorporate the ideas and approaches of qualified specialists and be published through sponsoring organizations such as the American Petroleum Institute or the American Association of Professors in Sanitary Engineering.

(10) The necessary survey and treatability information should be available before any industrial treatment facility is designed. In order to be reasonably certain that the pollution control system envisioned will satisfy the requirements of the industry and regulatory authority, it is necessary that a comprehensive survey and treatability program be required. This requirement can be satisfied through the proper establishment of guidelines by the regulatory agency and/or industry management.

(11) Wastewater surveys and treatability studies should reflect the possible influence of shift changes, week-end operations, seasonal variations, etc., on the wastewater characterization and its treatability. The more accurately these variations are documented in the preliminary phases, the greater the probability of optimum plant design being achieved.

(12) As many new instrumental approaches for analyzing wastewaters currently are being introduced, an evaluation of the instrument performance with respect to analytical efficiency, interferences, and general limitations should be documented by independent sources.

COMPANY INDEX

l'Air Liquide, SA pour l'Etude de l'Exploitation des Procedes Georges Claude - 65, 248
Allied Chemical Corporation - 259
American Waagner-Biro Company, Inc. - 222, 223
Asahi Kasei Kogyo KK - 70
Ashland Oil & Refining Co. - 220
The Badger Co., Inc. - 188, 195
Badische Anilin & Soda-Fabrik AG - 216
Charles River Foundation - 219
Chemische Werke Huls AG - 188
Ciba-Geigy - 250
Day & Zimmermann, Inc. - 256
The Dow Chemical Co. - 218
Dravo Corp. - 224
Dynamit Nobel AG - 253
E.I. du Pont de Nemours & Co. - 206
Farbwerke Hoechst AG - 154, 247
Firma Carl Still - 221, 222
The Gas Council - 221
Gewerkschaft Auguste Victoria - 253
Gulf Oil Corporation - 144
Gulf Research & Development Co. - 195
Hartung, Kuhn & Co. Maschinenfabrik GmbH - 222
Hercules Powder Company - 216
Inventa AG - 70
Knapsack AG - 240
Koppers Company, Inc. - 221, 225
Michigan Chemical Corporation - 216
Monsanto Company - 69, 70
Dr. C. Otto & Company GmbH - 223, 224
The Permutit Company, Ltd. - 244
Petro-Tex Chemical Corp. - 91
Phillips Petroleum Co. - 92, 93, 219
Polymer Corporation Ltd. - 219
Process Engineers Incorporated - 232
Pulp and Paper Research Institute of Canada - 221
Shell Oil Company - 161
Socony Mobil Oil Co., Inc. - 239
The Standard Oil Co. - 65
Stewarts & Lloyds Ltd. - 244
Sumikin Coke Co., Ltd. - 224
Sumitomo Shipbuilding & Machinery Co., Ltd. - 67
Texaco, Inc. - 161, 248, 249, 253
U.S. Secretary of the Air Force - 156
United States Steel Corp. - 255, 262
Universal Oil Products Co. - 248
Woodall-Duckham Construction Co., Ltd. - 244
Yawata Chemical Engineering Co., Ltd. - 218
John Zink Company - 219, 220

INVENTOR INDEX

Bertram, C.G. - 188
Bhatia, S.P. - 221
Brainard, A.H. - 65
Breitbach, F. - 222
Campbell, C.R. - 69, 70
Carr, N.L. - 195
Caskey, E.K. - 220
Cathala, M.E.J. - 218
Chlanda, F.P. - 259
Choulat, G. - 222
Cole, E.L. - 248
Collier, H.E., Jr. - 206
Cottle, J.E. - 93
Crommelynck, F. - 248
Cunningham, S.A. - 92
Czehovsky, G. - 253
Danly, D.F. - 70
Davis, J.B. - 239
Deselaers, K. - 253
Didycz, W.J. - 262
Dille, R.M. - 161, 249
Doering, H. - 253
Drebes, F.-W. - 223
Edgar, W.D. - 225
Errgang, K. - 223
Ezaki, S. - 218
Feder, E. - 253
Ferrari, D.C. - 188
Fitzgibbons, W.O. - 65
Flasche, K. - 253
Ford, W.W. - 219
Fujii, Y. - 67
Glassman, D. - 262
Goedicke, E. - 240
Goodnight, H.E. - 220
Gregor, H.P. - 259
Griehl, W. - 70
Grigsby, A.J. - 65
Grover, F.N. - 216
Guttmann, A.T. - 65
Hall, N.R. - 244
Helm, E.J. - 221
Hess, H.V. - 248
Hinton, R.A. - 93
Hoffmann, G. - 188
Hughes, R.E. - 195
Hutson, T., Jr. - 92
Ikeda, M. - 224
Iwashita, H. - 70
Jaekel, G. - 240
Jasinsky, V. - 219
Johnson, R. - 69
Jonnet, A.G. - 225
Karpiuk, R.S. - 218
Karrenbauer, K. - 240
Kaunert, P. - 253
Kelmar, J.J. - 222
Kemmetmueller, R. - 222, 223
Kent, R.D. - 253
Kirstein, D. - 240
Kivari, A.M. - 232
Knappstein, J. - 221
Konicek, M.G. - 65
Kube, A.M.L. - 219
Kubsch, W. - 222
Kunzer, W. - 216
Kurland, R.S. - 144
Lentz, C.W. - 216
Leutert, F. - 216
Liu, K.-J. - 259
Masciantonio, P.X. - 255
Mathews, M.J., III - 70
Mattia, M.M. - 256
McCoy, D.E. - 249
McEachern, R.M. - 249
Merrill, E.W. - 219
Miller, A.F. - 65
Miyake, T. - 70

Morrison, J.N., Jr. - 224
Nashan, G. - 221
Nicklin, T. - 221
Oshimi, T. - 67
Panzarella, J.E. - 217
Pauling, L.E. - 216
Pfeiff, S. - 253
Prahacs, S. - 221
Pries, E.E. - 223, 224
Probst, O. - 154
Raymond, R.L. - 239
Reed, R.D. - 219
Renner, A. - 250
Riechert, K. - 224
Riemenschneider, W. - 154
Ringel, E. - 247
Riter, R.E. - 92
Roe, W.A., Jr. - 195
Rosenwald, R.H. - 248
Salehar, M.L. - 65
Schwerko, E.M. - 65
Seebold, J.E. - 216
Seko, M. - 70
Siewers, H. - 253
Spindeler, H. - 223
Stauffer, H.C. - 195
Stetter, A. - 253
Strausser, J.R. - 144
Striebeck, A. - 188
Sugimoto, M. - 224
Suter-Homuth, C. - 70
Taylor, E.R., Jr. - 206
Ter Haar, L.W. - 161
Teske, W. - 247
Upfold, A.T. - 219
Urban, P. - 248
Visser, P. - 161
Williams, D.C. - 220
Williams, T.H. - 244
Woerner, R.C. - 91
Wohlers, H.C. - 216
Wolverton, B.C. - 156
Yomiyama, A. - 70
Zink, J.S. - 219, 220
Zumbrunn, J.-P. - 65, 248

U.S. PATENT NUMBER INDEX

2,545,314 - 216
2,730,194 - 216
2,795,544 - 244
2,868,325 - 218
2,876,863 - 232
2,881,054 - 218
2,978,063 - 219
2,982,645 - 161
3,036,418 - 216
3,096,156 - 253
3,148,210 - 69
3,152,983 - 239
3,267,029 - 70
3,308,061 - 206
3,328,265 - 154
3,336,414 - 91
3,337,452 - 247
3,370,400 - 188
3,373,085 - 255
3,468,624 - 65
3,491,518 - 220
3,507,782 - 144
3,512,911 - 219
3,520,109 - 220
3,523,064 - 70
3,540,189 - 253
3,557,238 - 92
3,576,607 - 65
3,589,104 - 217
3,589,864 - 218
3,593,496 - 219
3,624,984 - 188
3,625,886 - 256
3,630,852 - 221
3,642,620 - 248
3,646,239 - 92
3,647,636 - 221
3,649,169 - 221
3,661,507 - 222
3,661,719 - 222
3,667,408 - 219
3,669,847 - 253
3,672,835 - 248
3,673,068 - 70
3,675,400 - 222
3,679,764 - 93
3,691,020 - 195
3,694,355 - 161
3,697,381 - 222
3,697,567 - 206
3,701,824 - 221
3,711,402 - 248
3,715,282 - 223
3,715,309 - 65
3,716,483 - 250
3,721,609 - 223
3,725,269 - 156
3,728,230 - 223
3,732,085 - 195
3,734,943 - 65
3,746,626 - 224
3,748,235 - 224
3,749,546 - 220
3,752,742 - 240
3,752,749 - 259
3,754,376 - 253
3,756,947 - 67
3,761,409 - 249
3,766,018 - 224
3,767,536 - 224
3,785,933 - 225
3,790,448 - 262

NOTICE

POLLUTION DETECTION AND MONITORING HANDBOOK 1974

by Marshall Sittig

Environmental Technology Handbook No. 1

This handbook contains methods for the detection and monitoring of pollutants in industrial effluents, notably air and water.

Such methods are prerequisites for any kind of management of environmental quality. Of equal importance are the data handling systems that must be used to collect and interpret the measurements, regardless of whether manual or automated identifying and monitoring techniques are chosen.

Generally the most important single factor in determining whether manual or automated methods of analysis and level monitoring should be used, is the required frequency interval. In the manual approach costs vary almost directly with the measurement frequency. At monitoring intervals oftener than once a day, or when a continuous watch must be maintained, installation of automatic sensing is more economical, provided reliable automatic sensors are available.

For meaningful control and prevention of pollution the accepted standards and parameters for air and water quality must be known, whether they have been legislated or not. Acceptable levels for each type of impurity from methylmercury to dissolved chlorides or carbon particles must be known by the industrialist or public health official, and this book, with its 1633 references makes a massive attempt to communicate such levels.

As a guideline for the industrial user, data on the toxicity of the various pollutants are included. Such data on the deleterious effects were taken from various government publications. This book is not a treatise of pharmacology or industrial toxicology, but a guide to the accurate evaluation of pollutant levels in a given effluent or ambient substrate.

Whenever applicable, the name of the pollutant is followed by chemical and other information. After this come directions for sampling in air or water or both, as the case may be. Thereafter are given qualitative and quantitative analytical methods suitable for identification and measurement of the degree or level of pollution. This comprehensive information is followed by measurement techniques suitable for repeated or continuous monitoring. Preference is given to methods giving reproducible results in environmental quality control, as recommended by the EPA and other U.S. government agencies. This is followed in each case by many up-to-date references to government publications, patents, and journal articles.

Arrangement is encyclopedic. The book is thus a worthy companion to the author's POLLUTANT REMOVAL HANDBOOK available from the same publisher.

Pollution control is a serious business, and in the design of a monitoring program the specific objectives to be served must be clearly in mind from the very beginning. Otherwise the effort most probably will not result in an efficient and meaningful program. The length of the initial survey, selection of monitoring points, parameter coverage, and sampling frequency all depend on the specific objectives and the timeframe in which the objectives must be achieved. These variables also have a very significant impact on monitoring costs.

Monitoring for the purpose of long-term trend identification, evaluation of standards compliance, and total management of environmental quality must, of course, be a continuing program without end.

It is hoped sincerely that this book, together with its companion volume POLLUTANT REMOVAL HANDBOOK, will do much to establish and sustain such antipollution programs.

A partial and condensed table of contents follows here in which subheadings are indicated below for only the first few entries. The book contains a total of 88 subject entries arranged in an alphabetical and encyclopedic fashion. The subject name refers to the polluting substance, and the text underneath each entry tells how to measure and monitor pollution by said substance:

INTRODUCTION

DEFINITION OF PROBLEMS
- Terms Used

POLLUTANT EFFECTS
- On Health
- On the Environment

AMBIENT QUALITY STANDARDS

EMISSION STANDARDS
- For Stationary Sources
- For Mobile Sources

MONITORING

ANALYTICAL SYSTEMS
- Manual
- Automated (General)
- Automated (Air)
- Automated (Water)
- Remote
- Public

POLLUTION INSTRUMENTATION
Size of the Market
References

SAMPLING
Air and Gas Sampling
Water Sampling
References

SUBSTANCES AND MATERIALS

ACIDS AND ALKALIS
ALDEHYDES
ALUMINUM
AMMONIA
ANTIMONY
ARSENIC
ASBESTOS
AUTOMOTIVE EXHAUST
BACTERIA
BARIUM
BERYLLIUM
BISMUTH
BORON
BROMIDES
CADMIUM
CALCIUM
CARBOHYDRATES
CARBON
CARBON DIOXIDE
CARBON MONOXIDE
CHLORIDES
CHLORINATED HYDROCARBONS
CHLORINE
CHROMIUM
COBALT
COPPER
CYANIDES
ETHYLENE
FLUORIDES
HALOGENS
HALOGEN COMPOUNDS
HYDROCARBONS
HYDROGEN CHLORIDE
HYDROGEN SULFIDE
IODIDES
IRON
LEAD
LEAD ALKYLS
MAGNESIUM
MANGANESE
MERCAPTANS
MERCURY
MOLYBDENUM
NICKEL
NICKEL CARBONYL
NITRATES
NITRIC ACID
NITRITES
NITROGEN COMPOUNDS
NITROGEN OXIDES
ODOROUS COMPOUNDS
Measurement in Air
Sampling Methods
Qualitative Analyses
Quantitative Methods (Organoleptic)
Quantitative Methods (Instrumental)
Measurement in Water
References
OIL AND GREASE
ORGANICS
OXYGEN DEMAND
OZONE AND OXIDANTS
PARTICULATES
PEROXYACETYL NITRATE (PAN)
PESTICIDES
PHENOLS
PHOSGENE
PHOSPHATES
PHOSPHITES
PHOSPHORIC ACID
PHOSPHORUS
PHOSPHORUS COMPOUNDS (ORG.)
POLLEN
POLYNUCLEAR AROMATICS
POTASSIUM
RADIOACTIVE MATERIALS
SEDIMENTS
SELENIUM
SILVER
SODIUM
STRONTIUM
SULFATES
SULFIDES
SULFITES
SULFUR COMPOUNDS (ORG.)
SULFUR DIOXIDE
SULFUR TRIOXIDE
SULFURIC ACID
SURFACTANTS
TIN
TITANIUM
URANIUM
VANADIUM
ZINC
ZIRCONIUM

FUTURE TRENDS

This book should prove to be very useful as an overall reference volume. The organization is clear, and the text is especially well arranged. The tables and figures are extensive and will be of value to students, engineers, and pollution control authorities.

ISBN 0-8155-0529-9

401 pages

FURANS 1973

Synthesis and Applications

by Alec Williams

Chemical Technology Review No. 18

In large scale production processes, most furans are obtained from the key chemical furfural (2-furaldehyde) which is prepared commercially from cereal straws and brans, especially corncobs, oat hulls, cottonseed hulls, and rice hulls. Treatment with sulfuric acid hydrolyzes the pentosan components of these agricultural residues and dehydrates the freed pentose sugars to furfural.

In this sense, furans are products of agriculture, yielding an abundant harvest crop each year.

Tetrahydrofuran is an important solvent and intermediate in the manufacture of printing inks, lacquers, adhesives and other coating compositions. It is also used in polyurethane technology.

Benzofuran (coumarone) and reduced benzofurans are of considerable importance in pharmaceutical, insecticidal and photochromic applications.

Furan polymers and resins are inexpensive and have great potential use where products with their characteristics are required.

This book describes synthetic routes, products and end use applications of the furans. Over 100 different patent-based syntheses are provided covering the many derivatives of this family of commercially valuable materials.

A partial and condensed table of contents follows. Numbers in ( ) indicate the number of processes per topic. Chapter headings are given, followed by examples of important subtitles.

1. **FURAN [DIVINYLENE OXIDE] (6)**
   Liquid Phase Decarbonylation of Furfural

2. **SUBSTITUTED FURANS (22)**
   Furonitrile
   2-Fluoroalkyl-5-nitrofurans
   5-Imino-1-methylcarbamoyl-3-(5-nitro-2-furyl)-1,2,4-triazoline
   Halogenated Cyclopentadiene Diadducts
   3-Choloroallyl Furans
   Furanaliphatic Acids
   Dimethylfuran
   2-Methylfuran

3. **TETRAHYDROFURAN (9)**
   Hydrogenolysis of Furan Using Nickel Catalyst
   Purification of THF
   Separation of a Mixture of THF, Toluene, and Water

4. **SUBSTITUTED TETRAHYDROFURAN (9)**
   Nitroalkyl
   Trinitromethyl
   Polycyanotetrahydrofurans
   Tetrahydrofuranyl Derivatives of Benzomorphans

5. **COPOLYMERS (3)**
   Vulcanizable Products with Epoxies
   With Decaborane

6. **COMPLEXES (15)**
   Chromium
   Manganese
   Organoboron Cyclic Ethers
   With Hexafluoro-2-propanol

7. **STABILIZERS (3)**
   4,4'Thiobis (6-tert.-butyl-m-cresol)
   Alkyl Phosphites
   Sodium Borohydride-Phosphite Additives

8. **BENZOFURAN [COUMARONE] (3)**
   Catalytic Oxidative Dehydrocyclization

9. **SUBSTITUTED BENZOFURANS (30)**
   With Pharmacological Activity
   For Ultraviolet Screening
   For Dyestuffs and Other Uses

10. **DIBENZOFURANS (28)**
    2-Dibenzofuranacetohydroxamic Acid
    Dibenzofurans as Functional Fluids
    Dihydrobenzofurans with Insecticidal Activity

11. **DIHYDROFURANS (4)**
    2,3- and 2,5-Dihydrofurans
    Purification of 2,5-Dihydrofuran

12. **SUBSTITUTED DIHYDROFURANS (4)**
    Polycyano-2,5-dihydrofurans
    Hexahydro-9b-methylfuro[3,2-c] quinolines
    Naphthospirofurans

13. **FURANONES AS FLAVOR AGENTS (5)**
    Furanone Meat Flavor Compositions
    Furanone-Hydrogen Sulfide Reaction Products

14. **GRISEOFULVINS (8)**
    5'-Hydroxygriseofulvin by Fermentation
    Reduction Using Metal Hydrides
    Isolation of Crude Material
    Purification of Crude Material
    Tetrahydrofuranyl Ethers
    Pivaloyl Esters of Griseofulvol

ISBN 0-8155-0506-X

303 pages

ORGANIC PEROXIDE TECHNOLOGY 1973

by Louis F. Martin

Chemical Technology Review No. 15

Some 40 to 50 different peroxide compounds are commercially available at present, and many are used as initiators in the polymerization of polyethylene, polyvinyl chloride, polyester thermosets, styrene-butadiene rubber, and for in-process cross-linking of ethylene-propylene copolymers and silicone rubbers.

Benzoyl peroxide, used in polystyrene and polyester markets such as toys, furniture, automobiles, trucks, airplanes and ships, is by far the largest volume product. Methyl ethyl ketone peroxide is also used in large volumes for glass fiber reinforced plastics applications such as pleasure craft, shower stalls, tub components and sports equipment. The so-called per esters are growing more slowly, because of some substitution with the cheaper peroxydicarbonates and azo compounds.

Research efforts are toward high performance specialty compounds, offering added value at reasonable prices. The development of safer, easily handled, and more efficient initiators is a major goal.

This book describes some 140 different, patented processes relating to all aspects of peroxide manufacture, and over 600 specific, detailed examples are provided. The very significant interest the world over is reflected by the 40 processes developed in Europe and Japan. A partial and condensed table of contents follows. Numbers in ( ) indicate the number of processes per topic. Chapter headings are given, followed by examples of important subtitles.

1. HYDROPEROXIDES — AUTOOXIDATION PROCESSES (37)
- Cumene Catalysts
- Ethylbenzene
- Other Aralkyl Hydrocarbons
- Dihydroperoxides of Alkylated Tetrahydronaphthalene
- Terpene
- Cyclic Compounds
- 6-Hydroperoxyhexanoic Acid from Cyclohexane
- Peroxynitrates

2. OTHER HYDROPEROXIDE PRODUCTS AND PROCESSES (8)
- Hydrogen Peroxide Reactions
- 3,6-Dimethyloctane Dihydroperoxide
- 2,2-Dihydroperoxypropane
- Use of Quaternary NH_4-Salts
- Perfluoroalkyls
- Stabilized Compositions

3. TRUE PEROXIDES (46)
- Hydroperoxide-Alcohol Reactions
- Tris(peroxides) from 1,3,5-Triisopropylbenzene
- Hydroperoxide-Organic Chloride Reactions
- Use of Triphenylphosphine Oxide Catalyst
- Diaryl Peroxides Using Hypochlorites
- Ozone Reactions
- Polybrominated Peroxides
- Anhydrides of Diperoxydicarboxylic Acids
- Formulated Compositions
- Aqueous Keto Peroxides + H_2O_2
- Diallyl Phthalate Compositions
- Stabilizers
- Dipicolinic Acid
- Phytic Acid
- Enolizable Ketones

4. KETONE & ALDEHYDE-BASED PEROXIDES (15)
- Ethyl Isoamyl Ketone
- Dialkyl Ketones
- Diacetone Alcohol
- Difunctional Hydroperoxides
- Peroxyketals
- Alkylidene Diperoxides

5. DIACYL PEROXIDES (13)
- Benzoyl Peroxide
- Acetyl Peroxide
- Diacetyl Peroxide
- Acyl Peroxides with Epoxy Groups

6. PEROXY ESTERS — PEROXY CARBONATES — FLUORINE-CONTAINING PEROXIDES (18)
- Tertiary Alkyl Peresters
- Esters of 1,1,3,3-Tetramethylbutyl Hydroperoxide
- Aromatic Peroxy Esters
- Peracetal Reactions
- Pseudo Per Esters of Keto Acids
- 2,5-Dimethylhexane Bis(Peroxy Carbonates)
- Carbonato Derivatives of Acetylenic Hydroperoxides
- Bis(trifluoromethyl) Trioxide
- Carbonyl Fluoride + Difluoromethylene Bix(oxyfluoride)
- Carbonyl Fluoride + Oxygen Difluoride
- Perfluorodimethyl Peroxide
- Peroxy Acids

ISBN 0-8155-0499-3

275 pages